ANNALS *of* THE NEW YORK ACADEMY OF SCIENCES

EDITOR-IN-CHIEF
Douglas Braaten

ASSOCIATE EDITOR
Rebecca E. Cooney

PROJECT MANAGER
Steven E. Bohall

EDITORIAL ADMINISTRATOR
Daniel J. Becker

Artwork and design by Ash Ayman Shairzay

The New York Academy of Sciences
7 World Trade Center
250 Greenwich Street, 40th Floor
New York, NY 10007-2157

annals@nyas.org
www.nyas.org/annals

Published by Blackwell Publishing
On behalf of the New York Academy of Sciences

Boston, Massachusetts
2012

ANNALS *of* THE NEW YORK ACADEMY OF SCIENCES

VOLUME 1266

ISSUE

Hematopoietic Stem Cells VIII

ISSUE EDITORS

Lothar Kanz,[a] Claudia Lengerke,[a] Willem E. Fibbe,[b] and John E. Dick[c]

[a]University of Tübingen, Tübingen, Germany; [b]Leiden University Medical Center, Leiden, the Netherlands; [c]University of Toronto, Toronto, Ontario, Canada

TABLE OF CONTENTS

Ann. N.Y. Acad. Sci. ISSN 0077-8923

ANNALS OF THE NEW YORK ACADEMY OF SCIENCES
Issue: *Hematopoietic Stem Cells VIII*

Preface for *Hematopoietic Stem Cells VIII*

The Eight International Conference and Workshop on Hematopoietic Stem Cells was held on September 22–24, 2011 in Tübingen, Germany. The conference was hosted by the University of Tübingen and took place in the historical rooms of the Protestant Collegiate (*Evangelisches Stift*) that has been a part of the theological faculty of the University of Tübingen since the sixteenth century. Over the years, the Protestant Collegiate has hosted many famous scientists, philosophers, and poets, including Johannes Kepler, Georg Wilhelm Friedrich Hegel, and Friedrich Hölderlin. In this inspiring atmosphere, twenty-four invited speakers presented their data on basic and clinical research in the exciting and rapidly evolving field of stem cell biology. The following topics were included in the program: stem cell regulation, development, microenvironment, translational research, stem cell cycling and division, cancer stem cells, and reprogramming.

As in previous HSC conferences, priority was given to extended discussions following each presentation. This format was well received by the participants, who readily engaged in lively, highly productive scientific discussions. For the speakers it provided, moreover, the opportunity to present and critically discuss unpublished data—in many instances presented for the first time—in a constructive and friendly manner with designated experts in the field.

As organizers we are very thankful to all the speakers for sharing their data and opinions, again making this conference an outstanding success. We hope that readers will find these papers informative and that the exciting new research presented here will inspire ongoing study.

We thank the New York Academy of Sciences for publishing these proceedings. Our special thanks go to the editorial staff of *Annals of the New York Academy of Sciences* for their continuous support throughout the planning of this volume. Finally, we would like to express our very special gratitude to Amgen (Germany) for its generous contribution in support of the conference.

Claudia Lengerke
University of Tübingen, Tübingen, Germany
Willem E. Fibbe
Leiden University Medical Center, Leiden, the Netherlands
John E. Dick
University of Toronto, Toronto, Ontario, Canada
Lothar Kanz
University of Tübingen, Tübingen, Germany

doi: 10.1111/j.1749-6632.2012.06709.x

Ann. N.Y. Acad. Sci. ISSN 0077-8923

ANNALS OF THE NEW YORK ACADEMY OF SCIENCES
Issue: *Hematopoietic Stem Cells VIII*

Epigenetic differences between sister chromatids?

Peter M. Lansdorp,[1,2,3] Ester Falconer,[1] Jiang Tao,[1] Julie Brind'Amour,[1] and Ulrike Naumann[1]

[1]Terry Fox Laboratory, BC Cancer Agency, Vancouver, British Columbia, Canada. [2]Division of Hematology, Department of Medicine, University of British Columbia, Vancouver, British Columbia, Canada. [3]European Research Institute for the Biology of Ageing, University of Groningen, University Medical Centre Groningen, Groningen, the Netherlands

Address for correspondence: Dr. Peter M. Lansdorp, European Research Institute for the Biology of Ageing, University of Groningen, University Medical Centre Groningen, A. Deusinglaan 1, NL-9713, AV Groningen, the Netherlands. p.m.lansdorp@umcg.nl

Semi-conservative replication ensures that the DNA sequence of sister chromatids is identical except for replication errors and variation in the length of telomere repeats resulting from replicative losses and variable end processing. What happens with the various epigenetic marks during DNA replication is less clear. Many chromatin marks are likely to be copied onto both sister chromatids in conjunction with DNA replication, whereas others could be distributed randomly between sister chromatids. Epigenetic differences between sister chromatids could also emerge in a more predictable manner, for example, following processes that are associated with lagging strand DNA replication. The resulting epigenetic differences between sister chromatids could result in different gene expression patterns in daughter cells. This possibility has been difficult to test because techniques to distinguish between parental sister chromatids require analysis of single cells and are not obvious. Here, we briefly review the topic of sister chromatid epigenetics and discuss how the identification of sister chromatids in cells could change the way we think about asymmetric cell divisions and stochastic variation in gene expression between cells in general and paired daughter cells in particular.

Keywords: DNA replication; chromatin; epigenetic marks; sister chromatids

Although asymmetric inheritance of proteins and RNA are known to contribute to differences in daughter cell fate, the possibility that epigenetic differences between sister chromatids also contribute to differences in cell fate has been proposed[1–5] but has been difficult to study. Indeed, most studies assume that sister chromatids, being genetically identical, are also functionally identical and that segregation to daughter cells proceeds randomly. However, given the plethora of epigenetic and chromatin modifications that are known to play a role in regulating gene expression (Fig. 1), this assumption is perhaps an oversimplification for some genes (Fig. 2). Many epigenetic marks are very dynamic during the cell cycle, and some of these marks are known to be "erased" prior to mitosis. For example, heterochromatic protein 1 (HP1) binding to trimethylated lysine at position 9 of histone H3 (Ref. 6) is disrupted by the phosphorylation of serine 10 of histone H3 by Aurora B kinase prior to mitosis.[7] Conversely, at least some epigenetic marks, including methylated cytosines and certain histone modifications (reviewed in Ref. 8), transcription factors such as CTCF,[9] and Polycomb proteins[10] remain associated or bound to DNA during DNA replication and are present on mitotic chromosomes. Most likely, such epigenetic marks serve to nucleate chromatin and higher order nuclear structures following mitosis and thereby function as the epigenetic "memory" that enables setting up similar gene expression patterns in parental cells and daughter cells. Therefore, the regulation of epigenetic marks during DNA replication to alter or preserve gene expression at a given locus could be an additional mechanism for differentiation or self-renewal fate decisions.

Several studies have shown that sister chromatids containing "old" template DNA are selectively retained in certain stem or progenitor cells.[11–13] These previous observations have typically been explained

doi: 10.1111/j.1749-6632.2012.06505.x

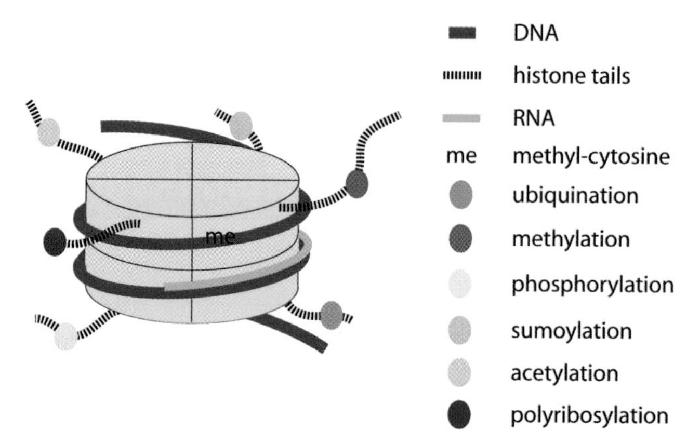

Figure 1. A plethora of histone modifications and other epigenetic marks such as (hydroxy-)methylation of cytosine and binding of specific RNA molecules are implicated in the regulation of gene expression.

by the "immortal strand" hypothesis, which proposes that stem cells retain template DNA in order to protect against the accumulation of mutations resulting from DNA replication.[14] Based on published data and theoretical reasoning,[4] we recently proposed that, if previous observations were not the result of technical artifacts, the primary function of selective segregation of DNA template strands could be the regulation of cell fate via epigenetic differences between sister chromatids (Fig. 3, the "silent sister" hypothesis[4]). Given the dynamic expression of some genes during the cell cycle[16] and the fidelity in which epigenetic marks appear to be copied,[17] epigenetic differences between sister chromatids could be limited to a select number of genes, for example, following processes restricted to lagging strand DNA replication.[18] Differences between sister chromatids are known to regulate the expression of the mating type–locus in fission yeast,[3] and it was recently shown that segregation of sister chromatids in *Escherichia coli* is not a random process.[19] Furthermore, another recent study showed that components of the yeast kinetochore, the protein complex that anchors chromosomes to the mitotic spindle, divide asymmetrically in a single postmeiotic lineage, suggesting a mechanism for the selective segregation of sister centromeres to daughter cells to establish different cell lineages or cell fates.[20]

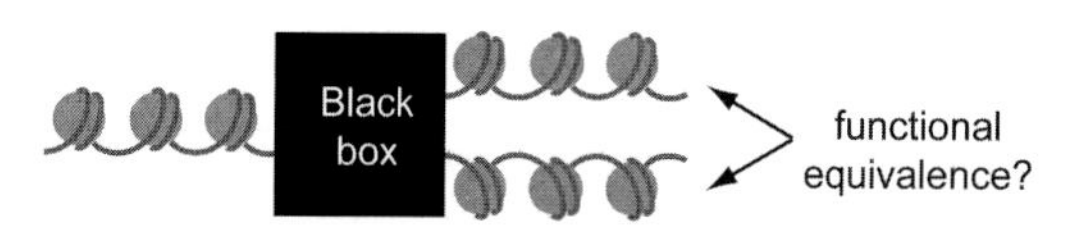

Figure 2. Given the challenge to faithfully copy the plethora of possible chromatin marks and epigenetic modifications (Fig. 1) onto both sister chromatids, it seems unlikely that sister chromatids can always be considered functionally equivalent.

In order to study sister chromatid segregation in relation to gene expression and cell fate, molecular features that can be used to tell sister chromatids apart are essential. Suitable features to distinguish sister chromatids are not obvious since both sisters are the product of semi-conservative DNA replication and, therefore, are expected to have exactly the same DNA sequence. However, we recently found that DNA template strands that are present *prior* to cell division provide a means to reproducibly distinguish and identify sister chromatids in daughter cells using chromosome orientation fluorescence *in situ* hybridization (CO-FISH).[21] Specifically, we found that murine chromosomes have an invariant orientation of pericentric major satellite DNA with respect to chromosome ends (Fig. 4). We exploited this polarity to differentially label and follow sister chromatid segregation in postmitotic cell pairs with probes specific for major satellite DNA. Importantly, we found that segregation of sister chromatids is nonrandom in a subset of murine colon cells.[21] Cell pairs exhibiting marked template strand asymmetry were found at different positions in the colon crypt, including at positions outside the crypt bottom where stem cells have been proposed to reside,[27] indicating that asymmetric segregation is not restricted to stem cells. Neither the mechanism nor the function of the observed asymmetric chromatid segregation is currently known. In order to allow nonrandom segregation, we presume that centrosomes and centromeres have properties that enable

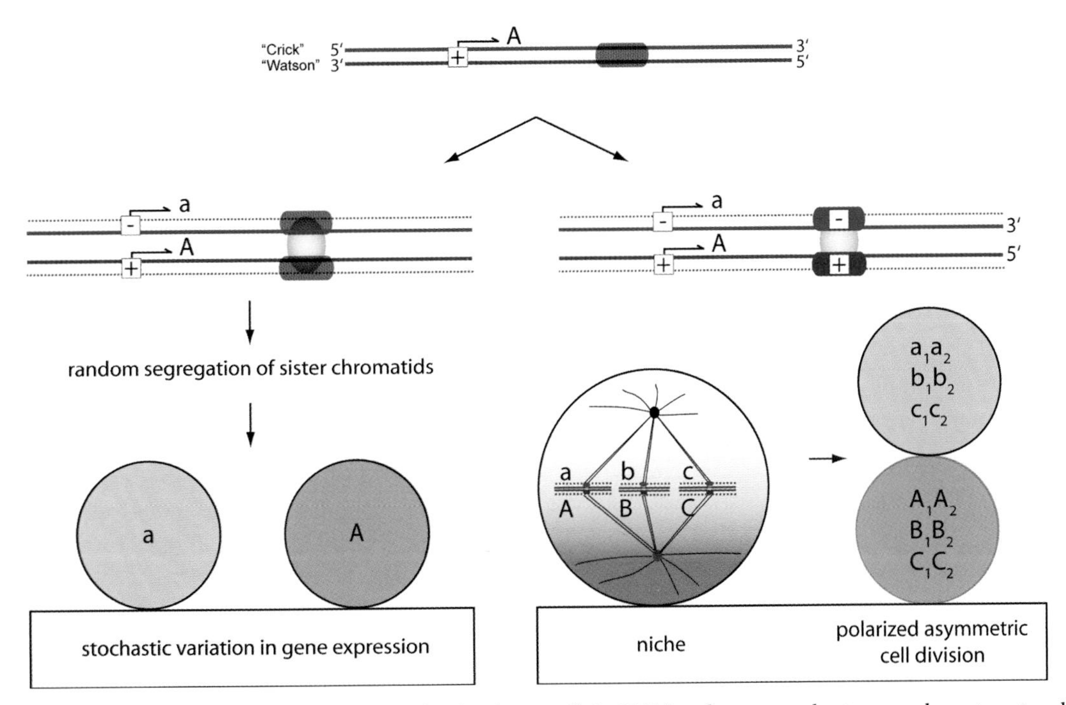

Figure 3. The silent sister hypothesis. Prior to replication (top panel) the DNA in a chromosome has two complementary strands: "Crick" (5′ to 3′, top strand, blue line) and "Watson" (3′ to 5′, bottom strand, red line). A single gene "A" is shown that is expressed as a result, of specific chromatin marks (+ sign in figure). Following DNA replication (middle panels), Watson and Crick DNA template strands are copied to yield two sister chromatids with identical DNA sequence. The silent sister hypothesis proposes that not all chromatin marks are copied onto both sister chromatids during or following DNA replication. As a result, only one sister chromatid will inherit the active chromatin mark (+) and the other sister chromatid (the "silent sister") will not (–). In the figure, active chromatin marks (+) follow the Watson DNA template strand, and the sister chromatid with the original Crick DNA template strand does not have this chromatin mark and therefore does not support expression of gene A (indicated by a small "a"). Following random segregation of sister chromatids (bottom left panel), the two daughter cells will show stochastic variation in the expression of gene A (A or a), which is predicted to follow the parental DNA template strand that was inherited. Note that failure to copy suppressive chromatin marks will result in similar stochastic variation in gene expression. If sister chromatids of specific chromosomes are furthermore specifically retained in one of the daughter cells (e.g., via specific chromatin marks at sister chromatid centromeres connecting microtubules to "mother" centrosomes),[15] sister chromatid asymmetry in chromatin marks at specific genes could directly regulate gene expression and cell fate as shown (bottom right panel).

specific kinetochore connections (Fig. 5). Epigenetic differences between sister chromatids could be directly involved in the regulation of cell fate if microtubules originating from "mother" centrosomes[15] were to prefer kinetochores present on one of the two sister chromatids of specific chromosomes (Fig. 5).

Conversely, it has been shown that DNA template strands are not retained in hematopoietic stem cells,[35] suggesting that asymmetric strand segregation is also not a general property of stem cells. It is possible that chromatin differences between sister chromatids do not exist in such cells, bypassing the need to selectively segregate sister chromatids to daughters. Alternatively, epigenetic differences between sister chromatids in these stem cells could be functionally insignificant or lead to stochastic differences in the expression of genes between daughter cells.

By combining sister chromatid information with gene expression data it should be possible to test the hypothesis that epigenetic differences between sister chromatids contribute to differences in gene expression. If the silent sister hypothesis is supported by further evidence, current models explaining stochastic variation in gene expression between cells and models of cell fate determinants in asymmetric cell divisions will have to be adjusted. In this case, broad implications are foreseen. If the silent sister hypothesis is disproven, an important

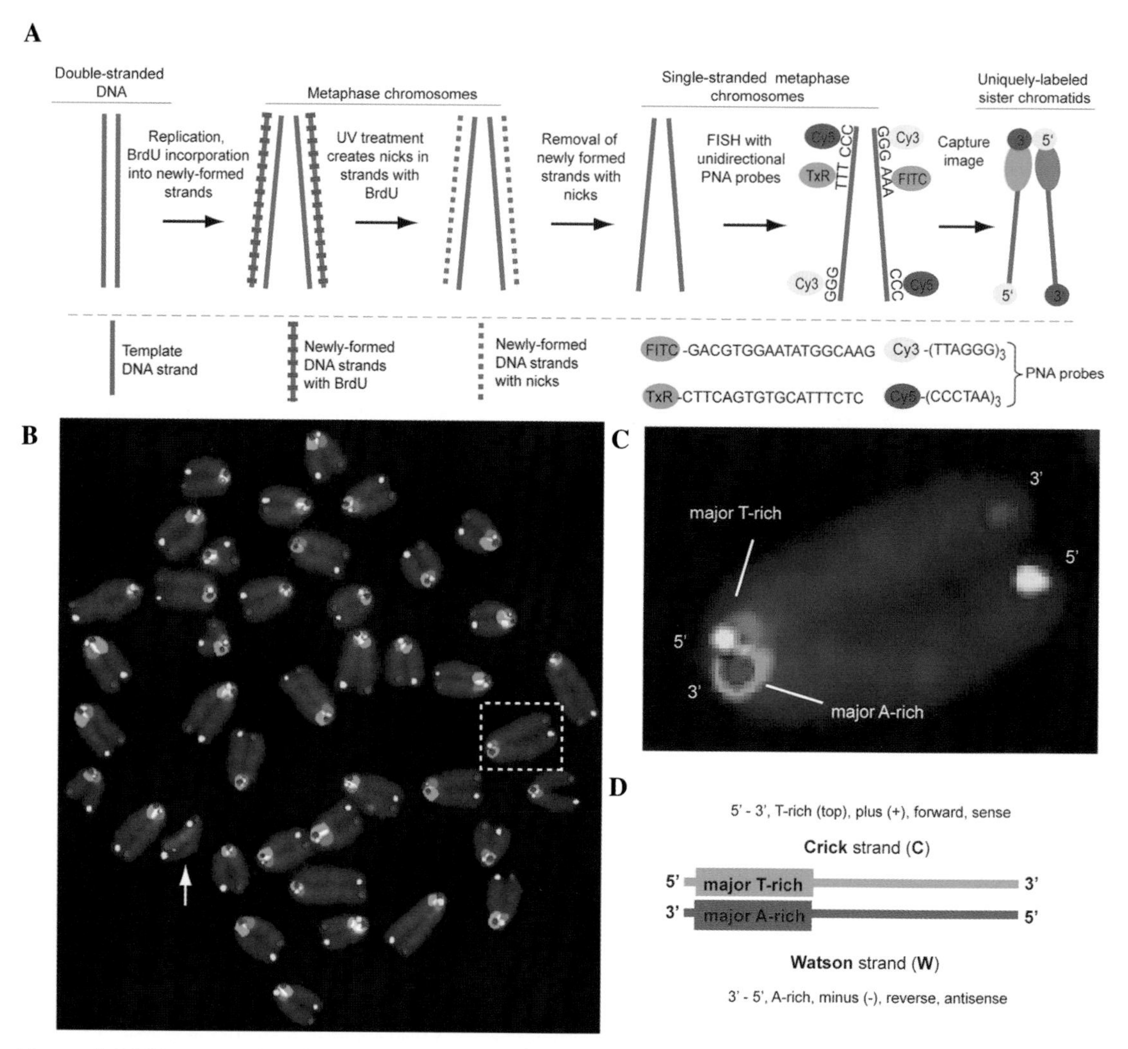

Figure 4. Highly conserved unidirectional orientation of major satellite DNA in murine chromosomes revealed by four color CO-FISH. (A) Schematic diagram of the CO-FISH procedure. (B) Pseudo-color CO-FISH image of murine chromosomes. The orientation of T-rich and A-rich major satellite sequences relative to the 5′ and 3′ end of chromosome ends is preserved in all chromosomes except the Y chromosome (arrow, no major satellite DNA). (C) Magnification of the boxed chromosome shown in (B). (D) Definition of DNA template strands based on the uniform orientation of repetitive DNA. The top strand containing T-rich major satellite DNA is designated as the Crick strand corresponds to the plus (+), 5′ to 3′, forward, sense strand in genome sequence databases. Figure 3 asymmetry at centrosomes[15] could result from differences in the timing or number of microtubules nucleating from each pole or from proteins enriched at a specific pole that can travel along microtubules (Fig. 5A). An example of the latter could be the adenomatous polyposis coli (APC) tumor suppressor protein, given its involvement in chromosome segregation and spindle assembly.[22–26] Distinguishing sister chromatids by a cell likely depends on differences in centromeric or pericentric chromatin. Asymmetry of kinetochore proteins was recently described in yeast,[20] and replication by either leading or lagging strand synthesis of (peri-)centric DNA could lead to asymmetric loading of chromatin proteins at centromeres.[18] Reprinted by permission from Macmillan Publishers (Ref. 21).

contribution to our understanding of biology will nevertheless have been made. Until recently, it was not possible to test the silent sister hypothesis because molecular features to reliably distinguish between sister chromatids had not been described. With our recently published data showing that sister chromatids can be identified using DNA template strand sequences,[21] it should be possible to test the silent sister hypothesis in the next few years.

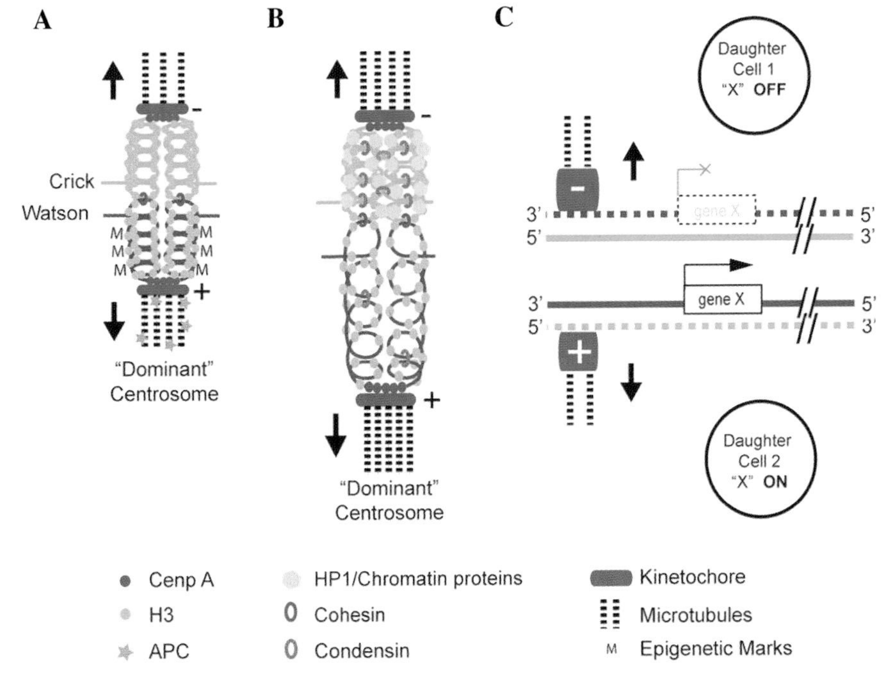

Figure 5. Models for asymmetric segregation of sister chromatids. Only the template strand of double-stranded DNA in sister chromatids prior to mitosis is shown (green and red uninterrupted lines). (A) The unidirectional orientation of repetitive DNA at centromeres could result in uneven distribution of epigenetic marks (M) between sister chromatids that foster nucleation or capture of microtubules coming from the dominant centrosome. (B) Differences in higher-order chromatin structure could alter the elastic properties of (peri-)centric chromatin allowing selection of sister chromatids via microtubules from the dominant centrosome. (C) Regulation of cell fate predicted by the silent sister hypothesis. Alternatively, strand-specific methylation[28] or strand-specific transcription of centromeric DNA[29–31] could help establish chromatin asymmetry at sister chromatid centromeres. Centromeric RNA has been implicated in centromere assembly[32] and opposite strands of major satellite DNA are differentially transcribed in specific cell types during murine development.[33] Such chromatin differences could be recognized by factors from asymmetric spindles or favor selective attachment to microtubules via differences in elastic properties (Fig. 5B).[34] Reprinted by permission from Macmillan Publishers (Ref. 21).

Acknowledgments

We thank members of the Lansdorp lab for helpful discussions. Work in the Lansdorp laboratory in Vancouver is supported by grants from the Canadian Institutes of Health Research (RMF-92093), the U.S. National Institutes of Health (R01GM094146), the Canadian Cancer Society, and the Terry Fox Foundation (Grants 018006 and 105265). Work in the Lansdorp laboratory in Groningen is supported by an Advanced Grant from the ERC.

Conflicts of interest

The authors declare no conflicts of interest.

References

1. Armakolas, A. & A.J. Klar. 2006. Cell type regulates selective segregation of mouse chromosome 7 DNA strands in mitosis. *Science* **311:** 1146–1149.
2. Bell, C.D. 2005. Is mitotic chromatid segregation random? *Histol. Histopathol.* **20:** 1313–1320.
3. Klar, A.J. 1987. Differentiated parental DNA strands confer developmental asymmetry on daughter cells in fission yeast. *Nature* **326:** 466–470.
4. Lansdorp, P.M. 2007. Immortal strands? Give me a break. *Cell* **129:** 1244–1247.
5. Rando, T.A. 2007. The immortal strand hypothesis: segregation and reconstruction. *Cell* **129:** 1239–1243.
6. Stewart, M.D., J. Li & J. Wong. 2005. Relationship between histone H3 lysine 9 methylation, transcription repression, and heterochromatin protein 1 recruitment. *Mol. Cell Biol.* **25:** 2525–2538.
7. Fischle, W. *et al.* 2005. Regulation of HP1-chromatin binding by histone H3 methylation and phosphorylation. *Nature* **438:** 1116–1122.
8. Probst, A.V., E. Dunleavy & G. Almouzni. 2009. Epigenetic inheritance during the cell cycle. *Nat. Rev. Mol. Cell Biol.* **10:** 192–206.
9. Burke, L.J. *et al.* 2005. CTCF binding and higher order chromatin structure of the H19 locus are maintained in mitotic chromatin. *EMBO J.* **24:** 3291–3300.

10. Francis, N.J. *et al.* 2009. Polycomb proteins remain bound to chromatin and DNA during DNA replication in vitro. *Cell* **137:** 110–122.
11. Potten, C.S., G. Owen & D. Booth. 2002. Intestinal stem cells protect their genome by selective segregation of template DNA strands. *J. Cell Sci.* **115:** 2381–2388.
12. Shinin, V. *et al.* 2006. Asymmetric division and cosegregation of template DNA strands in adult muscle satellite cells. *Nat. Cell Biol.* **8:** 677–687.
13. Smith, G.H. 2005. Label-retaining epithelial cells in mouse mammary gland divide asymmetrically and retain their template DNA strands. *Development* **132:** 681–687.
14. Potten, C.S. *et al.* 1978. The segregation of DNA in epithelial stem cells. *Cell* **15:** 899–906.
15. Yamashita, Y.M. *et al.* 2007. Asymmetric inheritance of mother versus daughter centrosome in stem cell division. *Science* **315:** 518–521.
16. Muramoto, T. *et al.* 2010. Methylation of H3K4 is required for inheritance of active transcriptional states. *Curr. Biol.* **20:** 397–406.
17. Laird, C.D. *et al.* 2004. Hairpin-bisulfite PCR: assessing epigenetic methylation patterns on complementary strands of individual DNA molecules. *Proc. Natl. Acad. Sci. U.S.A.* **101:** 204–209.
18. Lew, D.J., D.J. Burke & A. Dutta. 2008. The immortal strand hypothesis: how could it work? *Cell* **133:** 21–23.
19. White, M.A. *et al.* 2008. Non-random segregation of sister chromosomes in Escherichia coli. *Nature* **455:** 1248–1250.
20. Thorpe, P.H., J. Bruno & R. Rothstein. 2009. Kinetochore asymmetry defines a single yeast lineage. *Proc. Natl. Acad. Sci. US.A.* **106:** 6673–6678.
21. Falconer, E. *et al.* 2010. Identification of sister chromatids by DNA template strand sequences. *Nature* **463:** 93–97.
22. Etienne-Manneville, S. & A. Hall. 2003. Cdc42 regulates GSK-3beta and adenomatous polyposis coli to control cell polarity. *Nature* **421:** 753–756.
23. Hanson, C.A. & J.R. Miller. 2005. Non-traditional roles for the Adenomatous Polyposis Coli (APC) tumor suppressor protein. *Gene* **361:** 1–12.
24. Kaplan, K.B. *et al.* 2001. A role for the Adenomatous Polyposis Coli protein in chromosome segregation. *Nat. Cell Biol.* **3:** 429–432.
25. Kita, K. *et al.* 2006. Adenomatous polyposis coli on microtubule plus ends in cell extensions can promote microtubule net growth with or without EB1. *Mol. Biol. Cell* **17:** 2331–2345.
26. Yamashita, Y.M., D.L. Jones & M.T. Fuller. 2003. Orientation of asymmetric stem cell division by the APC tumor suppressor and centrosome. *Science* **301:** 1547–1550.
27. Barker, N. *et al.* 2007. Identification of stem cells in small intestine and colon by marker gene Lgr5. *Nature* **449:** 1003–1007.
28. Luo, S. & D. Preuss. 2003. Strand-biased DNA methylation associated with centromeric regions in Arabidopsis. *Proc. Natl. Acad. Sci. U.S.A.* **100:** 11133–11138.
29. Kanellopoulou, C. *et al.* 2005. Dicer-deficient mouse embryonic stem cells are defective in differentiation and centromeric silencing. *Genes Dev.* **19:** 489–501.
30. Moazed, D. *et al.* 2006. Studies on the mechanism of RNAi-dependent heterochromatin assembly. *Cold Spring Harb. Symp. Quant. Biol.* **71:** 461–471.
31. Murchison, E.P. *et al.* 2005. Characterization of Dicer-deficient murine embryonic stem cells. *Proc. Natl. Acad. Sci. U.S.A.* **102:** 12135–12140.
32. Bouzinba-Segard, H., A. Guais & C. Francastel. 2006. Accumulation of small murine minor satellite transcripts leads to impaired centromeric architecture and function. *Proc. Natl. Acad. Sci. U.S.A.* **103:** 8709–8714.
33. Rudert, F. *et al.* 1995. Transcripts from opposite strands of gamma satellite DNA are differentially expressed during mouse development. *Mamm. Genome* **6:** 76–83.
34. Bouck, D.C. & K. Bloom. 2007. Pericentric chromatin is an elastic component of the mitotic spindle. *Curr. Biol.* **17:** 741–748.
35. Kiel, M.J. *et al.* 2007. Haematopoietic stem cells do not asymmetrically segregate chromosomes or retain BrdU. *Nature* **449:** 238–242.

Ann. N.Y. Acad. Sci. ISSN 0077-8923

ANNALS OF THE NEW YORK ACADEMY OF SCIENCES
Issue: *Hematopoietic Stem Cells VIII*

Reprogramming cell fates: insights from combinatorial approaches

Carlos-Filipe Pereira, Ihor R. Lemischka, and Kateri Moore

Department of Developmental and Regenerative Biology, Black Family Stem Cell Institute, Mount Sinai School of Medicine, New York, New York

Address for correspondence: Carlos-Filipe Pereira and Kateri Moore, Department of Developmental and Regenerative Biology, Mount Sinai School of Medicine, 1 Gustave L. Levy Place, Box 1496, New York, NY 10029. filipe.pereira@mssm.edu and Kateri.Moore@mssm.edu

Epigenetic reprogramming can be achieved in different ways, including nuclear transfer, cell fusion, or the expression of transcription factors (TFs). Combinatorial overexpression provides an opportunity to define the minimal core network of TFs that instructs specific cell fates. This approach has been employed to induce mouse and human pluripotency and differentiated cell types from cells that can be also as distant as cells from different germ layers. This suggests the possibility that any specific cell type may be directly converted into another if the appropriate reprogramming TF core is determined. Herein, we review the factors used for reprogramming multiple cell identities and raise the question of whether there is a common underlying blueprint for reprogramming factors. In addition to the generation of human cell types of interest for cell-replacement therapies, we propose that the TF-mediated conversion of differentiated cell types, especially somatic stem cells, will have an impact on our understanding of their biological development.

Keywords: reprogramming; transcription factor; induced pluripotency; transdetermination; iPS; direct conversion

Introduction

Once committed, the differentiated state of a cell is normally stable and can be inherited through cell division. Under certain conditions, cell fate can, however, be modified or reversed.[1–3] Epigenetic reprogramming can be achieved experimentally in different ways, including through nuclear transfer, cell fusion, or the forced expression of transcription factors (TFs).[4]

The first evidence of successful nuclear reprogramming of somatic cells was provided by the live birth of animals cloned by injection of differentiated somatic nuclei into eggs.[5] This demonstrates that terminally differentiated cells retain cell plasticity and can be reprogrammed to produce an adult cloned animal.[5,6] It remained a formal possibility, however, that the donor nuclei that gave rise to the rarely observed clones were actually derived from tissue stem cells, which represent a small proportion of adult tissues. Subsequently, the successful generation of cloned mice from lymphocytes[7,8] unambiguously demonstrated that terminal differentiation does not restrict the potential of the nucleus to support development. Importantly, this established the principle that mechanisms underlying lineage restriction and cell identity are ultimately reversible.

The combination of two cell types through cell fusion to form somatic cell hybrids and heterokaryons yielded evidence that mammalian gene expression can be altered by diffusible *trans*-acting factors.[9] Cells derived from mesoderm, endoderm, and ectoderm could be dominantly reprogrammed into mouse myotubes that express muscle specific genes.[10] Cells that were more closely related to muscle—mesodermal derivatives—consistently expressed muscle genes sooner and to a greater extent than the cells of ectodermal or endodermal origin.[10] These studies established that differentiated cells do retain flexible lineage potential and that lineage conversion and gene activation can occur in absence of DNA replication.[11] The dominance of pluripotent cells over differentiated cells has also been shown in experimental hybrids and heterokaryons

doi: 10.1111/j.1749-6632.2012.06508.x

made between somatic cells and embryonic stem cells (ESCs).[12–14] The *trans*-acting factors mediating dominant reprogramming may reside in the nucleus or in the cytoplasm. This question was initially addressed by separating the nuclear compartment (karyoplast) from the cytoplasmic compartment (cytoplast) of an ESC; these elements were then individually fused with neuronal cells.[15] Only karyoplasts were able to reprogram upon cell fusion, suggesting that nuclear factors are essential for reprogramming. This conclusion is consistent with cloning experiments in amphibians[16] and mice,[17] which indicate that successful reprogramming depends on direct injection of nuclei into the germinal vesicle or into a metaphase oocyte, where nuclear factors are available in the cytoplasm. Interestingly, the overexpression of a single TF in somatic cells was unexpectedly found to modify or override lineage outcome, first in *Drosophila melanogaster*[18] and subsequently in mammals,[19] leading to remarkable changes in cell fate.

In 2006, Takahashi and Yamanaka reported the breakthrough discovery that a combination of defined TFs was sufficient to reprogram several somatic cell types to pluripotency (termed "iPS" for induced pluripotent stem).[20–22] In the initial study,[20] the authors sought to identify genes expressed in ESCs that would be sufficient to induce pluripotency. With a pioneer approach, 24 candidate genes were simultaneously expressed in fibroblasts using retroviral transduction. This resulted in the reprogramming of a small percentage of fibroblasts toward iPS cells. Genes were then systematically removed from the cocktail to define the minimal and optimal combinations of factors for inducing pluripotency. The same rationale has been more recently applied to induce differentiated cell types from cells that can be as distant as cells from different germ layers. Reprogrammed cells, by direct lineage conversion, include macrophages, brown fat cells, cardiomyocytes, neurons, hepatocytes, and beta cells.[23–28] Cell fate conversion to each of these cell types was accomplished by expression of a signature group of two to five TFs in unrelated cell types. Recently, the expression of miRNAs was also shown to either reprogram cells to pluripotency[29] or to assist in direct conversion toward the neural lineage in combination with TFs.[30] Because miRNAs generally target hundreds of mRNAs that coordinate expression of many different proteins, the underlying mechanism may involve the indirect activation of TFs. Collectively, these findings opened an avenue for combinatorial overexpression screening to define the minimal TF core required to induce any cell type of interest. Can this approach be applied to induce somatic stem cells? Is the induction of multipotency fundamentally different from direct lineage conversions? Do TFs share similar biochemical properties that confer the ability to trigger cellular reprogramming? Here, we review these questions with a focus on the *reprogramming factors*, TFs that are sufficient to instruct cell type conversion when overexpressed in unrelated somatic cells. We compiled data from TF overexpression studies that lead to induction of cell types and analyzed the TFs according to several features that may facilitate the selection of factors to induce additional cell fates. The establishment of cell type–specific TF minimal networks will make possible the direct generation of patient-specific cell types of interest for cell replacement or disease modeling. In addition, it also provides a novel approach to interrogate aspects of cell fate, stem cell function, and cellular differentiation.

Reprogramming cell fates by combinatorial expression of transcription factors

The induction of pluripotent stem cells was successfully achieved from mouse embryonic and adult fibroblasts after viral-mediated transduction of *Oct4*, *Sox2*, *c-Myc*, and *Klf4*. Importantly, iPS cells generated postnatal chimeras, contributed to the germ line,[31–33] and generated late-gestation embryos through tetraploid complementation[31] (the most stringent test for developmental potency). In addition, iPS cells were shown to be transcriptionally and epigenetically similar to ES cells. iPS cells showed reactivation of the somatically silenced X-chromosome, DNA demethylation of pluripotency-associated genes, and acquisition of chromatin structure that closely resemble ES cells.[31–33] Expression of the reprogramming factors in fibroblasts appears to initiate a cascade of events that leads to the conversion of a small percentage of cells. Similar to what is observed in nuclear transfer,[34] iPS reprogramming appears to be influenced by donor cell type. The same four TFs have been shown to drive mouse hepatocytes, stomach cells,[21] pancreatic beta cells,[35] and immature B lymphocytes to iPS, whereas adult mouse B lymphocytes[36] required silencing of the B cell–specific TF *Pax5* or the additional expression of *CEBPα*. Conversely, just two factors (*Oct4* and *Klf4*) or one factor (*Oct4*) were sufficient to reprogram adult neural stem cells and dermal papilla cells, which endogenously express *Sox2* and *c-Myc*.[37–41] In 2007, reprogramming of human fibroblasts was achieved in parallel approaches using *Oct4*, *Sox2*, and either *Nanog* plus *Lin28*[42] or *Klf4* plus *c-Myc*.[22,43,44] Lin28 is the only protein of the combination that is not a TF. Human cell types converted to iPS include fetal fibroblasts, adult dermal fibroblasts, bone marrow–derived mesenchymal cells, and keratinocytes. In order to reprogram keratinocytes a combination of five factors (*Oct4, Sox2, Klf4, c-Myc,* and *Nanog*) was used.[45]

Several combinations of factors have been used for iPS induction in mice and humans (Table 1, upper left panel), affecting mainly the efficiency and rapidity of reprogramming (reviewed in Ref. 46). iPS colonies were derived from mouse fibroblasts using the combination of *Oct4*, *Sox2*, and *Klf4*.[47] Similarly, human iPS can be generated with either *Oct4*, *Sox2*, and *Klf4* (no *c-Myc*) or *Oct4*, *Sox2*, and *Nanog* (no *Lin28*).[42,47] TFs that belong to the same family and are closely related can replace the function of some of the reprogramming factors; *Sox2* can be replaced by *Sox1* and *Sox3*, *Klf4* by *Klf2* and *Klf5*, and *c-Myc* by *N-Myc* and *L-Myc*.[42,47,48] TFs that belong to other families can also replace reprogramming factors, for example, the orphan nuclear receptor *Esrrb* can replace the function of both *Klf4* and *c-Myc*.[49] Interestingly, *Oct4* cannot be replaced by the closest homologues *Oct1* or *Oct6*.[47] Indeed, the only factor that has been reported to replace *Oct4* function in reprogramming is the nuclear receptor *Nr5a2*, previously shown to regulate *Oct4* levels.[50,51] These observations suggest that *Oct4* occupies an irreplaceable, yet mechanistically unidentified, role in induced pluripotency. This is supported by the observation that ESCs conditionally depleted for *Oct4* (before differentiation or loss of other reprogramming factors) lose dominant reprogramming capacity in heterokaryons, in contrast to *Sox2*.[13] Chromatin remodeling factors (*Brg1*, *Baf155*),[52] histone demethylation (*Jhdm1a/1b*),[53] nuclear receptors (*RAR-gamma*, *Nr5a2*),[54] additional TFs (*Tbx3*, *Glis1*, *Prdm14*),[55–57] and co-activators (*Utf1*),[58] have been shown to increase in the efficiency and rapidity of reprogramming when added in combination with *Oct4*, *Sox2*, *Klf4* (± *c-Myc*). Similar outcomes were achieved by ablation of factors that impede reprogramming: silencing of p53 and other regulators of senescence (acting as a major barrier to reprogramming by limiting cell cycle and inducing apoptosis[59]), repression of lineage-specifying TFs and inhibiting DNA methylation activity.[60,61] Together these reprogramming experiments suggest that iPS-inducing factors can be separated into two groups: core reprogramming factors that play a central role for induced pluripotency, including irreplaceable TFs (*Oct4*) and necessary TFs but interchangeable for closely related or functionally equivalent factors (*Sox2*, *Klf4*, *Nanog*, and *Esrrb*); and pro-programming factors,[57] which facilitate the reprogramming process resulting in increased efficiency and speed. The latter set can act on multiple cellular pathways such as cell cycle, chromatin remodeling, DNA demethylation, and mesenchymal-to-epithelial transition.[62–64]

It is also possible to induce cells directly into other cell fates using the appropriate factors (Table 1), a process that has been named transdifferentiation or direct conversion.[65] Conversion of fibroblasts into myoblasts by expressing *MyoD* was first demonstrated in mammals in 1987 by Davis and Weintraub.[19] However, ectopic expression of *MyoD* in ectoderm lineages *in vivo*[66] or in hepatocytes *in vitro*[67] does not initiate the complete conversion of these lineages into differentiated muscle. This suggests that *MyoD*, although a central player in the skeletal muscle program, needs cooperation from additional, yet unidentified factors to convert more distant cell types. Similarly, the TFs *CEBPα* and *CEBPβ* can induce committed mature B lymphocytes to become macrophages,[68] but an additional factor (*PU.1*) is required to convert fibroblasts.[23] Forced coexpression of *Prdm16* and *CEBPβ* was sufficient to convert mouse and human dermal fibroblasts into functional brown fat-like cells.[24]

More recently, the combinatorial overexpression approach was used to screen for induction of cell types emanating from different germ layers. After testing an initial pool of 19 neuronal lineage-specific TFs, a combination of three factors (*Ascl1*, *Brn2*, and *Mytl1*) was sufficient to induce neurons from fibroblasts, with almost 20% efficiency in less than one week.[26] The conversion of human fibroblasts to fully functional neurons appears to be greatly aided by the helix-loop-helix TFs *NeuroD1* and

Table 1. Reprogramming factors and their features

Factors to induce:	Mouse	Human	SS-TF	TF family/domains	Loss of function	Cell proliferation	Pioneer activity
Pluripotency	20 20 45 49 52 57 53 51 54 55	22 22 42 42 57 58 56					
Oct-4			Yes	Pou/Homeodomain	+	-	ND
Sox2			Yes	HMG-box/SOXp	+	-	ND
Klf4			Yes	Zn-finger	-*	+	ND
c-Myc			Yes	Helix-loop-helix/ Myc-N/Myc-LZ	+	+	ND
Esrrb			Yes	NR Zn-finger	+	-	ND
Nanog			Yes	Homeodomain	+	-	ND
Lin28			No	RNA-binding	-	-	ND
Tbx3			Yes	T-box	+	+	ND
Nr5a2			Yes	NR Zn-finger	+	+	ND
RAR-gamma			Yes	NR Zn-finger	-	-	ND
Utf1			No	Co-activator	-	-	ND
Brg1			No	Chromatin	+	-	ND
Baf155			No	remodelling		-	ND
Glis1			Yes	Zn-finger	ND	-	ND
Jhdm1a/1b			No	Histone demethylase	ND	-	ND
Prdm14			Yes	Zn-finger, SET	+	-	ND
Myocytes	19	97					
MyoD			Yes	Helix-loop-helix	-*	-	ND
Cardiomyocytes	25						
Tbx5		ND	Yes	T-box	+	-	ND
Mef2c			Yes	MADS	+	-	ND
Gata4			Yes	Zn-finger, Gata-type	+	-	ND
Neurons	26 70	69 71 70					
Ascl1	G D	G D D	Yes	Helix-loop-helix	+	-	ND
Brn2			Yes	Pou/Homeodomain	+	-	ND
Myt1l			Yes	Zn-finger	ND	-	ND
NeuroD1/2			Yes	Helix-loop-helix	+	-	ND
Lmx1a			Yes	Homeodomain	+	-	ND
Nurr1			Yes	NR Zn-finger	+	-	ND
Foxa2			Yes	Forkhead	+	-	+
Hepatocytes	27 72						
Gata4			Yes	Zn-finger, Gata-type	+	-	+
Hnf1α			Yes	Homeobox	+	-	ND
Hnf4α		ND	Yes	NR Zn-finger	+	-	ND
Foxa3			Yes	Forkhead	-*	-	+
Foxa2			Yes	Forkhead	+*	-	+
Foxa1			Yes	Forkhead		-	+
Macrophages	23, 68						
PU.1		ND	Yes	Ets	+	-	+
CEBPα/β			Yes	bZIP	+	-	ND
Brown fat cells	24	24					
Prdm16			Yes	Zn-finger, SET	+	-	ND
CEBPβ			Yes	bZIP	+	-	ND
Beta cells	28 28						
Ngn3			Yes	Helix-loop-helix	+	-	ND
NeuroD1		ND	Yes	Helix-loop-helix	+	-	ND
Pdx1			Yes	Homeobox	+	-	ND
Mafa			Yes	bZIP	+	-	ND

TF – Transcription factor. SS-TF – Sequence-specific transcription factor. NR – nuclear receptor. ND – not determined. G – Glutamatergic neurons. D – Dopaminergic neurons. *Reported compensatory effects by homologous TFs.

NeuroD2.[69] Two recent studies have reported that the addition of TFs that are involved in dopaminergic neuronal development can generate dopaminergic neurons from human and mouse fibroblasts that express tyrosine hydroxylase and release dopamine. Each study used a distinct group of TFs for conversion (*Ascl1, Brn2, Myt1l* + *Lmx1a,* and *Foxa2* versus *Ascl1, Nurr1,* and *Lmx1a*).[70,71] Interestingly, transduction of *Ascl1* alone was sufficient to induce some neural traits in fibroblasts, such as the expression of pan-neuronal proteins. This gene was present in all neural reprogramming cocktails and may be analogous to *Oct4* for the induction of pluripotency and *MyoD* for the induction of muscle. Leda *et al.* used an iterative process of elimination to define a minimal pool of three cardiac-specific TFs (*Gata4, Mef2c, Tbx5*) that directly induced mouse fibroblast transdifferentiation into cardiomyocytes.[25] Similar to the induction of neurons, the conversion of cardiomyocytes was rapid (one week) and efficient (>6% αMHC + cTnT + cells). Recently, two groups have demonstrated the conversion of mouse fibroblasts to hepatocyte-like cells by forced expression of *Gata4*, *Hnf1a*, and *Foxa3* and inactivation of *P19*^{Arf} or by expression of *Hnf4α* and *Foxa1*, *Foxa2*, or *Foxa3*.[27,72] Finally using a strategy of re-expressing key developmental regulators *in vivo*, Melton and colleagues identified a specific combination of three TFs (*Ngn3*, *Pdx1*, and *Mafa*) that converts differentiated pancreatic exocrine cells in adult mice into cells that closely resemble beta cells.[28] Interestingly, attempts to reprogram skeletal muscle and embryonic fibroblasts (mesodermal cell types) using this combination were unsuccessful.

In addition to these direct reprogramming studies, some reports suggest that overexpression of pluripotency-reprogramming factors in permissive culture conditions can generate cell types other than iPS. For example, the induction of hematopoietic progenitors was achieved by overexpression of *Oct4* in the presence of hematopoietic cytokines[73] and the induction of cardiomyocytes and neural progenitors by the induction of the *Oct4*, *Sox2*, *Klf4*, and *c-Myc*.[74,75] These results are unexpected because the generated cell types do not express the TFs used for induction. Future experiments may address the mechanisms underlying these reprogramming events that may involve transit through a partially reprogrammed state further instructed with culture conditions to differentiate into specific cell types.

In summary, seven cell types representative of the three germ layers were induced by overexpression of TF cocktails. With this approach, master regulators of cell fate induction have been allocated to these cell types. Reassuringly, the majority of these genes are essential for the generation of the target cell types during development. It remains an open question whether there are common features between reprogramming-competent TFs. Therefore, it is important to pay careful attention to those proteins and their molecular function.

Transcription factors with reprogramming abilities—are there common features?

Diverse arrays of proteins are crucial for successful transcription by RNA polymerase in eukaryotic cells. These proteins include general TFs, cofactors (coactivators and corepressors) and chromatin modifying proteins. In addition, sequence-specific DNA-binding TFs direct transcription initiation to specific promoters. The vast majority (>80%) of the reported reprogramming factors (Table 1) are annotated as sequence-specific TFs.[76] The exceptions include the RNA-binding protein Lin28, the chromatin remodelers Brg1 and Baf155, the histone demethylases Jhmd1a/1b and the coactivator Utf1. Interestingly, the exclusion of those factors from the reprogramming combinations only reduces the efficiency but does not abolish conversion, placing those genes in the pro-programming group and not in the minimal core of TFs to instruct cellular identity. DNA methyltransferases, chromatin remodelers, and chromatin modifiers were included in the initial pool of several combinatorial screens,[20,25,27] but none of those molecules were incorporated in the minimal combination to induce cell fates. Chromatin remodeling and DNA demethylation was shown to be required for converting the epigenome of one cell type to another.[60,61,77] However, with combinatorial overexpression approaches these molecules would lack the capacity to instruct. We suggest that factors with reprogramming abilities have proven or predicted sequence-specific DNA-binding TF activity.

There are currently 1,866 human proteins predicted to have sequence-specific TF activity in the gene ontology database.[76] A bioinformatic analysis of the TFs of the human genome[78] showed that in each tissue 150–300 TFs are expressed, and only a proportion (<35%) show a high degree of tissue-specificity (high expression restricted to up to three tissues). We used the publicly available BIOGPS database (www.biogps.org; GeneAtlas MOE430) to assess the expression profile of the known reprogramming factors across many tissues and cell lines (Fig. 1). The available gene expression data for the reprogramming factors to induce pluripotency, myoblasts, cardiomyocytes, neurons, hepatocytes, and macrophages were extracted from the database and hierarchical clustering was performed. The pattern of gene expression of these genes was sufficient to clearly cluster ESCs, muscle, heart, brain, liver, and macrophages. Interestingly, reprogramming factors show very restricted expression patterns in the cell type/tissue that they induce, especially *Oct4*, *MyoD*, *Tbx5*, and *Ascl1*. This implies that in addition to the comparison of the TF-transcriptome in the initial and target cell types, including expression in other cell types from other tissues may greatly increase the specificity of the analysis. Online databases such as BIOGPS or Unigene Differential Digital Display (www.ncbi.nlm.nih.gov/UniGene/ddd.cgi)

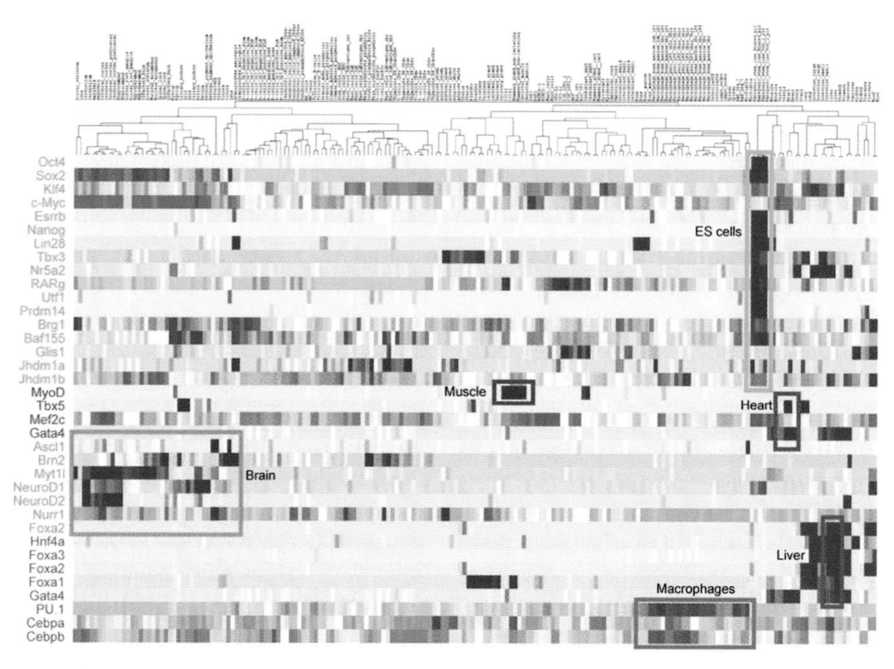

Figure 1. Heatmap representation of reprogramming factor expression across multiple tissues. Heatmap showing the gene expression of transcription factors (rows) in multiple tissues and cell lines (columns). Intersecting cells are shaded according to expression level (dark red for high expression and blue for low expression relative to the mean). Tissues are grouped according to hierarchical clustering based on the gene expression of 35 transcription factors. The boxes highlight high expression of reprogramming factors in induced tissue/cell types and the lack of broad expression on other tissues. Green, induced pluripotency; blue, induced myoblasts; gray, induced cardiomyocytes; brown, induced neurons; purple, induced hepatocytes; pink, induced macrophages. These data were extracted from the public online database BIOGPS (www.biogps.org; GeneAtlas MOE430).

provide good tools for broad gene expression comparisons.[20] This kind of analysis may aid the identification of lineage-restricted reprogramming factors.

The implication that a small set of lineage-restricted sequence-specific TFs will instruct cell fates predicts that: loss of function of those genes would affect the generation of the cell type during development, and gene expression may be tightly controlled in other cell types. We used the MGI database (www.informatics.jax.org) and PubMed (http://www.ncbi.nlm.nih.gov/pubmed/) to interrogate the loss of function phenotype (knockout or knockdown approaches) of the reprogramming factor deletion during tissue development/target cell type (Table 1). Interestingly the majority of the factors used for reprogramming have an impact either on the generation or maintenance of the target cell type (Table 1). The exceptions include factors whose function is compensated by homologous TFs on the *Klf*, *MyoD,* and the *Foxa* gene-families.[79] For example, *Klf4* depletion does not affect ESC self-renewal; however, simultaneous depletion of *Klf2*, *Klf4,* and *Klf5* leads to ESC differentiation.[80] Similarly *MyoD*-null mice are viable,[81] but when *Myf5*, another helix-loop-helix TF is deleted in combination with *MyoD*, there is a total absence of muscle fibers and precursor myoblasts.[82] In summary, reprogramming factors may lie at the intersection of sequence-specific TFs, factors that influence the generation of a target cell type during development, and tissue/cell type-restricted gene expression. These criteria can be used to rank candidates for combinatorial overexpression screens to determine TF core networks to induce cell fates. Further experimentation will be required to address the mechanisms of silencing of those genes in unrelated cell types and how they may differ from other TFs. In this regard, *Ascl1* is controlled not only by repressive histone modifications in ESCs but also replicates late during S-phase and associates with the nuclear periphery (temporal and spatial constraints associated with gene repression). Upon neural differentiation the *Ascl1* gene shifts to replicate early in S-phase and translocates to the nuclear interior where it is expressed.[83]

It is intriguing that some TFs are included in several different lineage-specific cocktails (Table 1). These factors include *NeuroD1* (neurons and beta-cells), *CEBPβ* (Brown fat cells and macrophages), *Gata4* (cardiomyocytes and hepatocytes), and *Foxa2* (hepatocytes and neurons). *Glis1* also enhanced the expression of *Foxa2*, indicating that this TF may also have a role in the promotion of iPS cell generation possibly by antagonizing the epithelial to mesenchymal transition.[57] This suggests that a small number of factors among the horde of sequence-specific TFs may have unique reprogramming features. For example, reprogramming factors may act as "pioneer factors," TFs found to be able to access their DNA target sites in silent chromatin when other factors cannot and continue to access the DNA prior to the time of transcriptional activation (reviewed by Refs. 84 and 85). Among the list of reprogramming factors (Table 1), five factors (*Gata4*, *Foxa1*, *Foxa2*, *Foxa3*, *PU.1*) have been implicated as having pioneer activity. In undifferentiated endoderm cells, the Foxa and Gata TFs are among the first to engage silent genes, such as the *Alb1* gene, helping to endow competence for cell type specification. Foxa proteins can bind their target sites in highly compacted chromatin and open up the local region for other factors to bind. In addition, Foxa1 protein remains bound to chromatin during mitosis.[86] PU.1 itself can expand the linker region between nucleosomes and promote local histone modifications, likely contributing to its ability to enhance binding of other factors.[87,88] It would be interesting to test the pioneer TF model and attempt to reprogram fibroblasts to beta cells by incorporating *Gata4*/*Foxa2* in the reported TF-core of *Pdx1*, *Mafa*, and *Ngn3*/*NeuroD1*.

Future experiments will be required to address whether there is a common underlying blueprint for reprogramming factors. Genome-wide location studies, *in vivo* footprinting, and determining their interaction partners may begin to shed light on the molecular function of these genes throughout various stages of development. This will also allow us to start addressing the mechanistic logic of TF-induced reprogramming by building the network of regulatory interactions. An additional important question is whether inducing pluripotent stem cells is fundamentally a different process from direct lineage conversion. Expanding combinatorial approaches to reprogram other cell identities, including somatic stem cells, will be crucial to define the nature of reprogramming.

Features of reprogramming toward pluripotency and somatic cells—perspectives to induce somatic stem cells

A comparison of the iPS cell reprogramming and direct lineage conversions reveal several distinct

Table 2. Reprogramming to pluripotency, somatic cells, and potentially somatic stem cells

	iPS cell reprogramming	Direct conversion of somatic cells	Somatic stem cell reprogramming
Target cell–type features			
Pluripotency	+	–	–
Multipotency	–	–	+
Self-renewal	+	–	+
Reprogramming features			
Reprogramming factors that regulate cell cycle	+	–	?
Reprogramming efficiency	Low	High	?
Reprogramming dynamics	Slow	Fast	?
Potential tumor risk	Low	High	?
Target cell generation	Difficult[a]	Easy	?
Cell scaling	Feasible	Limited	?
Gene correction	Feasible	Limited	?

[a]Protocols are lacking to efficiently generate some cell types, for example, hematopoietic stem cells.

features, although both involve TFs to trigger reprogramming (Table 2). Somatic lineage conversions are very rapid, with the first lineage reporters of the target cell type expressed just days after gene induction.[25,26] Inducing iPS cells typically takes 10–20 days and occurs with much lower efficiency.[20,22] Despite the early initiation of conversion, the activation of markers of mature cells, and the acquisition of functional properties appears to be delayed and to continue for several weeks; a process analogous to differentiation. Nonetheless, lineage conversion seems to take place without the generation of a progenitor cell type.[25,26,89] TF-mediated lineage reprogramming seems to be a direct phenotypic induction and not dedifferentiation followed by differentiation to an alternative fate.

Some direct cell fate conversion examples do not require cell division.[26,90] Conversely, cell proliferation increases the efficiency of generation of iPS,[91] and the reprogramming process is significantly delayed and less efficient in the absence of the *c-Myc* oncogene. We interrogated whether those differences are translated in the properties of TFs used for reprogramming, using gene ontology[76] to address their molecular function (Table 1). Interestingly, the factors that have been implicated in the control of the cell cycle or inducing proliferation are restricted to the induction of pluripotency (Tables 1 and 2). This can be either a feature of induced pluripotency or a broader determinant of self-renewal. The latter may have implications on future attempts to induce somatic stem cells, which also self-renew (Table 2). If this prediction is true, we would expect that to induce somatic stem cells, such as hematopoietic stem cells (HSCs), additional factors to induce self-renewal would be required in addition to the lineage-restricted reprogramming factors. On the other hand, triggering senescence/cell-cycle arrest has been shown to be a major barrier for iPS-reprogramming and also for the direct conversion to hepatocytes.[27,59]

One of the most pressing medical objectives is to obtain patient-specific cell types for cell replacement therapy. Although iPS cells have the clear advantage of unlimited growth and scalability, the potential risk for tumor formation is high. Directly converted somatic cells would be presumably less tumorigenic, but obtaining large numbers of cells and gene correction strategies will be challenging (Table 2). In addition, the differentiation of human ESCs and iPS cells is highly variable, cell-line dependent, and generates immature cells that differ from those found in mature organs *in vivo*.[92] Using the current protocols for hematopoietic differentiation, ESCs/iPS will readily give rise to differentiated hematopoietic cells as well as colony-forming cells.[93,94] However, HSCs capable of repopulating the hematopoietic system of lethally irradiated adult recipients are not generated. In this regard, differentiation recapitulates the development of the earliest embryonic hematopoietic tissue, the yolk sac.[95,96] Direct reprogramming provides the opportunity to induce human adult cells directly, possibly rendering homogenous populations. Inducing HSCs and other adult stem cells will potentially have the advantage to generate cells that self-renew, but since they are multipotent, their differentiation will be more straightforward. It remains an important question whether adult stem cells that retain self-renewal and multipotency properties are programmable.

To fully realize the potential of *in vitro* reprogrammed cells, we need to understand the molecular and epigenetic determinants that convert one cell type into another. Systematically defining the constellation of reprogramming factors provides the opportunity to allocate a master regulatory network for each cell type. The more the reprogramming factor combination resembles the target cell type regulatory network, the more efficient and accurate reprogramming would be. Taken together, these approaches will not only provide insights into development and disease but also a source of material to study and apply to both.

Conflicts of interest

The authors declare no conflicts of interest.

References

1. Simonsson, S. & J.B. Gurdon. 2005. Changing cell fate by nuclear reprogramming. *Cell Cycle* **4:** 513–515.
2. DiBerardino, M.A. 1988. Genomic multipotentiality of differentiated somatic cells. *Cell Differ. Dev.* **25**(Suppl): 129–136.
3. Surani, M.A. 2001. Reprogramming of genome function through epigenetic inheritance. *Nature* **414:** 122–128.
4. Yamanaka, S. & H.M. Blau. 2010. Nuclear reprogramming to a pluripotent state by three approaches. *Nature* **465:** 704–712.
5. Gurdon, J.B. 1962. Adult frogs derived from the nuclei of single somatic cells. *Dev. Biol.* **4:** 256–273.

6. Wilmut, I. *et al.* 1997. Viable offspring derived from fetal and adult mammalian cells. *Nature* **385:** 810–813.
7. Hochedlinger, K. & R. Jaenisch. 2002. Monoclonal mice generated by nuclear transfer from mature B and T donor cells. *Nature* **415:** 1035–1038.
8. Inoue, K. *et al.* 2005. Generation of cloned mice by direct nuclear transfer from natural killer T cells. *Curr. Biol.* **15:** 1114–1118.
9. Harris, H. *et al.* 1969. Suppression of malignancy by cell fusion. *Nature* **223:** 363–368.
10. Blau, H.M. *et al.* 1985. Plasticity of the differentiated state. *Science* **230:** 758–766.
11. Chiu, C.P. & H.M. Blau. 1984. Reprogramming cell differentiation in the absence of DNA synthesis. *Cell* **37:** 879–887.
12. Tada, M. *et al.* 2001. Nuclear reprogramming of somatic cells by in vitro hybridization with ES cells. *Curr. Biol.* **11:** 1553–1558.
13. Pereira, C.F. *et al.* 2008. Heterokaryon-based reprogramming of human B lymphocytes for pluripotency requires Oct4 but not Sox2. *PLoS Genet.* **4:** e1000170.
14. Cowan, C.A. *et al.* 2005. Nuclear reprogramming of somatic cells after fusion with human embryonic stem cells. *Science* **309:** 1369–1373.
15. Do, J.T. & H.R. Scholer. 2004. Nuclei of embryonic stem cells reprogram somatic cells. *Stem Cells* **22:** 941–949.
16. Byrne, J.A. *et al.* 2003. Nuclei of adult mammalian somatic cells are directly reprogrammed to oct-4 stem cell gene expression by amphibian oocytes. *Curr. Biol.* **13:** 1206–1213.
17. Wakayama, T. *et al.* 1998. Full-term development of mice from enucleated oocytes injected with cumulus cell nuclei. *Nature* **394:** 369–374.
18. Schneuwly, S., R. Klemenz & W.J. Gehring. 1987. Redesigning the body plan of Drosophila by ectopic expression of the homoeotic gene Antennapedia. *Nature* **325:** 816–818.
19. Davis, R.L., H. Weintraub & A.B. Lassar. 1987. Expression of a single transfected cDNA converts fibroblasts to myoblasts. *Cell* **51:** 987–1000.
20. Takahashi, K. & S. Yamanaka. 2006. Induction of pluripotent stem cells from mouse embryonic and adult fibroblast cultures by defined factors. *Cell* **126:** 663–676.
21. Aoi, T. *et al.* 2008. Generation of pluripotent stem cells from adult mouse liver and stomach cells. *Science* 321: 699–702, doi: 10.1126/science.1154884.
22. Takahashi, K. *et al.* 2007. Induction of pluripotent stem cells from adult human fibroblasts by defined factors. *Cell* **131:** 861–872.
23. Feng, R. *et al.* 2008. PU.1 and C/EBPalpha/beta convert fibroblasts into macrophage-like cells. *Proc. Natl. Acad. Sci. USA* **105:** 6057–6062.
24. Kajimura, S. *et al.* 2009. Initiation of myoblast to brown fat switch by a PRDM16-C/EBP-beta transcriptional complex. *Nature* **460:** 1154–1158.
25. Ieda, M. *et al.* 2010. Direct reprogramming of fibroblasts into functional cardiomyocytes by defined factors. *Cell* **142:** 375–386.
26. Vierbuchen, T. *et al.* 2010. Direct conversion of fibroblasts to functional neurons by defined factors. *Nature* **463:** 1035–1041.
27. Huang, P. *et al.* 2011. Induction of functional hepatocyte-like cells from mouse fibroblasts by defined factors. *Nature* **475:** 386–389.
28. Zhou, Q. *et al.* 2008. In vivo reprogramming of adult pancreatic exocrine cells to beta-cells. *Nature* **455:** 627–632.
29. Anokye-Danso, F. *et al.* 2011. Highly efficient miRNA-mediated reprogramming of mouse and human somatic cells to pluripotency. *Cell Stem Cell* **8:** 376–388.
30. Ambasudhan, R. *et al.* 2011. Direct reprogramming of adult human fibroblasts to functional neurons under defined conditions. *Cell Stem Cell* **9:** 113–118.
31. Wernig, M. *et al.* 2007. In vitro reprogramming of fibroblasts into a pluripotent ES-cell-like state. *Nature* **448:** 318–324.
32. Maherali, N. *et al.* 2007. Directly reprogrammed fibroblasts show global epigenetic remodeling and widespread tissue contribution. *Cell Stem Cell* **1:** 55–70.
33. Okita, K., T. Ichisaka & S. Yamanaka. 2007. Generation of germline-competent induced pluripotent stem cells. *Nature* **448:** 313–317.
34. Gurdon, J.B. & J.A. Byrne. 2003. The first half-century of nuclear transplantation. *Proc. Natl. Acad. Sci. USA* **100:** 8048–8052.
35. Stadtfeld, M., K. Brennand & K. Hochedlinger. 2008. Reprogramming of pancreatic Beta cells into induced pluripotent stem cells. *Curr. Biol.* **18:** 890–894.
36. Hanna, J. *et al.* 2008. Direct reprogramming of terminally differentiated mature B lymphocytes to pluripotency. *Cell* **133:** 250–264.
37. Kim, J.B. *et al.* 2008. Pluripotent stem cells induced from adult neural stem cells by reprogramming with two factors. *Nature* **454:** 646–650.
38. Eminli, S. *et al.* 2008. Reprogramming of neural progenitor cells into iPS cells in the absence of exogenous Sox2 expression. *Stem Cells* **26:** 2467–2474.
39. Di Stefano, B., A. Prigione & V. Broccoli. 2008. Efficient genetic reprogramming of unmodified somatic neural progenitors uncovers the essential requirement of Oct4 and Klf4. *Stem Cells Dev.* **18:** 707–716.
40. Tsai, S.Y. *et al.* 2011. Single transcription factor reprogramming of hair follicle dermal papilla cells to induced pluripotent stem cells. *Stem Cells* **29:** 964–971.
41. Tsai, S.Y. *et al.* 2010. Oct4 and klf4 reprogram dermal papilla cells into induced pluripotent stem cells *Stem Cells.* **28:** 221–228.
42. Yu, J. *et al.* 2007. Induced pluripotent stem cell lines derived from human somatic cells. *Science* **318:** 1917–1920.
43. Park, I.H. *et al.* 2008. Reprogramming of human somatic cells to pluripotency with defined factors. *Nature* **451:** 141–146.
44. Lowry, W.E. *et al.* 2008. Generation of human induced pluripotent stem cells from dermal fibroblasts. *Proc. Natl. Acad. Sci. USA* **105:** 2883–2888.
45. Maherali, N. *et al.* 2008. A high-efficiency system for the generation and study of human induced pluripotent stem cells. *Cell Stem Cell* **3:** 340–345.
46. Sterneckert, J., S. Hoing & H.R. Scholer. 2011. Concise review: oct4 and more: the reprogramming expressway. *Stem Cells* **30:** 15–21.

47. Nakagawa, M. *et al.* 2008. Generation of induced pluripotent stem cells without Myc from mouse and human fibroblasts. *Nat. Biotechnol.* **26:** 101–106.
48. Wernig, M. *et al.* 2008. c-Myc is dispensable for direct reprogramming of mouse fibroblasts. *Cell Stem Cell* **2:** 10–12.
49. Feng, B. *et al.* 2009. Reprogramming of fibroblasts into induced pluripotent stem cells with orphan nuclear receptor Esrrb. *Nat. Cell Biol.* **11:** 197–203.
50. Guo, G. & A. Smith. 2010. A genome-wide screen in EpiSCs identifies Nr5a nuclear receptors as potent inducers of ground state pluripotency. *Development* **137:** 3185–3192.
51. Heng, J.C. *et al.* 2010. The nuclear receptor Nr5a2 can replace Oct4 in the reprogramming of murine somatic cells to pluripotent cells *Cell Stem Cell* **6:** 167–174.
52. Singhal, N. *et al.* 2010. Chromatin-remodeling components of the BAF complex facilitate reprogramming. *Cell* **141:** 943–955.
53. Wang, T. *et al.* 2011. The histone demethylases jhdm1a/1b enhance somatic cell reprogramming in a vitamin-C-dependent manner. *Cell Stem Cell* **9:** 575–587.
54. Wang, W. *et al.* 2011. Rapid and efficient reprogramming of somatic cells to induced pluripotent stem cells by retinoic acid receptor gamma and liver receptor homolog 1. *Proc. Natl. Acad. Sci. USA* **108:** 18283–18288.
55. Han, J. *et al.* 2010. Tbx3 improves the germ-line competency of induced pluripotent stem cells. *Nature* **463:** 1096–1100.
56. Chia, N.Y. *et al.* 2010. A genome-wide RNAi screen reveals determinants of human embryonic stem cell identity. *Nature* **468:** 316–320.
57. Maekawa, M. *et al.* 2011. Direct reprogramming of somatic cells is promoted by maternal transcription factor Glis1. *Nature* **474:** 225–229.
58. Zhao, Y. *et al.* 2008. Two supporting factors greatly improve the efficiency of human iPSC generation. *Cell Stem Cell* **3:** 475–479.
59. Banito, A. *et al.* 2009. Senescence impairs successful reprogramming to pluripotent stem cells. *Genes Dev.* **23:** 2134–2139.
60. Mikkelsen, T.S. *et al.* 2008. Dissecting direct reprogramming through integrative genomic analysis. *Nature* **454:** 49–55.
61. Pereira, C.F. *et al.* 2010. ESCs require PRC2 to direct the successful reprogramming of differentiated cells toward pluripotency. *Cell Stem Cell* **6:** 547–556.
62. Samavarchi-Tehrani, P. *et al.* 2010. Functional genomics reveals a BMP-driven mesenchymal-to-epithelial transition in the initiation of somatic cell reprogramming. *Cell Stem Cell* **7:** 64–77.
63. Li, R. *et al.* 2010. A mesenchymal-to-epithelial transition initiates and is required for the nuclear reprogramming of mouse fibroblasts. *Cell Stem Cell* **7:** 51–63.
64. Egli, D., G. Birkhoff & K. Eggan. 2008. Mediators of reprogramming: transcription factors and transitions through mitosis. *Nat. Rev. Mol. Cell Biol.* **9:** 505–516.
65. Vierbuchen, T. & M. Wernig. 2011. Direct lineage conversions: unnatural but useful? *Nat. Biotechnol.* **29:** 892–907.
66. Faerman, A. *et al.* 1993. Ectopic expression of MyoD1 in mice causes prenatal lethalities. *Dev. Dyn.* **196:** 165–173.
67. Schafer, B.W. *et al.* 1990. Effect of cell history on response to helix-loop-helix family of myogenic regulators. *Nature* **344:** 454–458.
68. Xie, H. *et al.* 2004. Stepwise reprogramming of B cells into macrophages. *Cell* **117:** 663–676.
69. Pang, Z.P. *et al.* 2011. Induction of human neuronal cells by defined transcription factors. *Nature* **476:** 220–223.
70. Caiazzo, M. *et al.* 2011. Direct generation of functional dopaminergic neurons from mouse and human fibroblasts. *Nature* **476:** 224–227.
71. Pfisterer, U. *et al.* 2011. Direct conversion of human fibroblasts to dopaminergic neurons. *Proc. Natl. Acad. Sci. USA* **108:** 10343–10348.
72. Sekiya, S. & A. Suzuki. 2011. Direct conversion of mouse fibroblasts to hepatocyte-like cells by defined factors. *Nature* **475:** 390–393.
73. Szabo, E. *et al.* 2010. Direct conversion of human fibroblasts to multilineage blood progenitors. *Nature* **468:** 521–526.
74. Kim, J. *et al.* 2011. Direct reprogramming of mouse fibroblasts to neural progenitors. *Proc. Natl. Acad. Sci. USA* **108:** 7838–7843.
75. Efe, J.A. *et al.* 2011. Conversion of mouse fibroblasts into cardiomyocytes using a direct reprogramming strategy. *Nat. Cell Biol.* **13:** 215–222.
76. Ashburner, M. *et al.* 2000. Gene ontology: tool for the unification of biology. The Gene Ontology Consortium. *Nat. Genet.* **25:** 25–29.
77. Bhutani, N. *et al.* 2010. Reprogramming towards pluripotency requires AID-dependent DNA demethylation. *Nature* **463:** 1042–1047.
78. Vaquerizas, J.M. *et al.* 2009. A census of human transcription factors: function, expression and evolution. *Nat. Rev. Genet.* **10:** 252–263.
79. Hannenhalli, S. & K.H. Kaestner. 2009. The evolution of Fox genes and their role in development and disease. *Nat. Rev. Genet.* **10:** 233–240.
80. Jiang, J. *et al.* 2008. A core Klf circuitry regulates self-renewal of embryonic stem cells. *Nat. Cell Biol.* **10:** 353–360.
81. Rudnicki, M.A. *et al.* 1992. Inactivation of MyoD in mice leads to up-regulation of the myogenic HLH gene Myf-5 and results in apparently normal muscle development. *Cell* **71:** 383–390.
82. Rudnicki, M.A. *et al.* 1993. MyoD or Myf-5 is required for the formation of skeletal muscle. *Cell* **75:** 1351–1359.
83. Williams, R.R. *et al.* 2006. Neural induction promotes large-scale chromatin reorganisation of the Mash1 locus. *J. Cell Sci.* **119:** 132–140.
84. Zaret, K.S. *et al.* 2008. Pioneer factors, genetic competence, and inductive signaling: programming liver and pancreas progenitors from the endoderm. *Cold Spring Harb. Symp. Quant. Biol.* **73:** 119–126.
85. Zaret, K.S. & J.S. Carroll. 2011. Pioneer transcription factors: establishing competence for gene expression. *Genes Dev.* **25:** 2227–2241.
86. Yan, J. *et al.* 2006. The forkhead transcription factor FoxI1 remains bound to condensed mitotic chromosomes and stably remodels chromatin structure. *Mol. Cell Biol.* **26:** 155–168.
87. Ghisletti, S. *et al.* 2010. Identification and characterization of enhancers controlling the inflammatory gene expression program in macrophages. *Immunity* **32:** 317–328.

88. Heinz, S. *et al.* 2010. Simple combinations of lineage-determining transcription factors prime cis-regulatory elements required for macrophage and B cell identities. *Mol. Cell* **38:** 576–589.
89. Di Tullio, A. *et al.* 2011. CCAAT/enhancer binding protein alpha (C/EBP(alpha))-induced transdifferentiation of pre-B cells into macrophages involves no overt retrodifferentiation. *Proc. Natl. Acad. Sci. USA* **108:** 17016–17021.
90. Bussmann, L.H. *et al.* 2009. A robust and highly efficient immune cell reprogramming system. *Cell Stem Cell* **5:** 554–566.
91. Hanna, J. *et al.* 2009. Direct cell reprogramming is a stochastic process amenable to acceleration. *Nature* **462:** 595–601.
92. Murry, C.E. & G. Keller. 2008. Differentiation of embryonic stem cells to clinically relevant populations: lessons from embryonic development. *Cell* **132:** 661–680.
93. Doetschman, T.C. *et al.* 1985. The in vitro development of blastocyst-derived embryonic stem cell lines: formation of visceral yolk sac, blood islands and myocardium. *J. Embryol. Exp. Morphol.* **87:** 27–45.
94. Keller, G. *et al.* 1993. Hematopoietic commitment during embryonic stem cell differentiation in culture. *Mol. Cell Biol.* **13:** 473–486.
95. North, T. *et al.* 1999. Cbfa2 is required for the formation of intra-aortic hematopoietic clusters. *Development* **126:** 2563–2575.
96. Shalaby, F. *et al.* 1995. Failure of blood-island formation and vasculogenesis in Flk-1-deficient mice. *Nature* **376:** 62–66.
97. Weintraub, H. *et al.* 1989. Activation of muscle-specific genes in pigment, nerve, fat, liver, and fibroblast cell lines by forced expression of MyoD. *Proc. Natl. Acad. Sci. USA* **86:** 5434–5438.

Ann. N.Y. Acad. Sci. ISSN 0077-8923

ANNALS OF THE NEW YORK ACADEMY OF SCIENCES
Issue: *Hematopoietic Stem Cells VIII*

Molecular live cell bioimaging in stem cell research

Max Endele and Timm Schroeder

Research Unit Stem Cell Dynamics, Helmholtz Center Munich—German Research Center for Environmental Health, Neuherberg, Germany

Address for correspondence: Timm Schroeder, Research Unit Stem Cell Dynamics, Helmholtz Center Munich, Ingolstaedter Landstrasse 1, 85764 Neuherberg, Germany. timm.schroeder@helmholtz-muenchen.de

Functional heterogeneity within stem and progenitor cells has been shown to influence cell fate decisions. Similarly, intracellular signaling activated by external stimuli is highly heterogeneous and its spatiotemporal activity is linked to future cell behavior. To quantify these heterogeneous states and link them to future cell fates, it is important to observe cell populations continuously with single cell resolution. Live cell imaging in combination with fluorescent biosensors for signaling activity serves as a powerful tool to study cellular and molecular heterogeneity and the long-term biological effects of signaling. Here, we describe these methodologies, their advantages over classical approaches, and we illustrate how they could be applied to improve our understanding of the importance of heterogeneous cellular and molecular responses to external signaling cues.

Keywords: biosensors; cytokines; M-CSF; signaling; lineage choice; stem cell; imaging; time lapse

Introduction

Each second throughout our lives, millions of cells die within the human body. The regeneration of these cells is the task of tissue-specific multipotent adult stem cells. For instance, bone marrow hematopoietic stem cells (HSCs) are responsible for the lifelong replenishment of the blood system through a process termed hematopoiesis.[1] In the unperturbed adult, HSCs are a largely quiescent population that resides in specialized microenvironments (niches). These ill-defined niches are composed of skeletal lineage cells, blood vessels, and possibly other cellular and molecular components.[2,3] During hematopoietic differentiation, HSCs first differentiate into multipotent progenitors (MPPs), which have lost the ability to self-renew lifelong. MPPs, in turn, produce more lineage-restricted progenitors that ultimately give rise to all the differentiated mature cells of the blood system. This differentiation process is accompanied by lineage-specific epigenetic changes.[4]

Despite decades of research, the exact mechanisms controlling hematopoietic cell fate decisions still remain largely obscure. Stem cell fate decisions are controlled by the integration of cell-extrinsic signals and cell-intrinsic transcription factor networks.[5–7] The production of the correct numbers of different blood lineages is largely regulated by hematopoietic cytokines.[8] The prevalent model of lineage commitment assumed a permissive role of cytokines, inducing only the survival and proliferation of cells, which have intrinsically committed to one lineage.[9,10] However, there has also been recent evidence for an instructive role of cytokines on lineage choice.[11,12]

Obviously, it is important to purify cell types of interest when analyzing their molecular properties and control. Owing to the identification of novel surface markers in recent years, it is now possible to prospectively isolate hematopoietic stem and progenitor cells (HSPCs) to relatively high purity using fluorescent activated cell sorting (FACS).[13–20] However, single-cell transplantation studies have shown that even among highly enriched HSPC populations, subtypes exist that intrinsically behave differently regarding lineage commitment outcome, engraftment kinetics, or self-renewal capacity.[21–25] Similarly, single cells within a defined population can heterogeneously respond to the

doi: 10.1111/j.1749-6632.2012.06560.x

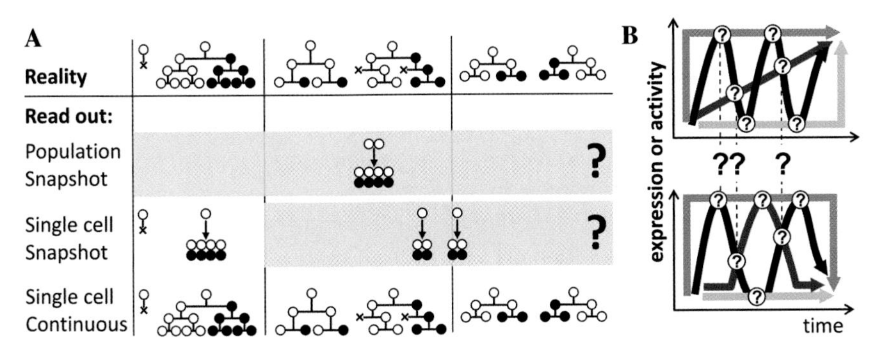

Figure 1. Necessity for continuous single cell quantification of cellular and molecular behavior. (A) Only continuous single cell analyses provide data allowing nonambiguous conclusions about cellular behavior. (B) Different molecular behaviors can lead to the same molecular phenotype. Snapshot analysis at given time points (indicated as question marks) cannot elucidate the true molecular behavior.

same external stimulus, leading to substantially different phenotypic outcomes.[26–29] Data from our laboratory have shown that HSPCs susceptible to the cytokines macrophage colony-stimulating factor (MCSF) and granulocyte colony-stimulating factor (GCSF) respond heterogeneously in regard to time to lineage commitment.[11]

Information on this important heterogeneity of populations is obscured when analyzed by classical experimental approaches that measure population averages. This is a particular problem when analyzing infrequent subpopulations such as stem and progenitor cells of a tissue. Furthermore, large numbers of cells are usually required for conventional molecular and cellular assays. Unfortunately, the numbers of available primary HSPCs is extremely low, making the applicability of these assays in HSPC research very tedious to impossible. Techniques such as flow cytometry or immunofluorescence allow analyses at the single cell level. However, most approaches provide only snapshot analyses of continuous processes, which can lead to ambiguous interpretation of cellular or molecular behavior, even when applied with single cell resolution. As depicted in Figure 1A, three scenarios starting with two single cells leading to an identical endpoint scenario (four white and four black differentiated cells) cannot be distinguished by snapshot population analysis. Snapshot analysis on the single cell level can only detect the true scenario in the special case in which one of the two starting cells produces all differentiated cells. Only continuous single cell analysis reveals the true differentiation process leading to the observed endpoint situation.

To experimentally quantify individual cell behaviors within a population over time, live cell imaging is required.[30–32] It allows the simultaneous quantification of several cellular parameters, such as morphology, motility, division kinetics, or survival rates of individual cells and their progeny. These features make continuous live cell imaging a powerful method allowing conclusions that would be obscured by noncontinuous or population analyses (Fig. 1).[11,33–35] In contrast to *in vivo* long-term imaging, whose applicability is thus far very limited due to technical restrictions (e.g., low sensitivity, low temporal and spatial resolution, limited time-span), *in vitro* imaging can be applied with high precision and temporal resolution over several weeks. Retrospective tracking and statistical analysis of the imaged cells is currently achieved manually, through self-made software, as commercially available products are too limited in their functions. Recent efforts in optimizing cell tracking include its automation and the implementation of automated cell segmentation, however none of these approaches is currently reliable enough for unsupervised tracking.[36]

Despite being a relatively new technique, *in vitro* single cell imaging has already contributed to a better understanding of HSPC behavior. For instance, one study could associate longer *in vitro* cell cycling times with *in vivo* HSC activity.[37] In another report, live-cell imaging and single-cell tracking was used to clarify the role of STAT5A/B in granulocyte-macrophage colony-stimulating factor (GMCSF)–mediated granulopoiesis.[38] Using a high-throughput single-cell array approach in combination with live imaging, Lecault *et al.* investigated the role of stem cell factor (SCF) concentrations on HSC survival and proliferation.[39]

Complementary to monitoring cellular behavior, there is also need to continuously quantify dynamic molecular properties, such as signaling activity. It is important to define the strength and timing of these intracellular processes in order to fully understand their functional consequence. Aforementioned biochemical approaches are unable to detect relationships between current molecular heterogeneity and future cell fates, as cells are either killed in the process of sample preparation, or their future identities lost after analysis. Moreover, as illustrated in Figure 1B, snapshot analyses cannot elucidate dynamic molecular processes, even when conducted with single cell resolution. Depending on the time point the snapshot is taken, different dynamic behaviors (indicated by colored lines) leading to different outcomes (upper and lower graph) are indistinguishable (indicated by dashed lines and question marks). Furthermore, the same dynamic behavior is scored differently at different time points (indicated by question marks along one dynamic pattern). To address these issues, fluorescent biosensors have been developed. In combination with long-term live cell imaging, they can allow to link dynamic signaling activity and future cell fates. Here, we discuss how these tools can be useful for investigating the spatiotemporal aspects of hematopoietic cytokine signaling.

Studying cytokine-mediated effects

Cytokines are secreted, soluble factors that bind to specific cytokine receptors activating multiple intracellular signaling pathways, thereby influencing the generation of blood lineages from HSPCs. Binding of a cytokine to its cognate receptor can have context-dependent pleiotropic biological effects on cells expressing the receptor. For example, cytokines can provide stimuli for survival and proliferation in early progenitors, but can additionally influence other cellular behavior, such as adherence or migration, in more mature cells. Cytokine effects can occur in a simultaneous or sequential fashion, as cells differentiate within a certain lineage. Furthermore, the influence of cytokines on cell fate can depend on dynamic intracellular molecular states or fluctuating levels of receptor expression, which can change HSPC responsiveness to the cytokine. It is therefore important to observe these effects on intrinsically heterogeneous HSPCs in the correct cell type, over time, and with single cell resolution.[11,40]

Macrophage colony-stimulating factor signaling and lineage choice

Macrophage colony-stimulating factor (M-CSF) is the cytokine principally regulating the production of monocytes/macrophages from hematopoietic progenitors.[41] Additionally, MCSF is important for the survival, adhesion, and motility of mature macrophages. We recently investigated whether M-CSF can only allow survival and proliferation of hematopoietic cells, or can also instruct the lineage choice of bone marrow–derived progenitor cells. The role of M-CSF in lineage decision has been controversial, mainly because it had been technically challenging to prove whether M-CSF can instruct uncommitted cells to a specific lineage (instructive model), or whether it only allows the survival and proliferation of spontaneously committed cells (selective model).[9,10,42] To exclude one of the models, it was essential to follow uncommitted cells at single cell resolution until their progeny adopted a new lineage fate. Using live cell imaging, we recently provided evidence that M-CSF can instruct the lineage choice of uncommitted bipotent granulocyte macrophage progenitors (GMPs).[11,18] Furthermore, this study showed that the time to lineage commitment of GMPs after M-CSF treatment is heterogeneous.

The identity and exact involvement of different signaling pathways downstream of M-CSF-mediated lineage instruction remains elusive. M-CSF signals through MCSFR, which belongs to the class III family of intrinsic tyrosine kinase growth factor receptors. The MCSFR has at least eight known functional tyrosine residues on its cytoplasmic domain that upon M-CSF binding and receptor dimerization are autophosphorylated. The phosphorylated tyrosines act as docking sites for numerous Src homology 2 (SH2)-containing adaptor proteins, activating a variety of different signaling pathways that mediate the pleiotropic actions of M-CSF.[41,43,44] The abundance of M-CSF-activated signaling cascades makes it difficult to functionally link specific pathways to specific biological effects. Previous studies investigating the association of phosphorylated tyrosine residues in the MCSFR with activated downstream signaling pathways have identified many of the early signaling events emanating from the activated MCSFR.[41,43] However, these studies frequently used ectopic expression of tyrosine-mutated receptors in transformed cells lacking endogenous MCSFR expression.[45–53] Other approaches involved the overexpression of chimeric mutant MCSFRs possessing a heterologous extracellular domain in myeloid cells that endogenously express the receptor.[54,55] Some of these studies led to conflicting results with regards to activated signaling pathways and concomitant influence on cell behavior, most likely due to the different cell types and experimental conditions used.[56]

Using a rat-derived fibroblast cell line (Rat-2) lacking endogenous MCSFR expression, it was shown that mutation of tyrosine residue 807 (Y807) resulted in a reduced proliferative response to M-CSF.[53] In contrast, the same mutant expressed in the MCSFR-deficient immature myeloid cell line FDC-P1 increased proliferation of these cells in the presence of M-CSF.[47] Yet another study analyzed the Y807 mutant in bone marrow–derived macrophages (BMDMs) using a receptor chimera consisting of the extracellular part of the erythropoietin receptor (EpoR) and the cytoplasmic part of the MCSFR. This study found that mutation of Y807 suppresses M-CSF-mediated macrophage proliferation.[54]

Analyzing the effects of mutated MCSFR tyrosine residue 559 (Y559) in the myeloblastic leukemia cell line M1, which is largely negative for endogenous MCSFR, suggested that this residue is not required for M-CSF–induced proliferation.[57] However, the identical mutant expressed in 32D cells—another myeloblast-like cell line devoid of endogenous MCSFR—led to a hyperproliferative M-CSF response.[58] In sharp contrast, overexpressing the Y559 mutant in BMDM using an EpoR-MCSFR chimera substantially reduced macrophage proliferation.[54]

Contradictory results have also been reported about involved signaling molecules downstream of the MCSFR. Phospholipase C gamma 2 (PLC-γ2) was initially described to bind the activated MCSFR on tyrosine residue 721 (Y721) as detected by yeast two hybrid screening and confirmed in FDC-P1 cells.[46] However, using an immortalized macrophage cell line, the binding of PLC-γ2 to Y721 has recently been questioned.[59] The major downstream signaling component of the PLC-γ pathway is protein kinase C (PKC). M-CSF has been described to activate PKC in human monocytes, while others did not find M-CSF–induced PKC activation in murine BMDMs.[60,61] PKC alpha (PKCα) has been implicated in the lineage determination and differentiation of bone marrow–derived bipotent granulocyte macrophage colony-forming cells.[62,63] Another study suggested PKC delta (PKCδ) to be involved in the M-CSF–induced differentiation of MCSFR-transduced FDC-P1 cells.[64] Using macrophage cell line BAC1.2F5, others did not find PKCδ activation upon M-CSF stimulation.[65]

These discrepancies exemplify the importance of studying cytokine signaling in the appropriate cellular context. Cytokine-induced effects can only be initiated if the cell has the cellular and molecular environment required to correctly interpret the external stimulus. Cells that do not endogenously express the MCSFR might not possess signaling molecules that are normally recruited by the receptor. Furthermore, molecular machineries exerting different cytokine effects may only be active or present during specific time windows of differentiation. Intriguingly, it has been shown that different expression levels of signaling molecules can lead to different biological responses.[66] Similarly, expression levels of cytokine receptors might affect cytokine-mediated cellular outcomes. Accordingly, studies involving the overexpression of mutated receptors have to be regarded with caution. Although studies in heterologous systems have contributed largely to our understanding of M-CSF signal transduction, it is for the above-mentioned reasons crucial to analyze, in a continuous manner, the biological effects of cytokines in cells physiologically responding to them (e.g., GMPs).

Novel tools to study cytokine signaling: biosensors

The detailed relationship between cytokine-activated signaling events and specific cell fates is only poorly understood. Classical biochemical studies have elucidated many of the underlying mechanisms of cytokine-induced signal transduction, albeit with poor temporal and spatial resolution. However, it becomes increasingly clear that it is crucial to quantify signaling activity continuously over time.[29,67–71] Signaling processes are generally not synchronous between cells and many of the intracellular changes that happen in a millisecond to second range. To obtain an improved understanding of cytokine-induced signal transduction, methods are required that also reveal the spatial distribution and dynamic aspect of these molecular processes. The development of genetically encoded fluorescent biosensors in combination with live cell imaging can allow addressing these points by directly visualizing and quantifying molecular events with high resolution within living cells. Importantly, this allows direct correlation of the spatial and dynamic properties of cell signaling with long-term cellular behavior. Biosensors can be categorized according to different working principles, of which two will be discussed in more detail.

Translocation-based biosensors

Many proteins within signaling cascades change their subcellular localization in response to external stimuli. Signaling proteins frequently bind specific second messengers on the plasma membrane that transiently accumulate upon activation of a specific pathway. Fusion of the full-length signaling protein to a fluorescent protein can serve as a reporter for its localization and visualize stimulus-induced translocation events.[72–74] However, it is often sufficient to fuse the fluorescent protein to the minimal protein binding domain responsible for the translocation (Fig. 2). For instance, pleckstrin homology (PH) domains of different proteins have been used to visualize and measure the accumulation and turnover of specific phospholipids in membranes.[75–78] In a similar fashion, the C1 domain of PKC fused to GFP has been shown to be an indicator for diacylglycerol accumulation, thereby indirectly detecting PLC-γ activity.[72,74] Furthermore, fusion of transcription factors to fluorescent proteins can serve as probes for cytosol-nucleus shuttling events. Applying live cell imaging combined with a biosensor for NF-κB (based on a p65-GFP fusion), the spatiotemporal coordination of NF-κB signaling in response to the inflammatory cytokine TNF-alpha (TNF-α) has recently been investigated.[79]

Translocation-based sensors are particularly useful to get insight into the dynamics of intracellular events. Owing to the wide range of fluorescent hues, multiple pathway activations can be monitored within a single cell.

FRET-based signaling biosensors

In cases of GTPases or kinases, it is more useful to measure protein activity, rather than reading out localization changes as indication for signaling activation. To this end, fluorescence resonance energy transfer (FRET)-based biosensors have been developed for a whole range of different signaling molecules.[80–88] FRET describes the process of

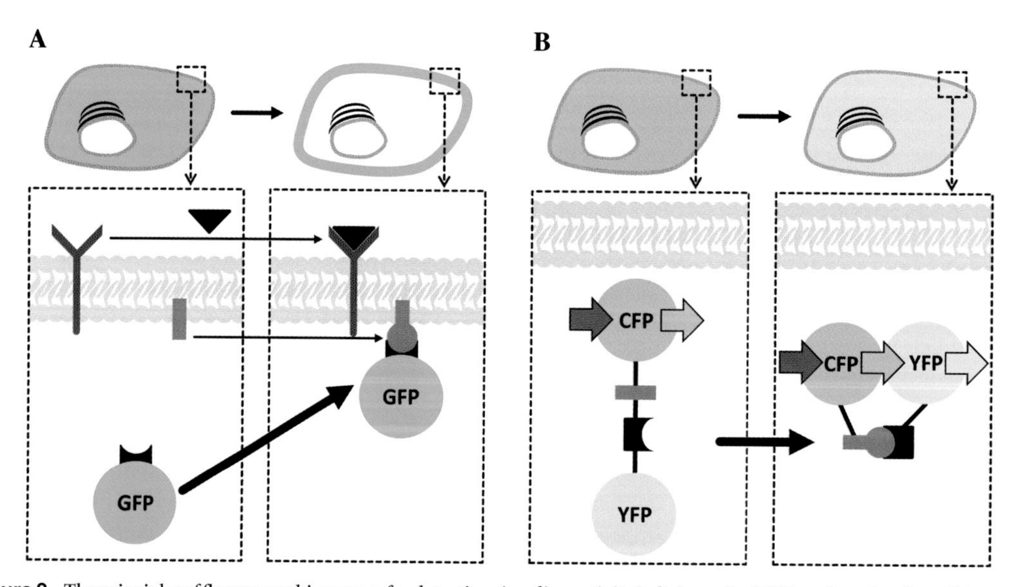

Figure 2. The principles of fluorescent biosensors for detecting signaling activity in living cells. (A) Translocation-based biosensors. External stimulation leads to the transient accumulation of a second messenger in a specific subcellular localization. The second messenger is bound by the biosensor, resulting in the translocation of the biosensor to this location. (B) FRET-based biosensors. FRET-based biosensors contain a sensing domain, which is modified by the signaling activity of interest. Modification of the sensing domain leads to a conformational change of the sensor inducing FRET. Colored arrows indicate the light of different wavelengths.

nonradioactive transfer of energy from an excited donor fluorophore to an acceptor fluorophore. For FRET to occur the emission spectrum of the donor has to overlap with the excitation spectrum of the acceptor and the two fluorophores have to be in close proximity (< 10 nm). Most commonly, FRET changes are measured by ratiometric imaging, which is based on monitoring changes in donor and acceptor fluorescence intensities due to FRET.

A typical FRET-based sensor for kinase activity consists of a kinase-specific substrate linked to a binding domain specific for the phosphorylated substrate. This sensing region is sandwiched between two FRET-capable fluorophores (typically CFP and YFP). If the substrate gets phosphorylated by the activated kinase, the binding domain binds the substrate, resulting in a conformational change that brings the two fluorophores in close proximity (Fig. 2B). Only then can FRET occur, providing a read out for kinase activity. Dephosphorylation of the substrate by specific phosphatases conversely leads to loss of FRET. As they are genetically encoded, biosensors can be targeted to distinct intracellular compartments via specific genetic localization sequences. This allows measuring signaling activity from functionally distinct organelles within the cell and has provided insight into how intracellular compartmentalization can affect signaling.[85,89]

Caveats of fluorescent biosensors

As illustrated above, fluorescent biosensors overcome many of the limitations associated with studying heterogeneous populations by classical biochemistry. They allow us to look at protein dynamics during signal transduction within the natural milieu of single cells in real time. Most importantly, the signaling events can be directly linked to the future fates of the cells. The generated data provides novel platforms for mathematical modeling, improving our understanding of these dynamic and compartmentalized biological processes.[90] However, there are some caveats involved in the development and application of biosensors that need to be carefully considered.

The most important issue when using biosensors is to ensure that cellular physiology is not or only minimally disrupted through potential competition of the sensor with natural ligands. This usually requires rigorous screening and characterization of several constructs during the development of a new sensor. Ideally, the binding target of a FRET-based sensor should be a molecule that does not naturally exist in the cell of interest and for that reason cannot interact with cellular ligands once phosphorylated. Another strategy is to modify the sensing domain to make it inert to cellular interaction partners.[91] The strength of biosensor expression in the cell should not exceed levels of endogenous products, while simultaneously being high enough to reliably detect it.

A second issue to be kept in mind is phototoxicity. In order to detect highly dynamic intracellular processes, temporal resolution (i.e., imaging frequency) often needs to be in the seconds to minutes range. In order to minimize photo toxicity, fluorophores having longer excitation wavelengths (thus lower energy) are generally preferred over blue or near-ultraviolet excited ones. With the development of monomeric variants of coral fluorescent proteins, the range of suitable colors for biosensing has increased.[92]

Another point concerns the simultaneous detection of several intracellular processes, which is so far limited by the spectral separation of the coimaged sensors. This particularly affects the simultaneous imaging of several FRET-based reporters. Yet, it will be important to track related processes in a single cell at the same time in order to correlate these events in the same context. For FRET-based sensors, CFP and YFP is still the most commonly used fluorophore pair. However, there has been increasing effort to pair novel fluorescent donors and

Table 1. **Available translocation-based biosensors**

Biosensor	Detection	Fluorophores
PH-Akt[77]	PI(3,4)P_2 and PI(3,4,5)P_3	ECFP, VENUS, tagRFP, tdTOMATO
PLCδ-PH[78]	PI(4,5)P_2 and IP_3	mRFP
PKCγ-C1[72]	Diacylglycerol	EGFP, mRFP
p65-GFP[79]	NFkB activity	EGFP
PKCδ[85]	PKC activity	VENUS, mRFP
PKCε[85]	PKC activity	YFP, mRFP
PKCα[85]	PKC activity	mRFP
PKCβ1[85]	PKC activity	mRFP
PKCβ2[85]	PKC activity	mRFP
PKCγ[85]	PKC activity	mRFP

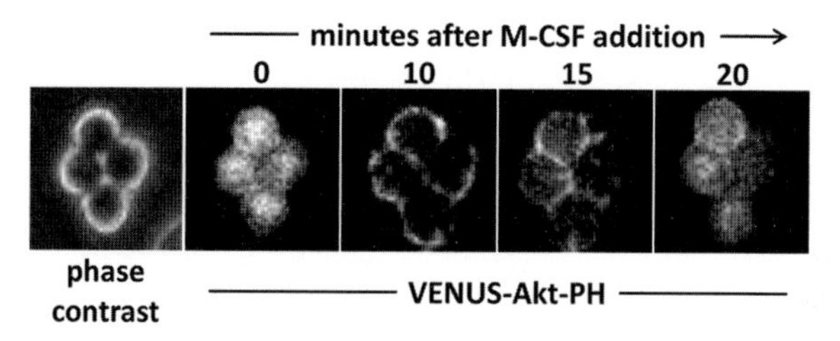

Figure 3. Rapid and transient activation of cytokine signaling. Time-lapse microscopy of live RAW 264 cells endogenously expressing the M-CSF receptor. Cells were transduced with a biosensor for PI3K activation. Stimulation with M-CSF (20 ng/mL) results in rapid and transient translocation of the probe from cytosol and nucleus to the plasma membrane. Time-lapse pictures were taken every 60 seconds.

acceptors.[93–97] The development of new FRET pairs for biosensors and their combination with other sensors has already enabled the visualization and quantification of several processes in parallel.[98]

Lastly, the application of biosensors to living mammalian organisms is hampered by the fact that most fluorescent proteins cannot be detected in deep tissues due to their short excitation and emission wavelengths and concomitant insufficient tissue penetration. This might be overcome in the future by the development of fluorescent proteins in the near-infrared to infrared emission range.[99,100]

Biosensors for studying cytokine signaling in mammalian stem and progenitor cells

To exploit the advantages of biosensors over classical cellular and molecular analyses, we have gathered genetically encoded biosensors detecting different signaling pathways typically activated in mammalian stem and progenitor cells (Table 1). In order to facilitate simultaneous detection of several pathways, the sensing domains of translocation-based biosensors were fused to fluorescent proteins spanning the available range of wavelengths. As primary stem and progenitor cells are often difficult to transfect using conventional techniques (e.g., lipofection), constructs were cloned into a third generation lentiviral backbone allowing efficient transgene delivery.

To test the general applicability of biosensors in combination with M-CSF, monocytic RAW 264 cells expressing MCSFR endogenously were lentivirally transduced with a PI3K sensor based on the PH domain of Akt. Stimulation with M-CSF led to a rapid and transient translocation of the sensor from the nucleus and cytoplasm to the plasma membrane (Fig. 3), demonstrating the single cell sensitivity and temporal resolution that can be achieved through this approach.

Unfortunately, the high similarity between the CFP and YFP genes typically used for FRET-based biosensors proved to make the use of lentiviruses extremely difficult. We were not able to produce virus particles with high titers harboring the correct FRET biosensors (Raichu-Cdc42, Raichu-RhoA, Akind, Miu2, CKAR, KCP-1[80,82–84,86,88]; data not shown). Most likely, the highly similar CFP and YFP sequences led to homologous recombination during reverse transcription, preventing their efficient use in difficult to transfect mammalian HSPCs. The advent of novel fluorescent proteins without sequence similarities to the GFP family of proteins could help to remedy this problem. Combinations of GFP family and novel fluorescent proteins as FRET donor/acceptor pairs could then allow the use also of FRET-based sensors in these highly interesting and clinically relevant cell types.

In conclusion, the combination of long-term single cell imaging with detection of live signaling via biosensors will yield new insights into cytokine signaling dynamics and kinetics, and help identify the individual functions of cytokine-triggered signaling cascades in controlling the fates of stem and progenitor cells in health and disease.

Conflicts of interest

The authors declare no conflicts of interest.

References

1. Orkin, S.H. & L.I. Zon. 2008. Hematopoiesis: an evolving paradigm for stem cell biology. *Cell* **132:** 631–644.
2. Wilson, A. *et al.* 2008. Hematopoietic stem cells reversibly switch from dormancy to self-renewal during homeostasis and repair. *Cell* **135:** 1118–1129.
3. Lo Celso, C. & D.T. Scadden. 2011. The haematopoietic stem cell niche at a glance. *J. Cell Sci.* **124:** 3529–3535.

4. Ji, H. *et al.* 2010. Comprehensive methylome map of lineage commitment from haematopoietic progenitors. *Nature* **467:** 338–342.
5. Zhu, J. & S.G. Emerson. 2002. Hematopoietic cytokines, transcription factors and lineage commitment. *Oncogene* **21:** 3295–3313.
6. Zhang, C.C. & F. Harvey. 2008. Cytokines regulating hematopoietic stem cell function. *Curr. Opin. Hematol.* **15:** 307–311.
7. Zon, L.I. 2008. Intrinsic and extrinsic control of haematopoietic stem-cell self-renewal. *Nature* **453:** 306–313.
8. Metcalf, D. 2008. Hematopoietic cytokines. *Blood* **111:** 485–491.
9. Enver, T. *et al.* 1998. Do stem cells play dice? *Blood* **92:** 348–352.
10. Cross, M.A. & T. Enver. 1997. The lineage commitment of haemopoietic progenitor cells. *Curr. Opin. Genet. Develop.* **7:** 609–613.
11. Rieger, M.A. *et al.* 2009. Hematopoietic cytokines can instruct lineage choice. *Science* **325:** 217–218.
12. Sarrazin, S. *et al.* 2009. MafB restricts M-CSF-dependent myeloid commitment divisions of hematopoietic stem cells. *Cell* **138:** 300–313.
13. Spangrude, G.J. *et al.* 1988. Purification and characterization of mouse hematopoietic stem cells. *Science* **241:** 58–62.
14. Osawa, M. *et al.* 1996. Long-term lymphohematopoietic reconstitution by a single CD34-low/negative hematopoietic stem cell. *Science* **273:** 242–245.
15. Kiel, M.J. *et al.* 2005. SLAM family receptors distinguish hematopoietic stem and progenitor cells and reveal endothelial niches for stem cells. *Cell* **121:** 1109–1121.
16. Benveniste, P. *et al.* 2010. Intermediate-term hematopoietic stem cells with extended but time-limited reconstitution potential. *Cell Stem Cell* **6:** 48–58.
17. Kent, D.G. *et al.* 2009. Prospective isolation and molecular characterization of hematopoietic stem cells with durable self-renewal potential. *Blood* **113:** 6342–6350.
18. Akashi, K. *et al.* 2000. A clonogenic common myeloid progenitor that gives rise to all myeloid lineages. *Nature* **404:** 193–197.
19. Doulatov, S. *et al.* 2010. Revised map of the human progenitor hierarchy shows the origin of macrophages and dendritic cells in early lymphoid development. *Nature Immunol.* **11:** 585–593.
20. Notta, F. *et al.* 2011. Isolation of single human hematopoietic stem cells capable of long-term multilineage engraftment. *Science* **333:** 218–221.
21. Dykstra, B. *et al.* 2007. Long-term propagation of distinct hematopoietic differentiation programs in vivo. *Cell Stem Cell* **1:** 218–229.
22. Müller-Sieburg, C.E. *et al.* 2002. Deterministic regulation of hematopoietic stem cell self-renewal and differentiation. *Blood* **100:** 1302–1309.
23. Morita, Y. *et al.* 2010. Heterogeneity and hierarchy within the most primitive hematopoietic stem cell compartment. *J. Exp. Med.* **207:** 1173–1182.
24. Challen, G.A. *et al.* 2010. Distinct hematopoietic stem cell subtypes are differentially regulated by TGF-beta1. *Cell Stem Cell* **6:** 265–278.
25. Schroeder, T. 2010. Hematopoietic stem cell heterogeneity: subtypes, not unpredictable behavior. *Cell Stem Cell* **6:** 203–207.
26. Balaban, N.Q. *et al.* 2004. Bacterial persistence as a phenotypic switch. *Science* **305:** 1622–1625.
27. Sprinzak, D. *et al.* 2010. Cis-interactions between Notch and Delta generate mutually exclusive signalling states. *Nature* **465:** 86–90.
28. Ashall, L. *et al.* 2009. Pulsatile stimulation determines timing and specificity of NF-kappaB-dependent transcription. *Science* **324:** 242–246.
29. Marshall, C.J. 1995. Specificity of receptor tyrosine kinase signaling: transient versus sustained extracellular signal-regulated kinase activation. *Cell* **80:** 179–185.
30. Schroeder, T. 2008. Imaging stem-cell-driven regeneration in mammals. *Nature* **453:** 345–351.
31. Schroeder, T. 2011. Long-term single-cell imaging of mammalian stem cells. *Nature Methods* **8:** S30–S35.
32. Schroeder, T. 2005. Tracking hematopoiesis at the single cell level. *Ann. N. Y. Acad. Sci.* **1044:** 201–209.
33. Eilken, H.M. *et al.* 2009. Continuous single-cell imaging of blood generation from haemogenic endothelium. *Nature* **457:** 896–900.
34. Kuang, S. *et al.* 2007. Asymmetric self-renewal and commitment of satellite stem cells in muscle. *Cell* **129:** 999–1010.
35. Cohen, A.R. *et al.* 2010. Computational prediction of neural progenitor cell fates. *Nature Methods* **7:** 213–218.
36. Scherf, N. *et al.* 2012. On the symmetry of siblings: automated single-cell tracking to quantify the behavior of hematopoietic stem cells in a biomimetic setup. *Exp. Hematol.* **40**:119–130.
37. Dykstra, B. *et al.* 2006. High-resolution video monitoring of hematopoietic stem cells cultured in single-cell arrays identifies new features of self-renewal. *Proc. Natl. Acad. Sci. USA* **103:** 8185–8190.
38. Kimura, A. *et al.* 2009. The transcription factors STAT5A/B regulate GM-CSF-mediated granulopoiesis. *Blood* **114:** 4721–4728.
39. Lecault, V. *et al.* 2011. High-throughput analysis of single hematopoietic stem cell proliferation in microfluidic cell culture arrays. *Nature Methods* **8:** 581–586.
40. Rieger, M.A. & T. Schroeder. 2009. Analyzing cell fate control by cytokines through continuous single cell biochemistry. *J. Cell. Biochem.* **108:** 343–352.
41. Pixley, F.J. & E.R. Stanley. 2004. CSF-1 regulation of the wandering macrophage: complexity in action. *Trends Cell Biol.* **14:** 628–638.
42. Metcalf, D. 1998. Lineage commitment and maturation in hematopoietic cells: the case for extrinsic regulation. *Blood* **92:** 345–347.
43. Bourette, R.P. & L.R. Rohrschneider. 2000. Early events in M-CSF receptor signaling. *Growth Factors* **17:** 155–166.
44. Yu, W. *et al.* 2008. CSF-1 receptor structure/function in MacCsf1r-/- macrophages: regulation of proliferation, differentiation, and morphology. *J. Leukocyte Biol.* **84:** 852–863.
45. Bourette, R.P. *et al.* 2001. Suppressor of cytokine signaling 1 interacts with the macrophage colony-stimulating factor receptor and negatively regulates its proliferation signal. *J. Biol. Chem.* **276:** 22133–22139.

46. Bourette, R.P. *et al.* 1997. Sequential activation of phoshatidylinositol 3-kinase and phospholipase C-gamma2 by the M-CSF receptor is necessary for differentiation signaling. *EMBO J.* **16:** 5880–5893.
47. Bourette, R. *et al.* 1995. Uncoupling of the proliferation and differentiation signals mediated by the murine macrophage colony-stimulating factor receptor expressed in myeloid FDC-P1 cells. *Cell Growth Differentiation* **6:** 631–645.
48. Bourgin, C. *et al.* 2002. Induced expression and association of the Mona/Gads adapter and Gab3 scaffolding protein during monocyte/macrophage differentiation. *Mol. Cell. Biol.* **22:** 3744–3756.
49. Liu, Y. *et al.* 2001. Scaffolding protein Gab2 mediates differentiation signaling downstream of Fms receptor tyrosine kinase. *Mol. Cell. Biol.* **21:** 3047–3056.
50. Mancini, A. *et al.* 1997. Identification of a second Grb2 binding site in the v-Fms tyrosine kinase. *Oncogene* **15:** 1565–1572.
51. Wilhelmsen, K. *et al.* 2002. C-Cbl binds the CSF-1 receptor at tyrosine 973, a novel phosphorylation site in the receptor's carboxy-terminus. *Oncogene* **21:** 1079–1089.
52. Wolf, I. *et al.* 2002. Gab3, an new DOS / Gab family member, facilitates macrophage differentiation. *Mol. Cell. Biol.* **22:** 231–244.
53. Geer, P.V.D. & T. Hunter. 1991. Tyrosine 706 and 807 phosphorylation site mutants in the murine colony-stimulating factor-1 receptor are unaffected in their ability to bind or phosphorylate phosphatidylinositol-3 kinase but show differential defects in their ability to induce early response gene transcription. *Mol. Cell. Biol.* **11:** 4698–4709.
54. Takeshita, S. *et al.* 2007. c-Fms tyrosine 559 is a major mediator of M-CSF-induced proliferation of primary macrophages. *J. Biol. Chem.* **282:** 18980–18990.
55. Faccio, R. *et al.* 2007. M-CSF regulates the cytoskeleton via recruitment of a multimeric signaling complex to c-Fms Tyr-559/697/721. *J. Biol. Chem.* **282:** 18991–18999.
56. Hamilton, J.A. 1997. CSF-1 signal transduction. *J. Leukocyte Biol.* **62:** 145–155.
57. Marks, D.C. *et al.* 1999. Expression of a Y559F mutant CSF-1 receptor in M1 myeloid cells: a role for Src kinases in CSF-1 receptor-mediated differentiation. *Mol. Cell Biol. Res. Commun.* **1:** 144–152.
58. Rohde, C.M. *et al.* 2004. A juxtamembrane tyrosine in the colony stimulating factor-1 receptor regulates ligand-induced Src association, receptor kinase function, and down-regulation. *J. Biol. Chem.* **279:** 43448–43461.
59. Sampaio, N.G. *et al.* 2011. Phosphorylation of CSF-1R Y721 mediates its association with PI3K to regulate macrophage motility and enhancement of tumor cell invasion. *J. Cell Sci.* **124:** 2021–2031.
60. Imamura, K. *et al.* 1990. Colony-stimulating factor 1 activates protein kinase C in human monocytes. *EMBO J.* **9:** 2423–2429.
61. Jaworowski, A. *et al.* 1994. Phospholipase D is activated by phorbol ester but not CSF-1 in murine bone marrow-derived macrophages. *Biochem. Biophys. Res. Commun.* **201:** 733–739.
62. Heyworth, C.M. *et al.* 1994. Cytokine-mediated protein kinase C activation is a signal for lineage determination in bipotential granulocyte macrophage colony-forming cells. *J. Cell Biol.* **125:** 651–659.
63. Pierce, A. *et al.* 1998. An activated protein kinase C alpha gives a differentiation signal for hematopoietic progenitor cells and mimicks macrophage colony-stimulating factor-stimulated signaling events. *J. Cell Biol.* **140:** 1511–1518.
64. Junttila, I. *et al.* 2003. M-CSF induced differentiation of myeloid precursor cells involves activation of PKC-δ and expression of Pkare. *J. Leukocyte Biol.* **73:** 281–288.
65. Xu, X. *et al.* 1993. Phosphatidylcholine hydrolysis and c-myc expression are in collaborating mitogenic pathways activated by colony-stimulating factor 1. *Mol. Cell. Biol.* **13:** 1522–1533.
66. Rossi, F. *et al.* 1996. Lineage commitment of transformed haematopoietic progenitors is determined by the level of PKC activity. *EMBO J.* **15:** 1894–1901.
67. Nakakuki, T. *et al.* 2010. Ligand-specific c-Fos expression emerges from the spatiotemporal control of ErbB network dynamics. *Cell* **141:** 884–896.
68. von Kriegsheim, A. *et al.* 2009. Cell fate decisions are specified by the dynamic ERK interactome. *Nature Cell Biol.* **11:** 1458–1464.
69. Rocks, O. *et al.* 2005. An acylation cycle regulates localization and activity of palmitoylated Ras isoforms. *Science* **307:** 1746–1752.
70. Nelson, D.E. *et al.* 2004. Oscillations in NF-kappaB signaling control the dynamics of gene expression. *Science* **306:** 704–708.
71. Shimojo, H. *et al.* 2008. Oscillations in notch signaling regulate maintenance of neural progenitors. *Neuron* **58:** 52–64.
72. Oancea, E. *et al.* 1998. Green fluorescent protein (GFP)-tagged cysteine-rich domains from protein kinase C as fluorescent indicators for diacylglycerol signaling in living cells. *J. Cell Biol.* **140:** 485–498.
73. Reither, G. *et al.* 2006. PKCalpha: a versatile key for decoding the cellular calcium toolkit. *J. Cell Biol.* **174:** 521–533.
74. Oancea, E. & T. Meyer. 1998. Protein kinase C as a molecular machine for decoding calcium and diacylglycerol signals. *Cell* **95:** 307–318.
75. Nishio, M. *et al.* 2007. Control of cell polarity and motility by the PtdIns(3,4,5)P3 phosphatase SHIP1. *Nature Cell Biol.* **9:** 36–44.
76. Várnai, P. & T. Balla. 2008. Live cell imaging of phosphoinositides with expressed inositide binding protein domains. *Methods* **46:** 167–176.
77. Haugh, J.M. *et al.* 2000. Spatial sensing in fibroblasts mediated by 3′ phosphoinositides. *J. Cell Biol.* **151:** 1269–1280.
78. Stauffer, T.P. *et al.* 1998. Receptor-induced transient reduction in plasma membrane PtdIns(4,5)P2 concentration monitored in living cells. *Curr. Biol.* **8:** 343–346.
79. Tay, S. *et al.* 2010. Single-cell NF-kappaB dynamics reveal digital activation and analogue information processing. *Nature* **466:** 267–271.
80. Schleifenbaum, A. *et al.* 2004. Genetically encoded FRET probe for PKC activity based on pleckstrin. *J. Am. Chem. Soc.* **126:** 11786–11787.
81. Brumbaugh, J. *et al.* 2006. A dual parameter FRET probe for measuring PKC and PKA activity in living cells. *J. Am. Chem. Soc.* **128:** 24–25.

82. Violin, J.D. *et al.* 2003. A genetically encoded fluorescent reporter reveals oscillatory phosphorylation by protein kinase C. *J. Cell Biol.* **161:** 899–909.
83. Fujioka, A. *et al.* 2006. Dynamics of the Ras/ERK MAPK cascade as monitored by fluorescent probes. *J. Biol. Chem.* **281:** 8917–8926.
84. Yoshizaki, H. *et al.* 2003. Activity of Rho-family GTPases during cell division as visualized with FRET-based probes. *J. Cell Biol.* **162:** 223–232.
85. Kajimoto, T. *et al.* 2010. Protein kinase C δ-specific activity reporter reveals agonist-evoked nuclear activity controlled by Src family of kinases. *J. Biol. Chem.* **285:** 41896–41910.
86. Yoshizaki, H. *et al.* 2007. Akt–PDK1 complex mediates epidermal growth factor-induced membrane protrusion through Ral activation. *Mol. Biol. Cell* **18:** 119–128.
87. Fosbrink, M. *et al.* 2010. Visualization of JNK activity dynamics with a genetically encoded fluorescent biosensor. *Proc. Nat. Acad. Sci. USA* **107:** 5459–5464.
88. Itoh, R.E. *et al.* 2002. Activation of Rac and Cdc42 video imaged by fluorescent resonance energy transfer-based single-molecule probes in the membrane of living cells. *Mol. Cell. Biol.* **22:** 6582–6591.
89. Gallegos, L.L. *et al.* 2006. Targeting protein kinase C activity reporter to discrete intracellular regions reveals spatiotemporal differences in agonist-dependent signaling. *J. Biol. Chem.* **281:** 30947–30956.
90. Megason, S.G. & S.E. Fraser. 2007. Imaging in systems biology. *Cell* **130:** 784–795.
91. Palmer, A.E. *et al.* 2006. Ca2+ indicators based on computationally redesigned calmodulin-peptide pairs. *Chem. Biol.* **13:** 521–530.
92. Shaner, N.C. *et al.* 2008. Improving the photostability of bright monomeric orange and red fluorescent proteins. *Nature Methods* **5:** 545–551.
93. van der Krogt, G.N.M. *et al.* 2008. A comparison of donor-acceptor pairs for genetically encoded FRET sensors: application to the Epac cAMP sensor as an example. *PloS One* **3:** e1916.
94. Grant, D.M. *et al.* 2008. Multiplexed FRET to image multiple signaling events in live cells. *Biophys. J.* **95:** L69–L71.
95. Merzlyak, E.M. *et al.* 2007. Bright monomeric red fluorescent protein with an extended fluorescence lifetime. *Nature Methods* **4:** 555–557.
96. Shcherbo, D. *et al.* 2009. Practical and reliable FRET/FLIM pair of fluorescent proteins. *BMC Biotechnology* **9:** 24–30.
97. Ai, H. *et al.* 2008. Fluorescent protein FRET pairs for ratiometric imaging of dual biosensors. *Nature Methods* **5:** 401–403.
98. Piljic, A. & C. Schultz. 2008. Simultaneous recording of multiple cellular events by FRET. *ACS Chem. Biol.* **3:** 156–160.
99. Shcherbo, D. *et al.* 2010. Near-infrared fluorescent proteins. *Nature Methods* **7:** 827–829.
100. Shu, X. *et al.* 2009. Mammalian expression of infrared fluorescent proteins engineered from a bacterial phytochrome. *Science* **324:** 804–807.

Ann. N.Y. Acad. Sci. ISSN 0077-8923

ANNALS OF THE NEW YORK ACADEMY OF SCIENCES
Issue: *Hematopoietic Stem Cells VIII*

The role of telomere shortening in somatic stem cells and tissue aging: lessons from telomerase model systems

Stefan Tümpel and K. Lenhard Rudolph
Institute of Molecular Medicine and Max-Planck-Research Department for Stem Cell Aging, Ulm, Germany

Address for correspondence: Stefan Tümpel, Institute of Molecular Medicine and Max-Planck-Research Department for Stem Cell Aging, Albert-Einstein-Allee 11, 89075 Ulm, Germany. Stefan.Tuempel@uni-ulm.de

The analysis of model systems has broadened our understanding of telomere-related aging processes. Telomerase-deficient mouse models have demonstrated that telomere dysfunction impairs tissue renewal capacity and shortens lifespan. Telomere shortening limits cell proliferation by activating checkpoints that induce replicative senescence or apoptosis. These checkpoints protect against an accumulation of genomically instable cells and cancer initiation. However, the induction of these checkpoints can also limit organ homeostasis, regeneration, and survival during aging and in the context of diseases. The decline in tissue regeneration in response to telomere shortening has been related to impairments in stem cell function. Telomere dysfunction impairs stem cell function by activation of cell-intrinsic checkpoints and by the induction of alterations in the micro- and macro-environment of stem cells. In this review, we discuss the current knowledge about the impact of telomere shortening on disease stages induced by replicative cell aging as indicated by studies on telomerase model systems.

Keywords: telomerase; telomere shortening; checkpoints; aging; stem cells

Introduction

The enzyme telomerase is a ribonucleoprotein (RNP), which elongates telomeres by synthesizing six-nucleotide repeats. So far, four telomerase subunits have been identified, including the telomerase RNA component (TERC), the telomerase reverse transcriptase (TERT), dyskerin, and the telomerase and Cajal body protein 1 (TCAB1).[1,2] The TERC serves as a template for the *de novo* synthesis of telomeric sequences at the chromosome ends. Dyskerin is a pseudouridine synthase and has a central function in assembly and stability of RNPs, including telomerase.[3–6] Mutational analysis and knockdown experiments show that dyskerin depletion impairs TERC and telomerase activity.[2,3] TCAB1 is involved in the localization of TERC into Cajal bodies and telomere synthesis.[2] It has been suggested, that Cajal bodies coordinate the distribution of telomerase to the telomeres and that TCAB1 functions as a molecular link in this process.[7]

Telomeres shorten with every cell division due to the end replication problem of DNA polymerase and to telomere processing during S-phase.[8] Upon reaching a critically short length, telomeres lose capping function and DNA damage checkpoints are activated. Therefore, telomerase is required to enhance the replicative capacity of regenerative cells, such as stem and progenitor cells.

In humans, telomerase activity is repressed in most somatic tissues due to the suppression of TERT expression.[9] It is active only in a subset of adult cells, including stem cells, progenitor cells, germ cells, and activated lymphocytes.[9–11] In contrast, most differentiated organ cells show a complete suppression of TERT expression and do not exhibit telomerase activity.[9] The inactivation of telomerase in somatic tissue functions as a tumor-suppressor mechanism inhibiting the immortal growth of transformed cells.[12,13]

There are several human diseases that have been associated with mutations in dyskerin, TERC,

doi: 10.1111/j.1749-6632.2012.06547.x

TCAB1, or TERT, including dyskeratosis congenita (DC), aplastic anemia, idiopathic pulmonary fibrosis, and liver cirrhosis.[14–20] DC patients carrying specific mutations show signs of bone marrow failure and tumor predisposition (reviewed in Refs. 21, 22). Mutations that result in DC have been identified in genes that directly interact with telomerase, the reverse transcriptase, or with telomere-binding proteins—a set of proteins that are required for telomere capping and that are also known as the shelterin complex.[23–25]

In addition to the association of human diseases with telomerase or shelterin complex mutations, several chronic diseases have been shown to associate with accelerated telomere shortening in the affected organs, including liver cirrhosis, hematopoietic stem cell–associated disorders, chronic HIV infection, myelodysplastic syndromes, and ulcerative colitis.[26–28] These studies imply a functional role of telomere shortening in disease progression and tissue aging. Because stem cells regulate tissue homeostasis, the impairment of stem cell function has been implicated as one of the factors contributing to aging and disease progression.[29] One of the molecular characteristics of stem cells is the retention of telomerase activity. However, telomerase activity in adult stem cells seems not to be sufficient to maintain stable telomeres during aging, and significant age-associated telomere shortening has been demonstrated in human $CD34^+$ hematopoietic stem cells.[30]

In contrast to humans, telomerase activity is found in most somatic tissues of laboratory mouse strains, and it is not restricted to stem and progenitor cell compartments.[31] Moreover, most laboratory strains have very long telomere reserves (~50–70 kb) compared to humans (~10 kb). Therefore, telomerase knockout mice with shortened telomeres are used to delineate molecular causes of stem cell dysfunction and the evolution of diseases that are associated with telomere dysfunction. As an additional model system to laboratory mouse strains, wild-derived castaneous mice are highly relevant for the study of age-related disease because their telomere length mimics those of humans. Here, we will summarize the current advances in understanding somatic stem cell and tissue aging based on studies of different mouse models.

Aging phenotypes of telomerase model systems

Mice in which the *Terc* gene (mTerc) is disrupted exhibit a decline in telomere length with each successive generation.[32] Due to the long telomere reserve in laboratory mice, the first generation of $mTerc^{-/-}$ mice (G1) of a mixed genetic background were healthy and exhibited no apparent phenotype.[32,33] Successive generations of $mTerc^{-/-}$ mice, however, displayed shorter telomeres, a shortened lifespan, and impaired maintenance of organ systems, with high rates of cell turnover.[33,34] Depending on the telomerase-deficient mouse strain, critical telomere shortening, telomere dysfunction, and accelerated tissue aging occur in the 3rd–6th generation (G3–G6)[35] (Table 1). Aged, late-generation telomere-dysfunctional mice show defects in the hematopoietic system and reduced lung, liver, and pancreas regeneration.[43–45,58] Moreover, in these mice strong atrophy of the intestinal epithelium has been described, coinciding with weight loss and premature aging.[33] In addition to intestinal atrophy, it is possible that telomere dysfunction induces impairments in mitochondria function that may contribute to impairments in organ maintenance, including heart function and β cell metabolism.[59,60]

Similar to mTerc-deficient mice, early-generation $mTert^{-/-}$ mice show no obvious phenotype,[61,62] but late-generation $mTert^{-/-}$ mice show defects, including a decrease in fertility, testis mass, and impaired hematopoietic stem cell (HSC) renewal.[53,62] These data indicate that telomere dysfunction induces an overlapping spectrum of phenotypes in mouse strains carrying mutations in different components of the telomerase holoenzyme.

Several studies showed that Tert has also a telomere length–independent function.[51,63,64] For example, conditional activation of TERT protein is able to induce proliferation of hair follicle stem cells in an $mTerc^{-/-}$ background.[63] Moreover, it has been demonstrated that TERT can modulate Wnt/β-catenin-regulated gene transcription.[65] However, studies on the potential telomere length–independent role of Tert are based on constitutive overexpression experiments. In fact, several studies in Tert-deficient mice argue against an additional functional role for TERT under physiological conditions.[66,67]

In vitro analysis in embryonic stem cells show that overexpression of mTert leads to enhanced proliferation, resistance to apoptosis and oxidative stress, and improved differentiation.[68] Further, TERT overexpression results in the stabilization of telomere length of hematopoietic cells during serial transplantation; however, it does not increase the self-renewal capacity.[55] Aged mice with constitutive telomerase overexpression are more sensitive to spontaneous and induced tumors.[69–71] But not all transgenic mice develop tumors, and those that remained tumor-free exhibit decreased incidence of degenerative diseases but increased lifespan. These data imply that telomere dysfunction may contribute to some aspects of aging of laboratory mouse strains. An alternative explanation for the beneficial effects of Tert-transgene expression indicates that TERT-dependent effects on Wnt/β-catenin signaling may have protective effects.

The dyskerin-null mutation is embryonically lethal.[72] A hypomorphic dyskerin mutant was generated that shows bone marrow failure, increased susceptibility to tumor formation, shortened telomeres at late generation, and impairment of ribosome function.[36] An additional dyskerin mouse model was established carrying a dyskerin mutation that mimics a mutation found in DC.[37] These mutant mice demonstrated impaired cell proliferation that was dependent on telomerase activity.[37] In addition, the proliferative defect paralleled an accumulation of DNA damage and reactive oxidative species (ROS).[37,38] Studies on TERC-deficient mice showed that telomere dysfunction induced ROS formation in a p21-dependent manner and that ROS formation contributed to the permanent induction of senescence.[73]

The breeding of wild-type and late-generation mTerc$^{-/-}$ mice creates healthy mTerc$^{+/-}$ offspring, indicating that the expression of one copy of Terc is sufficient to restore telomere dysfunction during embryogenesis and postnatal life in some mouse strains.[74,75] The reintroduction of one copy of mTerc results in the elongation of critically short telomeres, restoration of chromosomal stability, and rescues proliferative defects.[74] In addition, mice from this intergenerational cross have wild-type levels of germ cell apoptosis, normal weight of the testis, and normal lifespan compared with telomere-deficient mice.[76]

The continuous breeding of Tert heterozygous mice leads to progressive telomere shortening without any evidence of infertility, despite the fact that telomere length is equivalent to that of late-generation Terc$^{-/-}$ mice that have reached infertility.[53] These data emphasize that a limiting amount of telomerase protects the shortest telomeres against dysfunction, and that loss of tissue function in Terc$^{-/-}$ mice is not the consequence of the average telomere length but is triggered by critically short telomeres.[53,75]

In contrast to the rescue of telomere dysfunction by heterozygous expression of telomerase in mouse models, heterozygous mutations of TERC or TERT led to organ failure and premature death in humans (see earlier, Refs. 14, 15, 24, 25). A possible interpretation indicates that in species with short telomeres (humans have short telomeres compared with inbred strains of laboratory mice), heterozygous expression of telomerase subunits may not be sufficient to maintain telomere length and cellular functions. In line with this interpretation, studies on wild-derived castaneous mouse strains (CAST/EiJ) revealed that heterozygous mutation in mTert or mTerc induce defects in organ maintenance and a shortened life span in the first generation,[77,78] similar to human DC.[66,77,79] These data suggest that modifier genes may exist in mouse strains that regulate telomere elongation by telomerase. It is tempting to speculate that tissue-specific expression of such modifier genes could contribute to organ specificity in diseases of patients carrying germline mutations in telomerase or shelterin complex components.

The role of telomerase in somatic stem cells

In mice with dysfunctional telomeres, the maintenance of organ systems that constantly regenerate from stem and progenitor cell compartments was impaired, indicating that these stem cell compartments are telomerase-dependent.[34] Stem cells are one of the few cell types in adult humans that express telomerase.[10] However, telomerase activity is not sufficient to maintain stable telomeres in aging stem cell compartments in humans, and telomere shortening may contribute to the accumulation of DNA damage in adult stem cells during aging.[30]

Several studies have shown that stem cell function declines during aging.[29,80,81] A decrease in stem cell function may contribute to the impaired maintenance and function of some tissue during aging. Experiments on mTerc$^{-/-}$ mice with dysfunctional telomeres have provided the first experimental evidence that telomere shortening can impair the function of somatic and germline stem cells. Telomere dysfunction was found to impair stem cell maintenance and *in vitro* growth capacities in several other organ systems, including neural stem cells, epidermal stem cells, spermatogonial stem cells, intestinal stem cells, muscle stem cells, and HSCs.[37,45–47,56,82–85]

In the hematopoietic system, it was shown that telomere dysfunction impairs the self-renewal and repopulation capacity of HSCs by induction of cell-intrinsic checkpoints.[46,86] Moreover, it was demonstrated that telomere dysfunction induces alterations in the cell environment that contribute to impairments in the differentiation of HSCs into B- and T-lymphoid lineages. This telomere dysfunction–induced alteration in stem cell differentiation could contribute to the declines in lymphopoiesis and immune functions that occur during aging.[87] Telomere dysfunction–induced alteration in the systemic, circulatory environment also impaired the engraftment of HSCs (see later, Ref. 85). It is possible that these cell-extrinsic mechanisms would also limit therapeutic approaches aiming to improve the function of degenerative tissues by stem cell transplantation. Data from hematopoietic stem cell transplantations (HSCT) in patients have shown that telomeres in the recipients are in general shorter than in their donors and that several exposures are thought to affect telomere length after HSCT, including age, genetics, and environmental factors (reviewed in Ref. 88).

Checkpoint response and environmental alteration

Induction of cell cycle checkpoints in response to telomere dysfunction

Studies on telomerase-deficient mice have shown that telomere dysfunction can impair stem cell functions by induction of cell-intrinsic checkpoints. When telomeres reach a critically short length, they are recognized as DNA damage–inducing checkpoints that lead to cell cycle arrest (senescence) or cell death (crisis) in primary human fibroblasts. The senescence checkpoint represents the primary response to dysfunctional telomeres in human fibroblast and it was shown that p21 is mediating the induction of senescence cell cycle arrest.[89] When cells bypass the senescence checkpoint (e.g., in the absence of p53) they continue to proliferate despite the presence of dysfunctional telomeres. At a certain telomere length, a second checkpoint is activated that is characterized by severe chromosomal instability and cell death. This checkpoint is p53 independent and is also known as "crisis."[90] The molecular nature of the crisis checkpoint is not understood and may include a series of different molecular events induced by different chromosomal lesions accumulating in cells in crisis.

Several genetic components that induce checkpoints in response to DNA damage and telomere dysfunction have also been analyzed in telomere-dysfunctional mice including Exonuclease-1 (Exo1), ataxia telangiectasia mutated (ATM), CHK2, p53, p21, Ink4a (codeletion of p16 and p19ARF), PMS2, and Msh2 (Table 2).[46,91–97]

Exo1

Exo1 is a DNA 5′-3′exonuclease and functions in recombination and mismatch repair.[100,101] It has been shown that Exo1 mediates the induction of DNA damage checkpoints in yeast by processing dysfunctional telomeres.[102] *In vivo* studies on telomere-dysfunctional mice provided the first experimental evidence that Exo1 processes dysfunctional telomeres and DNA breaks in mammalian cells.[91] Specifically, Exo1 deletion impaired the induction of single-stranded DNA and ataxia telangiectasia and Rad3-related (ATR) activation in telomere-dysfunctional mice and in response to irradiation induced DNA breaks. Deletion of Exo1 prevented the induction of DNA damage signals and reduces activation of apoptosis and cell cycle arrest checkpoints in telomere-dysfunctional tissues. Deletion of Exo1 also leads to a decreased formation of chromosomal fusion in response to telomere dysfunction in tissues, thus providing a plausible explanation why chromosomal instability or cancer formation was not induced.[91] The functional role of Exo1 in processing of DNA breaks and in the induction of DNA damage signals was reconfirmed in other models of DNA damage induction in mammalian cells.[103]

p53 and p21

In human fibroblasts, p53 activation in response to DNA damage induces cell cycle arrest.[90,104,105]

Table 1. Age-related phenotypes in telomerase model systems

	Genotype of the telomerase model	Phenotype	References
Dyskerin (DKC1)			
Various tissues	Hypomorphic Dkc1 mutant ($Dkc1^{m}$)	Bone marrow failure, increased susceptibility to tumor formation, shortening of telomeres at late generation, impairment of ribosome function	36
Various tissues	$Dkc1^{\delta 15}$	Impairment of cell proliferation, accumulation of DNA damage and reactive oxidative species	37, 38
Telomerase and Cajal body protein 1 (TCAB1)			
Various cell lines	TCAB1 knockdown	Disruption of telomerase–telomere association, abrogation of telomere synthesis by telomerase	2
Telomerase RNA component (TERC)			
Overall phenotype	G3-G6 $mTerc^{-/-}$	Neural tube closure defects in a subset of embryos, increased incidence of skin lesions, alopecia and hair graying, decreased body weight, development of anemia and leucopenia, kyphosis	33, 39
Intestine	G3-G6 $mTerc^{-/-}$	Increased radio sensitivity, age-dependent atrophy of the intestinal epithelium	33, 40
Stress response	G3-G6 $mTerc^{-/-}$	Delayed wound healing, reduced hematopoiesis in response to chemotherapy, reduced liver, pancreas and liver regeneration	33; 41–44
Hematopoietic cells	G3-G6 $mTerc^{-/-}$	Reduced self-renewal, colony-forming capacity, and repopulation capacity, skewed differentiation (impaired lymphopoiesis, increased myelopoiesis)	34, 45–47
Reproductive system	G3-G6 $mTerc^{-/-}$	Impaired spermatogenesis	34, 48
Cancer	G3-G6 $mTerc^{-/-}$	Increased chromosomal instability and cancer initiation, impaired tumor progression	33, 41, 49–52

Continued

Table 1. *Continued*

	Genotype of the telomerase model	Phenotype	References
Telomerase reverse transcriptase (TERT)			
Various tissues	mTert$^{-/-}$	Reduced self-renewal of hematopoietic cells, decrease in fertility and testis mass, increased intestinal cell apoptosis	45, 53, 54
Hematopoietic cells	Overexpression of TERT	Stabilization of telomere attrition, self-renewal capacity unaffected	55
Epidermis	Overexpression of TERT (K5-Tert)	Increased mobilization of epidermal stem cells, enhanced hair growth, and skin hyperplasia	51, 56
Various tissues	mTert reactivation in telomere-dysfunctional mice	Increased stem cell function	57

In vivo, telomere dysfunction induces cell cycle arrest or apoptosis.[46,52] Removal of p53 function rescues some telomere dysfunction–induced defects, including growth arrest in cell culture, testicular atrophy, and germ cell apoptosis.[52] However, it was not possible to study the function of p53 in somatic stem cell compartments during aging, because germline deletion of p53 resulted in premature death from cancer formation.[52] To bypass this problem, a conditional p53 knockout mouse was used, in which p53 was deleted specifically in intestinal epithelium in a telomere-deficient background.[92] These double mutants did not exhibit an improvement of the intestinal epithelium, but showed an atrophy of the intestinal epithelium, increased apoptosis, and premature weight loss compared to telomere-dysfunctional mice with intact p53. The increased apoptosis in the intestine in the double mutants was associated with increased rates of DNA damage and chromosomal instability at the stem cell level. The study provided experimental evidence that p53 contributes to the depletion of chromosomal instable stem cells in somatic tissue and that the aberrant survival of such instable stem cells can lead to accelerated tissue aging (Fig. 1).[92]

One downstream target of the p53 pathway is the cyclin-dependent kinase inhibitor p21.[89] p21 induces a transient cell cycle arrest in response to irradiation and a permanent G1 cell cycle arrest in response to telomere dysfunction, which is known as replicative senescence.[89,106] The function of p21 has been studied in mice with dysfunctional telomeres. In addition, p21 expression is increased in aging mice with dysfunctional telomeres.[46] Deletion of p21 (encoded by *Cdkn1a*) prolonged the lifespan and increased body weight of the mice. This rescue was associated with improved stem cell function and prolonged maintenance of high-turnover organs including the intestinal epithelium, spleen, and hematopoietic system.[46] These results provided experimental proof that p21-dependent checkpoints can impair stem cell function and self-renewal in response to telomere dysfunction. However, p21 deletion did not affect apoptosis checkpoints, and there was no accumulation of chromosomal instable cells in aging telomere-dysfunctional tissues as it was observed in response to p53 deletion.[46,92] These results imply that p21-independent checkpoints contribute to the inhibition of chromosomal instability and transformation in the context of telomere dysfunction. The nature of these checkpoints remains to be defined. It is possible that these checkpoints include p53-dependent apoptosis. Studies on p53 mutant

Table 2. Role of DNA damage checkpoint genes in telomerase model systems. The table shows alterations in the phenotype of late-generation, Terc-deficient mice induced by the deletion of the indicated checkpoint genes

Genotype	Phenotype	Lifespan	References
mTerc$^{-/-}$ Exo1$^{-/-}$	Prevents induction of ATR/p53 and chromosomal fusion, no increase in chromosomal instability or cancer formation	Prolonged	91
mTerc$^{-/-}$ p53$^{-/-}$	Accelerated intestinal crypt atrophy, increased apoptosis and weight loss, increased cancer formation	Shortened	92, 98
mTerc$^{-/-}$ p21$^{-/-}$	Improved fitness, extended lifespan of mice with dysfunction telomeres, improvement of stem cell function and organ maintenance, no increase in cancer	Prolonged	46
mTerc$^{-/-}$ Atm	Increased telomere shortening and genomic instability, increased germ cell death, reduced tumorigenesis, and defects in stem cell function	Shortened	40, 99
mTerc$^{-/-}$ CHK2$^{-/-}$	No improvement in progenitor cell function or organ maintenance, no rescue in p53 induction	Unaffected	93
mTerc PMS2	Rescue of degenerative pathologies, reduced tumor formation	Prolonged	96
mTerc MSH2	Rescue of degenerative pathologies	Shortened	94
mTerc$^{-/-}$ Ink4a$^{-/-}$	Suppression of tumor initiation	Unaffected	50, 95

mice showed that p21 can suppress tumor formation when p53-dependent apoptosis responses are abrogated.[107,108] These data suggest that p53-dependent apoptosis and cell cycle arrest may represent compensatory checkpoints that contribute to impairments in stem cell function and tissue aging in the context of telomere dysfunction. Alternatively, other p53 targets may be involved, for example, autophagy or G2 cell cycle arrest.

Atm

The ataxia telangiectasia mutated gene (*Atm*) encodes for a kinase, which inhibits cell cycle progression in response to DNA damage. *Atm* has been identified in the autosomal recessive disease ataxia telangiectasia, which is characterized by sterility, immune defects, genomic instability, cancer predisposition, and neurodegeneration.[109] It has been shown that Atm is activated in response to telomere dysfunction.[110] Further, it was reported that Atm deficiency leads to telomere shortening in yeast and mammalian cells.[110,111] In Terc-deficient mice, *Atm* deletion accelerated telomere erosion, genomic instability, reduced tumorigenesis, and defects in stem cell function.[97,99] These results indicated that ATM-independent pathways can mediate checkpoint induction and tissue atrophy in response to telomere dysfunction possibly involving Exo1/ATR-dependent signaling (see earlier and Ref. 91). It remains to be delineated what mechanisms accelerate tissue aging in telomere-dysfunctional mice in response to *Atm* deletion; the role of ATM in telomere shortening could contribute, but the exact nature of this process remains to be defined. It was also observed that ATM deficiency results in increased levels of ROS inhibiting stem cell function.[112] Although under debate, it is possible that increasing ROS levels accelerate cellular and organismal aging in the context of telomere dysfunction.[113]

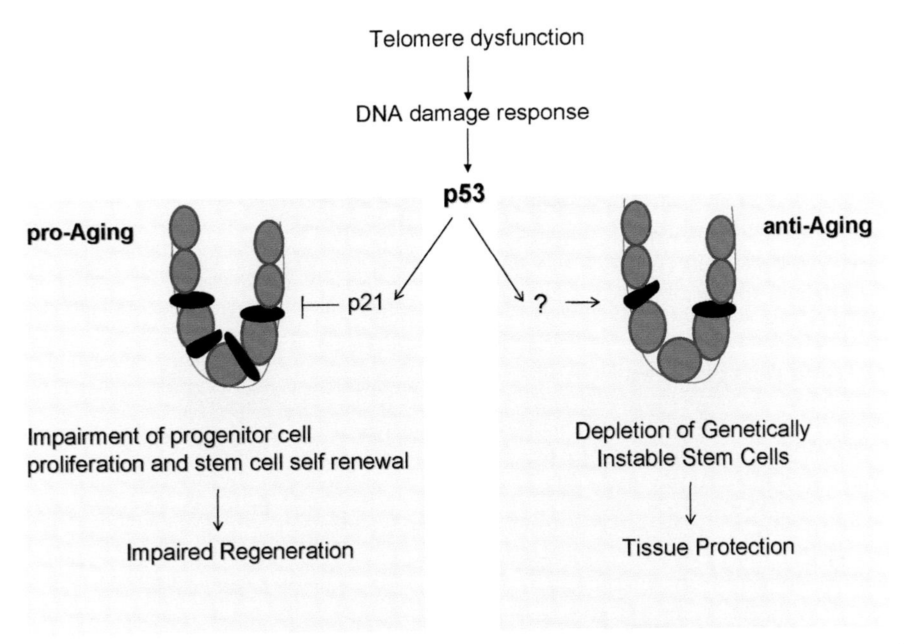

Figure 1. A dual role of p53 in telomere dysfunction–induced aging. Telomere dysfunction leads to the induction of p53, which impairs tissue maintenance by inhibiting self-renewal and proliferation of stem and progenitor cells. In contrast, p53 has a tissue-protective effect by eliminating chromosomal instable stem cells. The mediators of this checkpoint remain to be defined.

Chk2

Chk2 is activated in an ATM-dependent manner in response to DNA damage. Chk2 knockout mice were crossed with Terc$^{-/-}$ mice to study its role in the induction of telomere dysfunction checkpoints *in vivo*.[93] Chk2 was shown to mediate the induction of senescence of primary human fibroblasts in response to telomere shortening.[114] In contrast, Chk2 deletion did not rescue activation of p53, cell cycle arrest, or apoptosis in the intestinal stem cell compartment of telomere-dysfunctional mice. Chk2 deletion did not improve progenitor cell function, organ maintenance, or life span of telomere-dysfunctional mice. These results indicated that stem cells exhibit different checkpoint responses to dysfunctional telomeres compared to fibroblast cultures. Given the role of stem cells in tissue aging and cancer formation, these results could have implications for both processes and indicate the need to study checkpoint responses *in vivo* in stem cell compartments.

Telomere shortening and environmental alterations

Alterations of the environment can either change the macro-environment (systemic factors) or the micro-environment (niche).[29,115] One cell-intrinsic factor that is altered by environmental factors is telomere length.[116] The effect of oxidative stress on telomere length has been observed under chronic hyperoxia[110] and by exposure of various reagents.[117,118] Furthermore, fibroblasts from patients with Fanconia anemia, a disease that results in elevated level of oxidative stress, show accelerated telomere shortening.[119] As an additional extrinsic factor, hormones have been demonstrated to have an effect on telomere length regulation.[120] For example, *in vivo* studies revealed a link between estrogen and telomere shortening.[121]

Telomere dysfunction is a cell-intrinsic factor that leads to the induction of checkpoints and DNA repair. However, telomere dysfunction was also shown to induce environmental alterations that impair the function of HSCs.[85] HSC transplantation experiments revealed an influence of telomere dysfunction on the stem cell environment. One of the strong phenotypes of the hematopoietic system in telomere-dysfunctional mice is a skewing in lympho-/myelopoiesis. Telomere-dysfunctional mice show an impairment in B lymphocyte development and enhanced myeloid lineage differentiation.[47] The same phenotype occurs during human aging. HSC transplantation and bone transplantation experiments revealed that alteration in the circulatory environment contributes to impaired lymphopoiesis in response to telomere dysfunction.[47] It remains to be elucidated what factors contribute to this process and whether similar mechanisms contribute to human aging. Studies in telomerase-deficient mice revealed that increased levels of G-CSF could contribute to the observed skewing in hemato-lymphopoiesis.[85]

Conclusions and perspectives

The accumulation of DNA damage represents one important factor that can contribute to organismal aging and that affects aging stem cells. Telomere shortening was identified as a cell-intrinsic mechanism contributing to the accumulation of DNA damage in replicative cell aging, and it also affects aging stem cells. Association studies revealed that telomere shortening seems to contribute to organismal aging. Moreover, telomerase mutations were identified as a cause of impaired stem cell function and reduced organ maintenance in humans. To better understand telomere-dependent effects on aging and cancer, it is important to delineate molecular pathways that mediate cellular and organismal defects in response to telomere dysfunction. These studies could ultimately identify molecular targets for future therapies aiming to improve organ homeostasis and regeneration during aging and in chronic disease. Telomeres could also represent a target for cancer therapies aiming to destroy the immortality of cancer cells. The complexity of telomere-related biology in aging and tissue degeneration indicates the need to study these processes using *in vivo* models. Given the increasing evidence that aging of stem cells contributes to tissue aging and the development of disease, it seems to be of utmost importance to analyze checkpoint response at stem cell level. In the past few years, it became clear that telomere dysfunction can have an impact on stem cell function involving both cell-extrinsic and cell-intrinsic mechanisms. In addition, there is emerging evidence that telomere dysfunction can cooperate with other factors influencing cellular and organismal function during aging, for example, mitochondria dysfunction. The delineation of stem cell responses to telomere dysfunction will ultimately improve our understanding of disease pathologies, aging, and cancer and can lead to the development of new therapies aiming to prevent or delay the progression of these processes.

Conflicts of interest

The authors declare no conflicts of interest.

References

1. Cohen, S.B., M.E. Graham, G.O. Lovrecz, *et al.* 2007. Protein composition of catalytically active human telomerase from immortal cells. *Science* **315:** 1850–1853.
2. Venteicher, A.S., E.B. Abreu, Z. Meng, *et al.* 2009. A human telomerase holoenzyme protein required for Cajal body localization and telomere synthesis. *Science* **323:** 644–648.
3. Mitchell, J.R., E. Wood & K. Collins. 1999. A telomerase component is defective in the human disease dyskeratosis congenita. *Nature* **402:** 551–555.
4. Mochizuki, Y., J. He, S. Kulkarni, *et al.* 2004. Mouse dyskerin mutations affect accumulation of telomerase RNA and small nucleolar RNA, telomerase activity, and ribosomal RNA processing. *Proc. Natl. Acad. Sci. U. S. A.* **101:** 10756–10761.
5. Meier, U.T. 2005. The many facets of H/ACA ribonucleoproteins. *Chromosoma* **114:** 1–14.
6. Mitchell, J.R., J. Cheng & K. Collins. 1999. A box H/ACA small nucleolar RNA-like domain at the human telomerase RNA 3' end. *Mol. Cell Biol.* **19:** 567–576.
7. Venteicher, A.S. & S.E. Artandi. 2009. TCAB1: driving telomerase to Cajal bodies. *Cell Cycle* **8:** 1329–1331.
8. Rajaraman, S., J. Choi, P. Cheung, *et al.* 2007. Telomere uncapping in progenitor cells with critical telomere shortening is coupled to S-phase progression in vivo. *Proc. Natl. Acad. Sci. U. S. A.* **104:** 17747–17752.
9. Wright, W.E., M.A. Piatyszek, W.E. Rainey, *et al.* 1996. Telomerase activity in human germline and embryonic tissues and cells. *Dev. Genet.* **18:** 173–179.
10. Chiu, C.P., W. Dragowska, N.W. Kim, *et al.* 1996. Differential expression of telomerase activity in hematopoietic progenitors from adult human bone marrow. *Stem Cells* **14:** 239–248.
11. Martens, U.M., V. Brass, L. Sedlacek, *et al.* 2002. Telomere maintenance in human B lymphocytes. *Br. J. Haematol.* **119:** 810–818.
12. Wright, W.E. & J.W. Shay. 2001. Cellular senescence as a tumor-protection mechanism: the essential role of counting. *Curr. Opin. Genet. Dev.* **11:** 98–103.
13. Hahn, W.C., S.A. Stewart, M.W. Brooks, *et al.* 1999. Inhibition of telomerase limits the growth of human cancer cells. *Nat. Med.* **5:** 1164–1170.
14. Armanios, M.Y., J.J. Chen, J.D. Cogan, *et al.* 2007. Telomerase mutations in families with idiopathic pulmonary fibrosis. *N. Engl. J. Med.* **356:** 1317–1326.
15. Armanios, M., J.L. Chen, Y.P. Chang, *et al.* 2005. Haploinsufficiency of telomerase reverse transcriptase leads to anticipation in autosomal dominant dyskeratosis congenita. *Proc. Natl. Acad. Sci. U. S. A.* **102:** 15960–15964.
16. Calado, R.T., J.A. Regal, D.E. Kleiner, *et al.* 2009. A spectrum of severe familial liver disorders associate with telomerase mutations. *PLoS One* **4:** e7926.

17. Hartmann, D., U. Srivastava, M. Thaler, *et al.* 2011. Telomerase gene mutations are associated with cirrhosis formation. *Hepatology* **53:** 1608–1617.
18. Calado, R.T., J. Brudno, P. Mehta, *et al.* 2011. Constitutional telomerase mutations are genetic risk factors for cirrhosis. *Hepatology* **53:** 1600–1607.
19. Vulliamy, T.J., S.W. Knight, N.S. Heiss, *et al.* 1999. Dyskeratosis congenita caused by a 3' deletion: germline and somatic mosaicism in a female carrier. *Blood* **94:** 1254–1260.
20. Zhong, F., S.A. Savage, M. Shkreli, *et al.* 2011. Disruption of telomerase trafficking by TCAB1 mutation causes dyskeratosis congenita. *Genes. Dev.* **25:** 11–16.
21. Nishio, N. & S. Kojima. 2011. Recent progress in dyskeratosis congenita. *Int. J. Hematol.* **92:** 419–424.
22. Kirwan, M. & I. Dokal. 2009. Dyskeratosis congenita, stem cells and telomeres. *Biochim. Biophys. Acta.* **1792:** 371–379.
23. Savage, S.A., N. Giri, G.M. Baerlocher, *et al.* 2008. TINF2, a component of the shelterin telomere protection complex, is mutated in dyskeratosis congenita. *Am. J. Hum. Genet.* **82:** 501–509.
24. Vulliamy, T., A. Marrone, F. Goldman, *et al.* 2001. The RNA component of telomerase is mutated in autosomal dominant dyskeratosis congenita. *Nature* **413:** 432–435.
25. Vulliamy, T., A. Marrone, R. Szydlo, *et al.* 2004. Disease anticipation is associated with progressive telomere shortening in families with dyskeratosis congenita due to mutations in TERC. *Nat. Genet.* **36:** 447–449.
26. Wiemann, S.U., A. Satyanarayana, M. Tsahuridu, *et al.* 2002. Hepatocyte telomere shortening and senescence are general markers of human liver cirrhosis. *Faseb. J.* **16:** 935–942.
27. Brummendorf, T.H., J.P. Maciejewski, J. Mak, *et al.* 2001. Telomere length in leukocyte subpopulations of patients with aplastic anemia. *Blood* **97:** 895–900.
28. Kinouchi, Y., N. Hiwatashi, M. Chida, *et al.* 1998. Telomere shortening in the colonic mucosa of patients with ulcerative colitis. *J. Gastroenterol.* **33:** 343–348.
29. Rando, T.A. 2006. Stem cells, ageing and the quest for immortality. *Nature* **441:** 1080–1086.
30. Vaziri, H., W. Dragowska, R.C. Allsopp, *et al.* 1994. Evidence for a mitotic clock in human hematopoietic stem cells: loss of telomeric DNA with age. *Proc. Natl. Acad. Sci. U. S. A.* **91:** 9857–9860.
31. Prowse, K.R. & C.W. Greider. 1995. Developmental and tissue-specific regulation of mouse telomerase and telomere length. *Proc. Natl. Acad. Sci. U. S. A.* **92:** 4818–4822.
32. Blasco M.A., H.W. Lee, M.P. Hande, *et al.* 1997. Telomere shortening and tumor formation by mouse cells lacking telomerase RNA. *Cell* **91:** 25–34.
33. Rudolph, K.L., S. Chang, H.W. Lee, *et al.* 1999. Longevity, stress response, and cancer in aging telomerase-deficient mice. *Cell* **96:** 701–712.
34. Lee, H.W., M.A. Blasco, G.J. Gottlieb, *et al.* 1998. Essential role of mouse telomerase in highly proliferative organs. *Nature* **392:** 569–574.
35. Herrera, E., E. Samper, J. Martin-Caballero, *et al.* 1999. Disease states associated with telomerase deficiency appear earlier in mice with short telomeres. *Embo. J.* **18:** 2950–2960.
36. Ruggero, D., S. Grisendi, F. Piazza, *et al.* 2003. Dyskeratosis congenita and cancer in mice deficient in ribosomal RNA modification. *Science* **299:** 259–262.
37. Gu, B.W., M. Bessler & P.J. Mason. 2008. A pathogenic dyskerin mutation impairs proliferation and activates a DNA damage response independent of telomere length in mice. *Proc. Natl. Acad. Sci. U. S. A.* **105:** 10173–10178.
38. Gu, B.W., J.M. Fan, M. Bessler & P.J. Mason. 2011. Accelerated hematopoietic stem cell aging in a mouse model of dyskeratosis congenita responds to antioxidant treatment. *Aging Cell* **10:** 338–348.
39. Herrera, E., E. Samper & M.A. Blasco. 1999. Telomere shortening in mTR-/- embryos is associated with failure to close the neural tube. *Embo. J.* **18:** 1172–1181.
40. Wong, K.K., S. Chang, S.R. Weiler, *et al.* 2000. Telomere dysfunction impairs DNA repair and enhances sensitivity to ionizing radiation. *Nat. Genet.* **26:** 85–88.
41. Rudolph, K.L., M. Millard, M.W. Bosenberg & R.A. DePinho. 2001. Telomere dysfunction and evolution of intestinal carcinoma in mice and humans. *Nat. Genet.* **28:** 155–159.
42. Satyanarayana, A., S.U. Wiemann, J. Buer, *et al.* 2003. Telomere shortening impairs organ regeneration by inhibiting cell cycle re-entry of a subpopulation of cells. *Embo. J.* **22:** 4003–4013.
43. von Figura, G., M. Wagner, K. Nalapareddy, *et al.* 2011. Regeneration of the exocrine pancreas is delayed in telomere-dysfunctional mice. *PLoS One* **6:** e17122.
44. Alder, J.K., N. Guo, F. Kembou, *et al.* 2011. Telomere length is a determinant of emphysema susceptibility. *Am. J. Respir. Crit. Care Med* **184:** 904–912.
45. Allsopp, R.C., G.B. Morin, R. DePinho , *et al.* 2003. Telomerase is required to slow telomere shortening and extend replicative lifespan of HSCs during serial transplantation. *Blood* **102:** 517–520.
46. Choudhury, A.R., Z. Ju, M.W. Djojosubroto, *et al.* 2007. Cdkn1a deletion improves stem cell function and lifespan of mice with dysfunctional telomeres without accelerating cancer formation. *Nat. Genet.* **39:** 99–105.
47. Song, Z., J. Wang, L.M. Guachalla, *et al.* 2010. Alterations of the systemic environment are the primary cause of impaired B and T lymphopoiesis in telomere-dysfunctional mice. *Blood* **115:** 1481–1489.
48. Hemann, M.T., K.L. Rudolph, M.A. Strong, *et al.* 2001. Telomere dysfunction triggers developmentally regulated germ cell apoptosis. *Mol. Biol. Cell* **12:** 2023–2030.
49. Farazi, P.A., J. Glickman, S. Jiang, *et al.* 2003. Differential impact of telomere dysfunction on initiation and progression of hepatocellular carcinoma. *Cancer Res.* **63:** 5021–5027.
50. Greenberg, R.A., L. Chin, A. Femino, *et al.* 1999. Short dysfunctional telomeres impair tumorigenesis in the INK4a(delta2/3) cancer-prone mouse. *Cell* **97:** 515–525.
51. Artandi, S.E., S. Alson, M.K. Tietze, *et al.* 2002. Constitutive telomerase expression promotes mammary carcinomas in aging mice. *Proc. Natl. Acad. Sci. U. S. A.* **99:** 8191–8196.
52. Chin, L., S.E. Artandi, Q. Shen, *et al.* 1999. p53 deficiency rescues the adverse effects of telomere loss and cooperates with telomere dysfunction to accelerate carcinogenesis. *Cell* **97:** 527–538.
53. Chiang, Y.J., R.T. Calado, K.S. Hathcock, *et al.* 2010. Telomere length is inherited with resetting of the telomere setpoint. *Proc. Natl. Acad. Sci. U. S. A.* **107:** 10148–10153.

54. Meznikova, M., N. Erdmann, R. Allsopp & L.A. Harrington. 2009. Telomerase reverse transcriptase-dependent telomere equilibration mitigates tissue dysfunction in mTert heterozygotes. *Dis. Model Mech.* **2:** 620–626.
55. Allsopp, R.C., G.B. Morin, J.W. Horner, *et al.* 2003. Effect of TERT over-expression on the long-term transplantation capacity of hematopoietic stem cells. *Nat. Med.* **9:** 369–371.
56. Flores, I., M.L. Cayuela & M.A. Blasco. 2005. Effects of telomerase and telomere length on epidermal stem cell behavior. *Science* **309:** 1253–1256.
57. Jaskelioff, M., F.L. Muller, J.H. Paik, *et al.* 2011. Telomerase reactivation reverses tissue degeneration in aged telomerase-deficient mice. *Nature* **469:** 102–106.
58. Rudolph, K.L., S. Chang, M. Millard, *et al.* 2000. Inhibition of experimental liver cirrhosis in mice by telomerase gene delivery. *Science* **287:** 1253–1258.
59. Sahin, E., S. Colla, M. Liesa, *et al.* 2011. Telomere dysfunction induces metabolic and mitochondrial compromise. *Nature* **470:** 359–365.
60. Guo, N., E.M. Parry, L.S. Li, *et al.* 2011. Short telomeres compromise beta-cell signaling and survival. *PLoS One* **6:** e17858.
61. Liu, Y., B.E. Snow, M.P. Hande, *et al.* 2000. The telomerase reverse transcriptase is limiting and necessary for telomerase function in vivo. *Curr. Biol.* **10:** 1459–1462.
62. Erdmann, N., Y. Liu & L. Harrington. 2004. Distinct dosage requirements for the maintenance of long and short telomeres in mTert heterozygous mice. *Proc. Natl. Acad. Sci. U.S.A.* **101:** 6080–6085.
63. Sarin, K.Y., P. Cheung, D. Gilison, *et al.* 2005. Conditional telomerase induction causes proliferation of hair follicle stem cells. *Nature* **436:** 1048–1052.
64. Lee, J., Y.H. Sung, C. Cheong, *et al.* 2008. TERT promotes cellular and organismal survival independently of telomerase activity. *Oncogene* **27:** 3754–3760.
65. Park, J.I., A.S. Venteicher, J.Y. Hong, *et al.* 2009. Telomerase modulates Wnt signalling by association with target gene chromatin. *Nature* **460:** 66–72.
66. Strong, M.A., S.L. Vidal-Cardenas, B. Karim, *et al.* 2011. Phenotypes in mTERT/ and mTERT/ mice are due to short telomeres, not telomere-independent functions of telomerase reverse transcriptase. *Mol. Cell Biol.* **31:** 2369–2379.
67. Vidal-Cardenas, S.L. & C.W. Greider. 2010. Comparing effects of mTR and mTERT deletion on gene expression and DNA damage response: a critical examination of telomere length maintenance-independent roles of telomerase. *Nucleic Acids Res.* **38:** 60–71.
68. Armstrong, L., G. Saretzki, H. Peters, *et al.* 2005. Overexpression of telomerase confers growth advantage, stress resistance, and enhanced differentiation of ESCs toward the hematopoietic lineage. *Stem Cells* **23:** 516–529.
69. Gonzalez-Suarez, E., E. Samper, A. Ramirez, *et al.* 2001. Increased epidermal tumors and increased skin wound healing in transgenic mice overexpressing the catalytic subunit of telomerase, mTERT, in basal keratinocytes. *Embo. J.* **20:** 2619–2630.
70. Gonzalez-Suarez, E., C. Geserick, J.M. Flores & M.A. Blasco. 2005. Antagonistic effects of telomerase on cancer and aging in K5-mTert transgenic mice. *Oncogene* **24:** 2256–2270.
71. Gonzalez-Suarez, E., J.M. Flores & M.A. Blasco. 2002. Cooperation between p53 mutation and high telomerase transgenic expression in spontaneous cancer development. *Mol Cell. Biol.* **22:** 7291–7301.
72. He, J. S., Navarrete, M. Jasinski, *et al.* 2002. Targeted disruption of Dkc1, the gene mutated in X-linked dyskeratosis congenita, causes embryonic lethality in mice. *Oncogene* **21:** 7740–7744.
73. Passos, J.F., G. Nelson, C. Wang, *et al.* 2010. Feedback between p21 and reactive oxygen production is necessary for cell senescence. *Mol. Syst. Biol.* **6:** 347.
74. Samper, E., J.M. Flores & M.A. Blasco. 2001. Restoration of telomerase activity rescues chromosomal instability and premature aging in Terc$^{-/-}$ mice with short telomeres. *EMBO Rep.* **2:** 800–807.
75. Hemann, M.T., M.A. Strong, L.Y. Hao & C.W. Greider. 2001. The shortest telomere, not average telomere length, is critical for cell viability and chromosome stability. *Cell* **107:** 67–77.
76. Gonzalez-Suarez, E., E. Samper, J.M. Flores & M.A. Blasco. 2000. Telomerase-deficient mice with short telomeres are resistant to skin tumorigenesis. *Nat. Genet.* **26:** 114–117.
77. Hao, L.Y., M. Armanios, M.A. Strong, *et al.* 2005. Short telomeres, even in the presence of telomerase, limit tissue renewal capacity. *Cell* **123:** 1121–1131.
78. Hemann, M.T. & C.W. Greider. 2000. Wild-derived inbred mouse strains have short telomeres. *Nucl. Acids Res.* **28:** 4474–4478.
79. Armanios, M., J.K. Alder, E.M. Parry, *et al.* 2009. Short telomeres are sufficient to cause the degenerative defects associated with aging. *Am. J. Hum. Genet.* **85:** 823–832.
80. Van Zant, G. & Y. Liang. 2003. The role of stem cells in aging. *Exp. Hematol.* **31:** 659–672.
81. Rossi, D.J., D. Bryder & I.L. Weissman. 2007. Hematopoietic stem cell aging: mechanism and consequence. *Exp. Gerontol.* **42:** 385–390.
82. Rossi, D.J., D. Bryder, J. Seita, *et al.* 2007. Deficiencies in DNA damage repair limit the function of haematopoietic stem cells with age. *Nature* **447:** 725–729.
83. Ferron, S., H. Mira, S. Franco, *et al.* 2004. Telomere shortening and chromosomal instability abrogates proliferation of adult but not embryonic neural stem cells. *Development* **131:** 4059–4070.
84. Sacco, A., F. Mourkioti, R. Tran, *et al.* 2010. Short telomeres and stem cell exhaustion model Duchenne muscular dystrophy in mdx/mTR mice. *Cell* **143:** 1059–1071.
85. Ju, Z., H. Jiang, M. Jaworski, *et al.* 2007. Telomere dysfunction induces environmental alterations limiting hematopoietic stem cell function and engraftment. *Nat. Med.* **13:** 742–747.
86. Allsopp, R.C., S. Cheshier & I.L. Weissman. 2001. Telomere shortening accompanies increased cell cycle activity during serial transplantation of hematopoietic stem cells. *J. Exp. Med.* **193:** 917–924.
87. Wang, J., H. Geiger & K.L. Rudolph. 2011. Immunoaging induced by hematopoietic stem cell aging. *Curr. Opin. Immunol.* **23:** 532–536.
88. Gadalla, S.M. & S.A. Savage. 2011. Telomere biology in hematopoiesis and stem cell transplantation. *Blood Rev.* **25:** 261–269.

89. Brown, J.P., W. Wei & J.M. Sedivy. 1997. Bypass of senescence after disruption of p21CIP1/WAF1 gene in normal diploid human fibroblasts. *Science* **277:** 831–834.
90. Wright, W.E. & J.W. Shay. 1992. The two-stage mechanism controlling cellular senescence and immortalization. *Exp. Gerontol.* **27:** 383–389.
91. Schaetzlein, S., N.R. Kodandaramireddy, Z. Ju, *et al.* 2007. Exonuclease-1 deletion impairs DNA damage signaling and prolongs lifespan of telomere-dysfunctional mice. *Cell* **130:** 863–877.
92. Begus-Nahrmann, Y., A. Lechel, A.C. Obenauf, *et al.* 2009. p53 deletion impairs clearance of chromosomal-instable stem cells in aging telomere-dysfunctional mice. *Nat. Genet.* **41:** 1138–1143.
93. Nalapareddy, K., A.R. Choudhury, A. Gompf, *et al.* 2010. CHK2-independent induction of telomere dysfunction checkpoints in stem and progenitor cells. *EMBO Rep.* **11:** 619–625.
94. Martinez, P., I. Siegl-Cachedenier, J.M. Flores & M.A. Blasco. 2009. MSH2 deficiency abolishes the anticancer and pro-aging activity of short telomeres. *Aging Cell* **8:** 2–17.
95. Khoo, C.M., D.R. Carrasco, M.W. Bosenberg, *et al.* 2007. Ink4a/Arf tumor suppressor does not modulate the degenerative conditions or tumor spectrum of the telomerase-deficient mouse. *Proc. Natl. Acad. Sci. U. S. A.* **104:** 3931–3936.
96. Siegl-Cachedenier, I., P. Munoz, J.M. Flores, *et al.* 2007. Deficient mismatch repair improves organismal fitness and survival of mice with dysfunctional telomeres. *Genes. Dev.* **21:** 2234–2247.
97. Wong, K.K., R.S. Maser, R.M. Bachoo, *et al.* 2003. Telomere dysfunction and Atm deficiency compromises organ homeostasis and accelerates ageing. *Nature* **421:** 643–648.
98. Artandi, S.E., S. Chang, S.L. Lee, *et al.* 2000. Telomere dysfunction promotes non-reciprocal translocations and epithelial cancers in mice. *Nature* **406:** 641–645.
99. Qi, L., M.A. Strong, B.O. Karim, *et al.* 2003. Short telomeres and ataxia-telangiectasia mutated deficiency cooperatively increase telomere dysfunction and suppress tumorigenesis. *Cancer Res.* **63:** 8188–8196.
100. Tishkoff, D.X., A.L. Boerger, P. Bertrand, *et al.* 1997. Identification and characterization of Saccharomyces cerevisiae EXO1, a gene encoding an exonuclease that interacts with MSH2. *Proc. Natl. Acad. Sci. U. S. A.* **94:** 7487–7492.
101. Qiu, J., Y. Qian, V. Chen, *et al.* 1999. Human exonuclease 1 functionally complements its yeast homologues in DNA recombination, RNA primer removal, and mutation avoidance. *J. Biol. Chem.* **274:** 17893–17900.
102. Hackett, J.A. & C.W. Greider. 2003. End resection initiates genomic instability in the absence of telomerase. *Mol. Cell. Biol.* **23:** 8450–8461.
103. Morin, I., H.P. Ngo, A. Greenall, *et al.* 2008. Checkpoint-dependent phosphorylation of Exo1 modulates the DNA damage response. *Embo. J.* **27:** 2400–2410.
104. d'Adda di Fagagna, F., P.M. Reaper, L. Clay-Farrace, *et al.* 2003. A DNA damage checkpoint response in telomere-initiated senescence. *Nature* **426:** 194–198.
105. Vaziri, H. & S. Benchimol. 1996. From telomere loss to p53 induction and activation of a DNA-damage pathway at senescence: the telomere loss/DNA damage model of cell aging. *Exp. Gerontol.* **31:** 295–301.
106. Herbig, U., W.A. Jobling, B.P. Chen, *et al.* 2004. Telomere shortening triggers senescence of human cells through a pathway involving ATM, p53, and p21(CIP1), but not p16(INK4a). *Mol. Cell* **14:** 501–513.
107. Feldser, D.M. & C.W. Greider. 2007. Short telomeres limit tumor progression in vivo by inducing senescence. *Cancer Cell* **11:** 461–469.
108. Cosme-Blanco, W., M.F. Shen, A.J. Lazar, *et al.* 2007. Telomere dysfunction suppresses spontaneous tumorigenesis in vivo by initiating p53-dependent cellular senescence. *EMBO Rep.* **8:** 497–503.
109. McKinnon, P.J. 2004. ATM and ataxia telangiectasia. *EMBO Rep.* **5:** 772–776.
110. Vaziri, H., M.D. West, R.C. Allsopp, *et al.* 1997. ATM-dependent telomere loss in aging human diploid fibroblasts and DNA damage lead to the post-translational activation of p53 protein involving poly(ADP-ribose) polymerase. *Embo. J.* **16:** 6018–6033.
111. Naito, T., A. Matsuura & F. Ishikawa. 1998. Circular chromosome formation in a fission yeast mutant defective in two ATM homologues. *Nat. Genet.* **20:** 203–206.
112. Ito, K., A. Hirao, F. Arai, *et al.* 2004. Regulation of oxidative stress by ATM is required for self-renewal of haematopoietic stem cells. *Nature* **431:** 997–1002.
113. Saretzki, G. & T. Von Zglinicki. 2002. Replicative aging, telomeres, and oxidative stress. *Ann. N. Y. Acad. Sci.* **959:** 24–29.
114. Gire, V., P. Roux, D. Wynford-Thomas, *et al.* 2004. DNA damage checkpoint kinase Chk2 triggers replicative senescence. *Embo. J.* **23:** 2554–2563.
115. Adams, G.B. & D.T. Scadden. 2006. The hematopoietic stem cell in its place. *Nat. Immunol.* **7:** 333–337.
116. von Zglinicki, T. 2002. Oxidative stress shortens telomeres. *Trends Biochem. Sci.* **27:** 339–344.
117. Dumont, P., M. Burton, Q.M. Chen, *et al.* 2000. Induction of replicative senescence biomarkers by sublethal oxidative stresses in normal human fibroblast. *Free Radic. Biol. Med.* **28:** 361–373.
118. von Zglinicki, T., R. Pilger & N. Sitte. 2000. Accumulation of single-strand breaks is the major cause of telomere shortening in human fibroblasts. *Free Radic. Biol. Med.* **28:** 64–74.
119. Adelfalk, C., M. Lorenz, V. Serra, *et al.* 2001. Accelerated telomere shortening in Fanconi anemia fibroblasts–a longitudinal study. *FEBS Lett.* **506:** 22–26.
120. Emmerson, E. & M.J. Hardman. 2012. The role of estrogen deficiency in skin ageing and wound healing. *Biogerontology* **13:** 3–20.
121. Bayne, S., H. Li, M.E. Jones, *et al.* 2011. Estrogen deficiency reversibly induces telomere shortening in mouse granulosa cells and ovarian aging in vivo. *Protein Cell* **2:** 333–346.

Ann. N.Y. Acad. Sci. ISSN 0077-8923

ANNALS OF THE NEW YORK ACADEMY OF SCIENCES
Issue: *Hematopoietic Stem Cells VIII*

In vivo divisional tracking of hematopoietic stem cells

Hitoshi Takizawa and Markus G. Manz
Division of Hematology, University Hospital Zurich, Zurich, Switzerland

Address for correspondence: Hitoshi Takizawa, Division of Hematology, University Hospital Zurich, CH-8091 Zurich, Switzerland. hitoshi.takizawa@usz.ch

Hematopoietic stem cell (HSC) division leads to self-renewal, differentiation, or death of HSCs, and adequate balance of this process results in sustained, lifelong, high-throughput hematopoiesis. Despite their contribution to hematopoietic cell production, the majority of cells within the HSC population are quiescent at any given time. Recent studies have tackled the questions of how often HSCs divide, how divisional history relates to repopulating potential, and how many HSCs contribute to hematopoiesis. Here, we summarize these recent findings on HSC turnover from different experimental systems and discuss hypothetical models for HSC cycling and maintenance in steady-state and upon hematopoietic challenge.

Keywords: hematopoietic stem cell division; divisional history; labeling technique; quiescence; active cycling

Introduction

Every second, millions of mature blood cells die and are replenished by newly generated ones over the lifetime of an individual. Homeostatic levels of blood cells are ensured by a highly organized hematopoietic hierarchy, where self-renewing hematopoietic stem cells (HSCs) in the bone marrow (BM) give rise to highly proliferating intermediate progenitors with limited self-renewal and restricted lineage potential, which subsequently undergo a stepwise differentiation program toward terminal maturation. Given immunesystem compatibility, HSCs are transplantable from one individual to another, sustaining in both lifelong hematopoiesis, a demonstration of functional excess of HSCs. Additionally, single HSCs possess a high BM-homing and multilineage repopulating capacity, as evidenced by *in vivo* transplantation experiments demonstrating that a single HSC can, lifelong, reconstitute the hemato-lymphoid system of lethally irradiated recipients.[1–5] The high accessibility, mobility, and robust regenerative potential of HSCs has been used for clinical HSC transplantation to cure inborn genetic disorders and acquired hematopoietic malignancies.

Besides self-renewal activity and multilineage blood-forming potential, a characteristic of HSCs is thought to be the relative quiescent cell cycle status of a majority of HSCs. Although recently debated,[6–9] HSC quiescence might lead to protection of metabolic toxicity and genotoxic events, and might be a reservoir for emergency hematopoiesis, for example, upon infection or massive blood loss.[10] However, it remains to be determined how often HSCs divide, to what extent their divisional history affects their function (i.e, their self-renewing and repopulating ability), and how many HSC clones contribute to hematopoiesis at any given time. To determine aspects of these issues *in vivo*, several groups have established different experimental techniques to track HSC division and differentiation.

Here, we summarize technical aspects of the different *in vivo* HSC-tracking systems and integrate experimental findings from those approaches to propose and discuss theoretical models for steady-state and demand-adapted HSC turnover.

Experimental approaches to determine *in vivo* steady-state HSC division

DNA and DNA complex labeling

Many studies have analyzed HSC cell cycle state with DNA-staining dyes, such as 7AAD (7-amino-actinomycin D), PI (propidium iodide), DAPI (4′, 6-diamidino-2-phenylindole), and Hoechst, partly

doi: 10.1111/j.1749-6632.2012.06500.x

in combination with the RNA-staining dye Pyronin Y or with the proliferation marker protein Ki67, a nuclear protein present during the active cell cycle phase except G0.[11] However, these analyses have technical limitations, as they can only give a snapshot of information on the status of the cell cycle. Thus, although these are powerful methods to analyze the effects of gene function, of treatment/challenge on HSC activation, and of biological functions of cell populations in different phases of cell cycle, these methods do not provide information on divisional history—that is, the number of cellular divisions that a cell has undergone during the period of analysis.[12,13]

Pioneering work on steady-state HSC turnover came from *in vivo* studies of incorporation of BrdU (5-bromo-2′-deoxyuridine), a DNA analogue that can be exited and detected by fluorescent-activated cell sorting (FACS). BrdU is given via oral or intraperitoneal injection, and the kinetics of BrdU incorporation into DNA provides information about cellular division. It has been estimated that approximately 8% of a population enriched by the phenotype for HSCs (lineage marker negative, c-Kit$^+$, Sca-1$^+$ (LKS) and Thy-1$^-$) randomly enter cell cycle per day, and almost all HSCs enter cell cycle on average every 57 days.[14] Similar results were obtained upon short-term BrdU pulse and analysis of highly enriched HSC populations (LKS CD48$^-$ CD41$^-$ CD150$^+$ cells),[15] half of which are functional HSCs, as shown by single cell transplantation.[1]

However, BrdU-mediated DNA labeling requires cell fixation and permeabilization for analysis, thus leading to cell death, which prohibits testing of the functionality of labeled cells subsequent to their biological *in vivo* HSC readout (the readout being transplantation into lethally irradiated animals followed by contribution to long-term hematopoiesis). To overcome this limitation, two groups employed transgenic animal models that conditionally express histone 2B protein fused with green fluorescent protein (H2B-GFP), under the control of a tetracycline regulatory element.[16] Upon treatment with doxycycline, expression of H2B-GFP is turned on or off (depending on the model and experimental setup) and GFP-retaining cells can subsequently be chased over time.[17,18] Interestingly, the experimental kinetics of GFP label retention did not fit a mathematical model assuming uniform HSC cycling, as previously reported with BrdU incorporation.[14] In contrast, a model assuming two populations fit the experimental data better and predicted a larger population cycling at 5–11% per day and another population cycling at 1–2% per day.[17] In other words, within the phenotypically defined HSCs, one population divided fast (every 9–18 days), whereas the other divided slow (every 55–125 days). Furthermore, in functional readouts (serial transplantation), the fast-cycling HSC fraction (LKS CD48$^-$ CD150$^+$) showed less sustained multilineage repopulating capacity compared with the slow-dividing HSCs. In line with these findings, the second model revealed that 15% of phenotypically defined HSCs actively cycle every 36 days, while 85% of HSCs divide only every 145 days, and the slow-dividing population contained HSCs with self-renewal and repopulation potential over serial transplantation.[18] On the basis of the assumption of fixed-cycling kinetics and a mathematical model constructed from data of BrdU incorporation for two weeks and subsequent follow-up of label-retaining cells, it was estimated that some HSCs divide only five times during the two-year life span of an average laboratory mouse.[18]

However, the results from H2B-GFP transgenic animals need to be interpreted with caution, as it has been reported that animals can express GFP even when not crossed with mice carrying a tetracycline responsive element, indicating promiscuous transgene expression.[19] Furthermore, it remains unclear if the phenotypically selected cells contained HSCs or if enough numbers of cells were transplanted to detect limiting amounts of HSCs in the population of tested cells.[3]

Cell surface and cytoplasmic protein labeling

Recently, alternatives to DNA and DNA complex labeling have been employed. Biotin, a natural vitamin of the B complex family, strongly binds avidin. *In vivo* injection of the esterized form of biotin allows cell surface protein labeling of almost all cells in different tissues within minutes and then subsequent detection of labeled cells with fluorescent label–conjugated streptavidin, a tetramer of avidin.[20] As cell division dilutes cell membrane–bound biotin, biotin-retaining cells have no or low divisional activity.

Interestingly, when the biotin-avidin method was used to label HSCs, biotin-retaining cells were mainly found in the bone marrow (BM) after one week of chase, and primitive hematopoietic progenitor cells showed higher biotin label intensity than other mature hematopoietic cells. In contrast to expectations from the previous findings discussed above, transplantation with biotin label–retaining HSC candidates (LKS CD150$^+$ cells isolated after three weeks of labeling) showed that HSC activity was surprisingly not contained in label-retaining, slow-cycling populations but in the relatively fast-cycling populations. However, as indicated by the authors of the study, low divisional resolution of biotin labeling, which can distinguish maximally to only four divisions, limits the assay for long-term *in vivo* tracking.[20]

We have recently modified a robust and non-invasive cellular labeling technique used largely in immunology research: labeling with 5(6)-carboxyfluorescein diacetate *N*-succinimidyl ester (CSFE),[3] a cell-permeable dye that covalently binds to intracellular proteins and equally distributes to daughter cells upon cellular division.[21,22] FACS-sorted LKS cells were labeled with CFSE *ex vivo* and intravenously transferred into nonirradiated recipients to follow steady-state HSC turnover kinetics. To validate CFSE dilution-based tracking, *in vivo* CFSE dilution was compared with CFSE-labeled nondividing T cells and with BrdU incorporation into CFSE-labeled LKS cells. Mice were treated with BrdU starting one week after transfer of CFSE-labeled LKS cells. Consistent with previous findings,[1,20] BrdU treatment induced cellular division, indicating a mitogenic effect of BrdU. Importantly, BrdU incorporation did not linearly correlate with cell divisions, nor did all divided cells incorporate BrdU; and, naturally, CFSE-retaining quiescent LKS cells remained unlabeled with BrdU. These data demonstrated that CFSE dilution provides higher divisional resolution (maximum seven divisions for less than four-week chase, and five divisions for more than four-week chase) than BrdU, that BrdU labeling changes the original cell cycle program, that BrdU labeling might miss some non-dividing cells, and that a small fraction of nondividing cells—in this context the most interesting HSC population—are not included in the respective analysis.

Weekly BM analysis to determine retention kinetics of CFSE labeling revealed that BM, but not other tissues, contained label-retaining cells with the same CFSE intensity as naive T cells for up to 21 weeks of chase, and these long-term label-retaining cells maintained a HSC immunophenotype. Similar results were obtained when highly HSC-enriched populations (LKS CD34$^-$ CD150$^+$) were labeled and transferred, suggesting that non-divided cells are likely enriched for HSCs. Resorting of labeled cells and limiting dilution, single cell, and serial transplantation with nondivided (dormant) versus >5 times–divided (fast-cycling) LKS cells demonstrated that, in contrast to previous findings, both dormant and fast-cycling HSCs retained long-term repopulating capacity.[3] This demonstrated that some HSCs are dormant over an extended time, whereas others cycle relatively frequently. More surprisingly, in recipients of limit-diluted and transplanted HSCs, donor-derived chimerism fluctuated over the serial transplantation time frame, irrespective of HSC cycling activity in the primary recipient. These findings led us to hypothesize that the turnover rate of HSCs is not permanently fixed, even in steady state, and that therefore contribution to blood formation from single HSCs varies over time. To test this experimentally, fast-cycling (i.e., the CFSE-diluted LKS population) HSCs were isolated from the primary host, relabeled with CFSE, and transferred into nonconditioned secondary recipients. BM analysis six weeks after secondary steady-state transplantation showed that some of the fast-cycling cells slowed down their division rate and regained quiescence, with *in vivo* biological HSC readout in subsequent transplantation to lethally irradiated animals. Consistent with these findings, mathematical modeling based on the CFSE dilution data gave no evidence for a dichotomy within the HSC population or of populations with distinct proliferating kinetics; modeling was consistent with one HSC population with broadly variable cell cycle activity. Furthermore, based on these data, we estimated that HSCs divide on average every 39 days, leading to about 18 divisions during a typical two-year mouse lifetime.

The methods, advantages, and disadvantages of the *in vivo* tracking systems for observation of HSC cycling discussed earlier are summarized in Table 1.

Discussion

Long-standing and intriguing questions on lifelong, sustained blood formation remain, including how many stem cell divisions are required to maintain hematopoiesis for a lifetime, and how many

Table 1. Advantages and disadvantages of *in vivo* steady-state HSC turnover tracking methods

Label	Labeling method	Labeling period	Longest period of chase	Label detection	Advantage of assay	Disadvantage of assay	Refs.
BrdU	i.p. + drinking water i.p. + drinking water i.p. + drinking water	180 days 10 days 13 days	180 days 120 days 306 days	Intracellular stain with anti-BrdU	• Simple and efficient labeling • No alteration in peripheral blood count and cell cycle of HSC observed	• No possibility to functionally test labeled cells because of cell fixation/ permeabilization for detection • Low divisional resolution • Mitogenic effect • Insensitive and nonspecific assay for HSCs	14 15 18
H2B-GFP (Tet-off)	Drinking water with DOX Drinking water with DOX	Until DOX treatment (timing not specified) Until DOX treatment (timing not specified)	240 days 84 days	Direct fluorescence emission	• Functional test of labeled cells possible	• Inconsistent labeling efficacy reported in different papers • Promiscuous and dim H2B-GFP expression • Poor detection of GFP^{+} cells in microscopic tissue analysis	18 19
H2B-GFP (Tet-on)	Drinking water with DOX	42 days (4–8-week-old mice)	500 days		• Efficient labeling regardless of cell cycle stage	• Unknown divisional resolution although seven–eight divisions assumed	17
Esterized biotin	i.v.	Single dose injection	21 days	Cell surface stain with biotin-streptavidin reaction	• Efficient labeling • Functional test of labeled cells possible	• Low divisional resolution (no more than four divisions)	20
CFSE	*Ex vivo* incubation	7 minutes	147 days	Direct fluorescence emission	• High divisional resolution (seven divisions) • Efficient labeling • No obvious cell toxicity • Functional test of labeled cells possible	• i.v. transfer of *ex vivo*–labeled cells • Fluorescence bleaching • Limited engraftment of cells in nonconditioned recipients	3

Tet-on, tetracycline-on system; Tet-off, tetracycline-off system; i.p., intraperitoneal injection; i.v., intravenous injection; DOX, doxycycline.

HSCs engage in blood cell production at any given time.[23]

To address these questions, clonal analysis is required. This has been achieved, by many studies, using HSC marking via *ex vivo* transduction with retro- or lentiviruses, followed by transplantation of the HSCs into irradiated animals to reconstitute the host blood system. Because the viral genome is randomly integrated into the host genome (though not in an entirely unbiased way) each HSC should have a unique integration site(s). Previous work demonstrated that as few as 20 integration sites were distinguishable, and among those, some HSC clones gave rise to blood cells in reconstituted tissues.[24–28] Interestingly, different classes of HSC clones were detected with distinct kinetics of hematopoietic contribution: most clones produced blood cells rapidly within several weeks after irradiation, and then decreased their contribution to hematopoiesis; other clones were inactive at the beginning of reconstitution, but later contributed persistently to blood production. In more recent studies, the HSC barcoding technique was combined with high-throughput sequencing, which allowed detection of more than 500 HSC clones, thus extending and confirming the limited use of HSC clones to sustain hematopoietic cell production at any given time.[29] However, these findings were based on viral infection of *ex vivo* HSCs that had been cultured from hours to days followed by hematopoietic system reconstitution after irradiation, two conditions that might change HSC cycling properties. It remains to be determined if the same holds true for steady-state hematopoiesis.

The clonal maintenance model, originally proposed by retrovirus-mediated marking studies and irradiated animals, proposes that in the steady state all HSCs divide continuously to produce blood cells and contribute equally to blood formation (Fig. 1A).[14,15,20,25,27] In contrast, the clonal succession model postulates that some HSCs frequently divide and give rise to more differentiated progeny until terminal differentiation or death, thus requiring subsequent replacement by HSCs that are activated from quiescence (Fig. 1B).[17,18,23,24] Our findings with CFSE divisional tracking do not fit either of the above models but instead propose a dynamic repetition model in which some HSCs divide and dominate blood production for a period of time, whereas other HSCs are activated to enter the cell cycle and take over blood production, followed by repetition of same cycle (Fig. 1C). The dynamic repetition model also indicates that the same HSC clone will reappear in future blood formation, as suggested by some rare findings in HSC marking studies[26] and in line with predictions from mathematical simulations.[30] These observations also suggest that fluctuating HSC cycling will result in rather homogenous turnover (divisional history) of all HSCs at end of life, which is in line with linear telomere shortening in the myeloid blood cell compartment, observed in cross-sectional studies in aging humans.[31] Equal use of all HSC resources over the lifetime of an animal, as suggested in the dynamic repetition model, would also, at the same time, help to protect HSCs from functional exhaustion, while still providing a dormant haven for the at-anytime quiescent HSC population to be protected from toxic metabolic or genotoxic events.

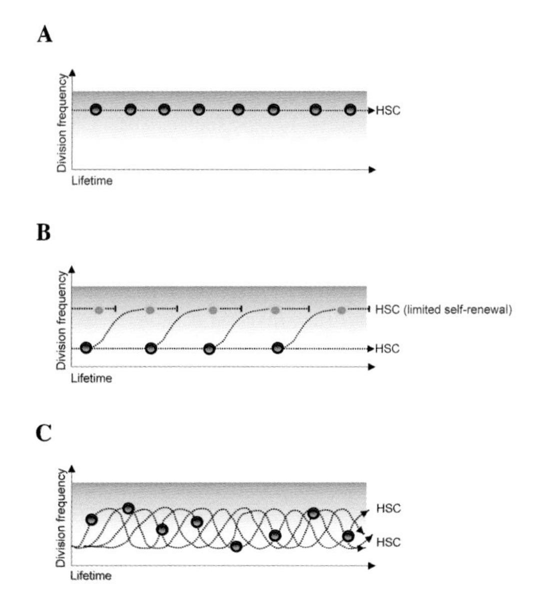

Figure 1. Hypothetical models for steady-state HSC cycling and hematopoiesis. (A) Clonal maintenance model: all HSCs continuously divide and equally contribute to hematopoiesis. (B) Clonal succession model: quiescent HSCs are recruited to cycle and produce mature blood cells until they die or differentiate. Subsequently, other quiescent HSCs follow the same fate. (C) Dynamic repetition model: some HSCs divide frequently and dominate blood formation for a certain period of time. Subsequently, they slow down division and become quiescent, whereas other fractions enter cell cycle and take over contribution to hematopoiesis. These cycles are repeated. Red color indicates contribution to blood formation. Adapted from Ref. 3.

However, as of now, little is known about how the two HSC cell cycle activities, quiescence and cycling, are regulated in the steady state—a state in which blood loss and active infection or inflammation is absent and blood values are within normal range (as defined by a Gaussian distribution in a healthy population). In contrast, it has been shown that upon hematopoietic challenge with irradiation or chemotherapy, almost all HSCs are recruited to a proliferative state, likely via increased inflammatory cytokines and chemokines that act on respective cell surface receptors expressed on HSCs.[32] Recent studies have demonstrated that infections might activate HSCs from dormancy by direct interferon-mediated signaling in HSCs.[33–35]

Moreover, our data provide evidence that *in vivo* lipopolysaccharide (LPS) challenge, which resembles/mimics Gram-negative bacterial infection, drives dormant HSCs into cycle.[3] Whether LPS-mediated HSC activation is direct, indirect, or both needs to be determined. Interestingly, primitive hematopoietic stem and progenitor cells express functional Toll-like receptors (TLRs), and their activation has been shown to have an impact at least on migration and differentiation.[36–38]

Although activation of almost all HSCs into proliferation and demand-adapted blood formation upon hematopoietic challenges might have delivered an evolutionary advantage, continuous activation of HSCs could increase the risk for accumulating genetic alterations that might trigger tumorigenesis. To avoid this outcome, HSCs seem to be equipped with a mechanism that secures their loss or functional exhaustion upon chronic stimulation.[33,39] An additional mechanism for HSC protection would be an intrinsic program that drives frequently divided HSCs into quiescence. In fact, we found that HSCs with extensive divisional history—that is, HSCs from aged mice, or HSCs that were forced into multiple divisions by limiting-dilution transplantation into irradiated recipients—tend to revert to quiescence in a permissive environment, suggesting intrinsic divisional history–dependent signals that determine HSC cell cycle.[3,40]

It is intriguing to speculate that the basic principles of HSC biology that have been unraveled in recent years will also hold true for other somatic stem cell systems of the body. Indeed, demand-adapted regulation of the stem cell cycle and coexistence of actively cycling and quiescent stem cells are observed in other stem cell–sustained organs, such as skin and gut.[41] Delineation of the detailed molecular mechanisms for regulation of HSC cycling will not only be important for understanding basic stem cell biology but also carries the potential to develop new therapies to efficiently target and control stem cells as well as their pathologic correlates—cancer-initiating cells.

Acknowledgments

This work was in part supported by a Postdoctoral Fellowship of the Japanese Society for the Promotion of Science for Research Abroad to H.T., the Swiss National Science Foundation (310030_131088/1), the Promedica Foundation (Chur, CH), and the Marlis Geiser-Lemken Stiftung (Zurich, CH) to M.G.M.

Conflicts of interest

The authors declare no conflicts of interest.

References

1. Kiel, M.J., O.H. Yilmaz, T. Iwashita, *et al.* 2005. SLAM family receptors distinguish hematopoietic stem and progenitor cells and reveal endothelial niches for stem cells. *Cell* **121:** 1109–1121.
2. Osawa, M., K. Hanada, H. Hamada & H. Nakauchi. 1996. Long-term lymphohematopoietic reconstitution by a single CD34-low/negative hematopoietic stem cell. *Science* **273:** 242–245.
3. Takizawa, H., R.R. Regoes, C.S. Boddupalli, *et al.* 2011. Dynamic variation in cycling of hematopoietic stem cells in steady state and inflammation. *J. Exp. Med.* **208:** 273–284.
4. Notta, F., S. Doulatov, E. Laurenti, *et al.* 2011. Isolation of single human hematopoietic stem cells capable of long-term multilineage engraftment. *Science* **333:** 218–221.
5. Dykstra, B., D. Kent, M. Bowie, *et al.* 2007. Long-term propagation of distinct hematopoietic differentiation programs in vivo. *Cell Stem Cell* **1:** 218–229.
6. Milyavsky, M., O.I. Gan, M. Trottier, *et al.* 2010. A distinctive DNA damage response in human hematopoietic stem cells reveals an apoptosis-independent role for p53 in self-renewal. *Cell Stem Cell* **7:** 186–197.
7. Mohrin, M., E. Bourke, D. Alexander, *et al.* 2010. Hematopoietic stem cell quiescence promotes error-prone DNA repair and mutagenesis. *Cell Stem Cell* **7:** 174–185.
8. Lane, A.A. & D.T. Scadden. 2010. Stem cells and DNA damage: persist or perish? *Cell* **142:** 360–362.
9. Seita, J., D.J. Rossi & I.L. Weissman. 2010. Differential DNA damage response in stem and progenitor cells. *Cell Stem Cell* **7:** 145–147.
10. Wilson, A. & A. Trumpp. 2006. Bone-marrow haematopoietic-stem-cell niches. *Nat. Rev. Immunol.* **6:** 93–106.
11. Scholzen, T. & J. Gerdes. 2000. The Ki-67 protein: from the known and the unknown. *J. Cell Physiol.* **182:** 311–322.
12. Passegue, E., A.J. Wagers, S. Giuriato, *et al.* 2005. Global analysis of proliferation and cell cycle gene expression in the regulation of hematopoietic stem and progenitor cell fates. *J. Exp. Med.* **202:** 1599–1611.
13. Bowie, M.B., K.D. McKnight, D.G. Kent, *et al.* 2006. Hematopoietic stem cells proliferate until after birth and show a reversible phase-specific engraftment defect. *J. Clin. Invest.* **116:** 2808–2816.
14. Cheshier, S.H., S.J. Morrison, X. Liao & I.L. Weissman. 1999. In vivo proliferation and cell cycle kinetics of long-term self-renewing hematopoietic stem cells. *Proc. Natl. Acad. Sci. U.S.A.* **96:** 3120–3125.
15. Kiel, M.J., S. He, R. Ashkenazi, *et al.* 2007. Haematopoietic stem cells do not asymmetrically segregate chromosomes or retain BrdU. *Nature* **449:** 238–242.

16. Tumbar, T., G. Guasch, V. Greco, *et al.* 2004. Defining the epithelial stem cell niche in skin. *Science* **303:** 359–363.
17. Foudi, A., K. Hochedlinger, D. Van Buren, *et al.* 2009. Analysis of histone 2B-GFP retention reveals slowly cycling hematopoietic stem cells. *Nat. Biotechnol.* **27:** 84–90.
18. Wilson, A., E. Laurenti, G. Oser, *et al.* 2008. Hematopoietic stem cells reversibly switch from dormancy to self-renewal during homeostasis and repair. *Cell* **135:** 1118–1129.
19. Challen, G.A. & M.A. Goodell. 2008. Promiscuous expression of H2B-GFP transgene in hematopoietic stem cells. *PLoS One* **3:** e2357.
20. Nygren, J.M. & D. Bryder. 2008. A novel assay to trace proliferation history in vivo reveals that enhanced divisional kinetics accompany loss of hematopoietic stem cell self-renewal. *PLoS One* **3:** e3710.
21. Lyons, A.B. & C.R. Parish. 1994. Determination of lymphocyte division by flow cytometry. *J. Immunol. Methods* **171:** 131–137.
22. Weston, S.A. & C.R. Parish. 1990. New fluorescent dyes for lymphocyte migration studies. Analysis by flow cytometry and fluorescence microscopy. *J. Immunol. Methods* **133:** 87–97.
23. Kay, H.E. 1965. How many cell-generations? *Lancet* **2:** 418–419.
24. Drize, N.J., J.R. Keller & J.L. Chertkov. 1996. Local clonal analysis of the hematopoietic system shows that multiple small short-living clones maintain life-long hematopoiesis in reconstituted mice. *Blood* **88:** 2927–2938.
25. Jordan, C.T. & I.R. Lemischka. 1990. Clonal and systemic analysis of long-term hematopoiesis in the mouse. *Genes Dev.* **4:** 220–232.
26. Lemischka, I.R., D.H. Raulet & R.C. Mulligan. 1986. Developmental potential and dynamic behavior of hematopoietic stem cells. *Cell* **45:** 917–927.
27. McKenzie, J.L., O.I. Gan, M. Doedens, *et al.* 2006. Individual stem cells with highly variable proliferation and self-renewal properties comprise the human hematopoietic stem cell compartment. *Nat. Immunol.* **7:** 1225–1233.
28. Guenechea, G., O.I. Gan, C. Dorrell & J.E. Dick. 2001. Distinct classes of human stem cells that differ in proliferative and self-renewal potential. *Nat. Immunol.* **2:** 75–82.
29. Lu, R., N.F. Neff, S.R. Quake & I.L. Weissman. 2011. Tracking single hematopoietic stem cells in vivo using high-throughput sequencing in conjunction with viral genetic barcoding. *Nat. Biotechnol.* **29:** 928–933.
30. Glauche, I., K. Moore, L. Thielecke, *et al.* 2009. Stem cell proliferation and quiescence—two sides of the same coin. *PLoS Comput. Biol.* **5:** e1000447.
31. Rufer, N., T.H. Brummendorf, S. Kolvraa, *et al.* 1999. Telomere fluorescence measurements in granulocytes and T lymphocyte subsets point to a high turnover of hematopoietic stem cells and memory T cells in early childhood. *J. Exp. Med.* **190:** 157–167.
32. Trumpp, A., M. Essers & A. Wilson. 2010. Awakening dormant haematopoietic stem cells. *Nat. Rev. Immunol.* **10:** 201–209.
33. Baldridge, M.T., K.Y. King, N.C. Boles, *et al.* 2010. Quiescent haematopoietic stem cells are activated by IFN-gamma in response to chronic infection. *Nature* **465:** 793–797.
34. Essers, M.A., S. Offner, W.E. Blanco-Bose, *et al.* 2009. IFN alpha activates dormant haematopoietic stem cells in vivo. *Nature* **458:** 904–908.
35. Sato, T., N. Onai, H. Yoshihara, *et al.* 2009. Interferon regulatory factor-2 protects quiescent hematopoietic stem cells from type I interferon-dependent exhaustion. *Nat. Med.* **15:** 696–700.
36. Massberg, S., P. Schaerli, I. Knezevic-Maramica, *et al.* 2007. Immunosurveillance by hematopoietic progenitor cells trafficking through blood, lymph, and peripheral tissues. *Cell* **131:** 994–1008.
37. Nagai, Y., K.P. Garrett, S. Ohta, *et al.* 2006. Toll-like receptors on hematopoietic progenitor cells stimulate innate immune system replenishment. *Immunity* **24:** 801–812.
38. Schmid, M.A., H. Takizawa, D.R. Baumjohann, *et al.* 2011. Bone marrow dendritic cell progenitors sense pathogens via Toll-like receptors and subsequently migrate to inflamed lymph nodes. *Blood* **118:** 4829–4840.
39. Esplin, B.L., T. Shimazu, R.S. Welner, *et al.* 2011. Chronic exposure to a TLR ligand injures hematopoietic stem cells. *J. Immunol.* **186:** 5367–5375.
40. Takizawa, H. & M.G. Manz. 2011. Dynamic regulation of hematopoietic stem cell cycling. *Cell Cycle* **10:** 2246–2247.
41. Li, L. & H. Clevers. 2010. Coexistence of quiescent and active adult stem cells in mammals. *Science* **327:** 542–545.

Ann. N.Y. Acad. Sci. ISSN 0077-8923

ANNALS OF THE NEW YORK ACADEMY OF SCIENCES
Issue: *Hematopoietic Stem Cells VIII*

Caudal genes in blood development and leukemia

Claudia Lengerke[1] and George Q. Daley[2]

[1]University of Tübingen Medical Center—Hematology & Oncology, Tübingen, Germany. [2]Children's Hospital Boston and Dana Farber Cancer Institute—Stem Cell Transplantation Program, Boston, Massachusetts

Address for correspondence: Dr. George Daley, Children's Hospital Boston and Dana Farber Cancer Institute—Stem Cell Transplantation Program, 300 Longwood Avenue, Boston, MA 02115. george.daley@childrens.harvard.edu; or Dr. Claudia Lengerke, University of Tübingen Medical Center—Hematology & Oncology, Otfried-Mueller-Strasse 10, 72076 Tübingen, Germany. claudia.lengerke@med.uni-tuebingen.de

Members of the caudal gene family (in mice and humans: Cdx1, Cdx2, and Cdx4) have been studied during early development as regulators of axial elongation and anteroposterior patterning. In the adult, Cdx1 and Cdx2, but not Cdx4, have been intensively explored for their function in intestinal tissue homeostasis and the pathogenesis of gastrointestinal cancers. Involvement in embryonic hematopoiesis was first demonstrated in zebrafish, where *cdx* genes render posterior lateral plate mesoderm competent to respond to genes specifying hematopoietic fate, and compound mutations in *cdx* genes thus result in a bloodless phenotype. Parallel studies performed in zebrafish embryos and murine embryonic stem cells (ESCs) delineate conserved pathways between fish and mammals, corroborating a BMP/Wnt-Cdx-Hox axis during blood development that can be employed to augment derivation of blood progenitors from pluripotent stem cells *in vitro*. The molecular regulation of *Cdx* genes appears complex, as more recent data suggest involvement of non-*Hox*–related mechanisms and the existence of auto- and cross-regulatory loops governed by morphogens. Here, we will review the role of *Cdx* genes during hematopoietic development by comparing effects in zebrafish and mice and discuss their participation in malignant blood diseases.

Keywords: Cdx; hematopoiesis; leukemia; Hox; blood development

Introduction

The caudal (Cdx) family of DNA-binding proteins was originally identified in *Drosophila*, but homologues with conserved molecular structure and function have been described in several organisms. The three members of the Cdx family in mice and humans, *Cdx1*, *Cdx2,* and *Cdx4*, have been intensively studied for their involvement in axial elongation and early anteroposterior patterning via *Hox* gene regulation (reviewed by Young and Deschamps[1]). While *Cdx* single- and compound-deficient mice exhibit overt defects in vertebrae and limbs, studies from zebrafish and murine embryonic stem cells (ESCs) indicate additional *Cdx*-driven patterning effects during the development of other mesoderm derivatives, such as blood,[2–4] kidney,[5] and cardiac[6] cells. Moreover, *cdx* genes regulate the development of neural (e.g., neural tube and spinal cord) and endodermal tissues (e.g., gut) and play important roles in adult intestinal tissue homeostasis and the pathogenesis of gastrointestinal cancers (reviewed by Guo, Suh, and Lynch[7]).

The role of *Cdx* genes in the blood system is less well elucidated. Data from zebrafish models demonstrate that *cdx* genes regulate embryonic hematopoiesis through activation of downstream *hox* genes. While overexpression studies performed with murine ESCs confirm these data, corresponding *in vivo* loss-of-function studies in mice are complicated by functional redundancy among the three *Cdx* family members. Consistent with the notion that reactivated developmental pathways can contribute to oncogenesis, emerging data indicate expression and functional roles of *Cdx2* in leukemia. In this review, we discuss the role of *Cdx* genes during hematopoietic development and their involvement in malignant blood disease.

Insights from knockout mouse models

During early development, *Cdx* genes follow a similar expression pattern to the developmentally

doi: 10.1111/j.1749-6632.2012.06625.x

related *Hox* genes, conferring positional identity to developing mesodermal tissues. In mice, expression is detected in the posterior epiblast and the overlying mesoderm at the posterior end of the primitive streak.[1] During their development in the posterior growth zone, anterior trunk tissues are exposed to *Cdx* genes but, as cells move anteriorly, *Cdx* transcripts decay.[8–10] Persistence of *Cdx* in the posterior region of the embryo and expression of more posterior *Hox* genes enable the development of posterior trunk mesoderm and tail tissues. The instructive function of *Cdx* and *Hox* genes strongly varies with the developmental stage. As such, overexpression of *Hox* genes in the epiblast alters the contribution of cells to the mesoderm[11] and overexpression at the mesoderm stage profoundly affects morphogenesis of developing tissues such as vertebrae. However, later overexpression in already formed somites shows no effect.[12]

Cdx mutant mice show posterior body truncations involving the axial skeleton, the neuraxis, and caudal urorectal structures. The severity of the phenotype depends on the individual *Cdx* gene and, consistent with the notion of redundancy, is more pronounced in compound mutants.[13,14] Studies on triple knockout mice are complicated by the essential role of *Cdx2* during placenta development, resulting in lethality of the $Cdx2^{-/-}$ genotype at 3.5 days postcoitum (dpc).[15] More recently, inactivation of *Cdx2* at postimplantation stages by a tamoxifen inducible *Cre*-system confirmed the axial truncation phenotype and the incomplete urorectal septation in *Cdx2*-deficient animals.[16,17] Next to anterior homeotic shifts of the axial skeleton, polyp-like lesions with proximal endoderm have been described in the coecum of $Cdx2^{+/-}$, suggesting anterior homeotic shifts in the intestinal mucosa.[18] Indeed *Cdx1* and *Cdx2* are expressed in a second wave starting with day 12.5 pc in elements of the developing gut and play important roles not only during gut formation but also in adult tissue homeostasis and carcinogenesis.[19] The murine *Cdx4* gene appears less potent than *Cdx1* and *Cdx2*. Analyses of *Cdx2/Cdx4* compound mutants revealed roles for *Cdx4* during placenta development and confirmed redundant roles with *Cdx2* during axial elongation. However, $Cdx4^{-/-}$ mice are born healthy and appear morphologically normal.[20]

In mice, the first hematopoietic cells arise in the yolk sac around 7.5 dpc, representing the primitive wave of hematopoiesis. The second wave of definitive hematopoiesis follows around 9 dpc from hematopoietic stem cells (HSCs), which are formed in the aorto-gonado-mesonephros region and then relocate to other anatomic sites including the yolk sac, the fetal liver and, shortly before birth, the bone marrow as the main site of adult hematopoiesis (reviewed by Lengerke and Daley[21]). Single- and compound-*Cdx*–deficient mice do not present overt hematopoietic phenotypes. However, careful analysis has revealed subtle defects, such as reduced numbers of yolk sac-derived erythroid colonies in *Cdx4*-deficient versus wild-type mice.[22] Functional redundancy between individual *Cdx* genes may mask effects in single- or double-knockout mice and targeted triple knockouts have not yet been analyzed.

The *cdx-hox* axis regulates embryonic hematopoiesis

The first link between *cdx* genes and developmental hematopoiesis was made in zebrafish. Homozygous zebrafish embryos carrying the autosomal recessive mutation *kugelig* (*kgg*), initially identified because of their tail defect were found to die early during development (day 5–10 postfertilization) and to exhibit pronounced anemia. In 2003, Davidson *et al.* identified loss of function mutations in the *cdx4* gene as the causative mutation for the hematopoietic phenotype of the *kgg* zebrafish embryo.[2] Zebrafish *cdx4* expression occurs in the early gastrula and becomes restricted to the posterior-most cells during gastrulation and early somitogenesis, preceding the expression of hematopoietic markers.[2] *In vivo* injection of *cdx4* mRNA rescues hematopoiesis in *kgg* mutants and induces a "posteriorized phenotype" with ectopic expression of hematopoietic markers in wild-type fish. During development, hematopoietic cells share common progenitors with endothelial cells and arise from so-called scl^{+} hemangioblasts. Up to the five somite stage, *cdx4* coexpresses with *scl* in the posterior blood islands, suggesting a role for *cdx4* in regulation of *scl*.[2] However, while in wild-type zebrafish embryos hematopoietic cells in the posterior lateral plate mesoderm were expanded by scl overexpression,[23] no rescue of hematopoiesis occurred in *kgg* embryo injected with *scl*-mRNA, indicating that *cdx4* is not a direct inducer of *scl*, or that other cofactors are needed to compensate for the loss of *cdx4*. Moreover, *cdx4* mutations induce a

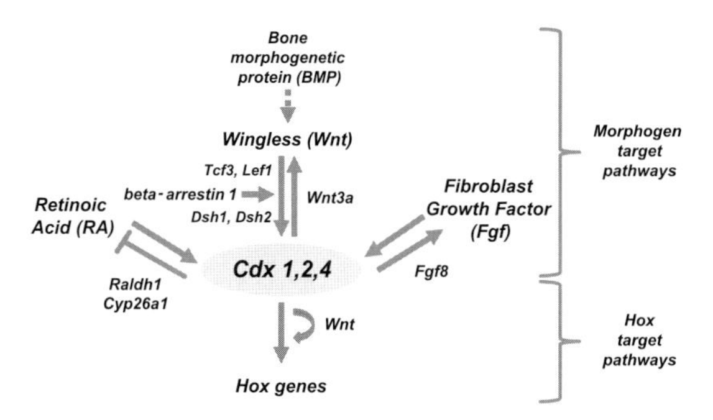

Figure 1. Schematic view of the molecular interactions between Cdx and morphogen pathways.

posterior shift in the boundary between anteriorly localized *scl*$^+$ cells giving rise to endothelial cells, and the more posterior population of cells, which display hemangioblastic properties. Thus, *cdx4* disruption inhibits the formation of blood but not endothelial cells.[2]

Several developmental studies demonstrate that *Cdx* genes act as master regulators of *Hox* genes.[13,24,25] Consistently, *kgg* mutants display profound alterations in *hox* expression domains, with almost complete absence of intermediate and more posterior *hox* genes such as *hoxb6b* and *hoxa9a*, while ectopic *cdx4* can restore *hox* expression patterns. Moreover, overexpression of individual target *hox* genes (e.g., *hoxb7a*, *hoxa9a*) also rescues the formation of *gata1*$^+$ hematopoietic cells in *kgg* mutants, corroborating the role of a *cdx-hox* axis during blood development. Notably, disruption of the developmental blood program cannot be achieved by targeted inhibition of single *hox* genes, indicating redundant functions of target *hox* genes during blood development.[2] Redundancy has also been observed between zebrafish *cdx* genes. In zebrafish, the *cdx4* mutation causes a severe but not complete loss of embryonic blood formation. Additional suppression of *cdx1a* in *kgg* mutants, however, induces a complete failure to specify blood and enhances the severity of *hox* gene deregulation.[3]

Shared upstream and downstream regulatory pathways

Individual *Cdx* genes display not only high amounts of functional redundancy but also auto- and cross-regulatory molecular loops through direct binding to promoter sites[26] or via regulation of molecular pathways that can act both up- and downstream (e.g., Wnt signaling) (Fig. 1).[4,14,27]

Interactions with the retinoic acid pathway

Classically, retinoic acid (RA) has been demonstrated as an upstream regulator of *Cdx1* expression.[28,29] *Cdx1* responsiveness to excess RA has been documented *in vivo* and functional RA-responsive elements have been identified in the *Cdx1* gene.[8,28] In the zebrafish embryo, RA modulates the formation of *gata1*$^+$ hematopoietic cells.[30] Interestingly, treatment with RA inhibitors rescues hematopoiesis in *cdx* mutants, suggesting that cdx proteins act by modulating the RA pathway.[30] During kidney development, *cdx*-dependent modulation of expression boundaries of the RA synthesizing and degrading *raldh2* and *cyp26a1* regulate the formation of distal tubule segments.[5] In the posterior growth zone, concomitant regulation of *cyp26a1* expression restraining RA signaling has been demonstrated to participate in the *Cdx-Hox*–orchestrated trunk-to-tail transition.[14] Furthermore, *cdx* inhibition expands the formation of *tbx5a*$^+$ anterior lateral plate cardiogenic mesoderm but effective differentiation of *tbx5a*$^+$ cells into *nkx2.5*$^+$ cardiac precursor cells requires simultaneous suppression of the retinoic acid pathway, supporting the notion that the *cdx* and RA pathways closely interact during development.[6]

Regulation by Wnt, BMP, and FGF

In murine embryoid bodies, BMP and Wnt are necessary for patterning hematopoietic fate from mesoderm. In detail, activation of BMP signaling induces Wnt3a and the canonical Wnt pathway, thereby activating the *Cdx-Hox* pathway.[4] Consistently, mice engineered for loss of responsive elements in the Wnt effector *Lef* display phenotypic effects resembling *Cdx1*$^{-/-}$ mice[29] and *Wnt3a* hypomorph mutants share phenotype with *Cdx2/4* mutants. On the molecular level, direct interactions between *Lef1* and *Cdx1/4* have been demonstrated,[4,31,32] but more complex molecular mechanisms also take place. As such, Wnt proteins have been recently shown to stimulate phosphorylation of the T cell factor (TCF) family member 3 (TCF3), thereby inducing its dissociation from the *Cdx4* promoter and relieving inhibitory effects on *Cdx4* expression.[33,34] Surprisingly, in *Cdx* mutants, *Cdx* expression can be restored and phenotypic posterior defects corrected by posterior gain of function of *Lef1*,[16] suggesting that Wnt signaling also acts downstream of Cdx proteins. This notion is supported by studies in

ESCs, where overexpression of *Cdx1* and *Cdx4* was shown to induce Wnt3a.[4]

Next to RA and Wnt, the Fgf pathway has been involved in activation of *Cdx* genes during development.[35] More recently, Fgf molecules have also been implicated as downstream targets of *Cdx*. For example, Fgf8, next to Wnt3a, T, and Cyp26a1, is downregulated in *Cdx2* conditional knockout mice and has been shown to directly respond to *Cdx2*.[17]

Intersection with other transcriptional pathways regulating developmental hematopoiesis

More recently, the signal transduction protein and nuclear transcription regulator *beta-arrestin 1* has been shown to regulate zebrafish hematopoiesis by binding and sequestration of the suppressive polycomb group recruiter *YY1*. Interestingly, overexpression of *cdx4*, *hoxa9a*, or *hoxb4a* was able to rescue hematopoiesis in *beta-arrestin 1* suppressed fish, suggesting that *beta-arrestin 1-YY1* effects are mediated by the *cdx-hox* axis.[36] Arrestins are known mediators of several developmental pathways important for developmental hematopoiesis such as Wnt, Hedgehog, Notch, and TGF-β.[37] In detail, beta-arrestin1 has been shown to interact with phosphorylated dishevelled proteins and thereby to modulate LEF-mediated transcriptional activity.[38]

Other data from zebrafish models suggest a linear activation of *cdx4* driven by the TATA-box-binding-protein (TBP)–related factor 3, *trf3* (or *tbp2*) through binding of *mespa*. Inhibition of *trf3* during zebrafish development induces multiple defects, including depletion of *cdx4* and a failure to undergo hematopoiesis, which can be rescued by *mespa* expression. Molecularly, *mespa* is a direct target of *trf3* and itself directly activates *cdx4*. Consistently, ectopic *mespa* can rescue the developmental defects of *trf3* suppressed zebrafish embryo, suggesting an ordered *trf3-mespa-cdx4* molecular axis during zebrafish blood development.[39] In mice, the *Mesp1* gene has been reported as one of the earliest markers of cardiac development,[40] but its involvement in hematopoiesis remains unclear. Interestingly, augmentation of *Mesp1* levels can be achieved in differentiating mouse ESCs by supplementation with high-dose BMP4, a condition associated with enhanced blood formation (Grauer and Lengerke, unpublished). The molecular interactions between *trf3* and classical morphogen pathways reported to induce *cdx* expression (Wnt, RA, and Fgf) are to our knowledge largely unknown.

To identify additional pathways interacting with *cdx4* during primitive hematopoiesis, Paik *et al.* conducted a chemical screen for compounds that increase *gata1* expression in *cdx4* mutant zebrafish.[41] Only 2 of 2,640 compounds performed a rescue, both belonging to the psoralen family and showing anteroposterior patterning effects similar to DEAB, a known inhibitor of Raldh enzymes. Further analyses are required to discern the molecular mechanisms by which psoralens influence embryonic hematopoiesis, and whether they act downstream or parallel to the *cdx4-hox* pathway.

Cdx proteins direct blood development from pluripotent stem cells

In vitro, differentiating pluripotent stem cells (PSCs) model early stages of development[21,42] and insights from developmental models can be used to modulate their efficient differentiation into tissues of interest. After exit from the undifferentiated stage, PSCs initiate gastrulation and form primitive-streak like cells. During this process, AP patterning can be imposed by exogenous supplementation with morphogens. For example, BMP4 can stimulate the generation of posterior mesoderm cells and afterward direct their differentiation into blood cells by activating a Wnt-Cdx-Hox pathway.[4] Accordingly, *Cdx* genes are expressed in waves, peaking during the developmental window of hemangioblast specification in murine and human differentiating PSCs.[43–45] In detail, BMP4 has been shown to induce Wnt3a and activate the *Cdx-Hox* pathway through the Wnt effector molecule Lef1.[4] Overexpression of *Cdx1* or *Cdx4* fully rescues hematopoiesis in the presence of BMP and Wnt inhibition, while overexpression of *Hoxa9* exhibits a partial rescue.[4] Taken together, these data indicate the conserved functions of the BMP-Wnt-Cdx-Hox pathway during blood formation in vertebrates, from zebrafish to mammals.[4]

Single *Cdx* gene–deficient mice present surprisingly modest hematopoietic phenotypes. Since overexpression studies indicate redundancy among *Cdx* genes, loss-of-function studies have been performed not only on single but also on compound-*Cdx*–deficient ESCs. As expected, *in vitro* differentiation of $Cdx4^{-/-}$, $Cdx1^{-/-}$, and $Cdx2^{-/-}$ ESCs reveal only subtle reductions in numbers of blood progenitors obtained in colony-forming assays, primarily involving erythroid and mixed progenitor colonies, and a slight decrease in hematopoietic expansion in OP9-coculture assays. However, more profound suppression of hematopoiesis is achieved by knockdown of *Cdx1* and/or *Cdx2* by RNA interference in the background of $Cdx4^{-/-}$ ESCs, culminating in an almost bloodless phenotype in $Cdx4^{-/-}$ cells treated with both *Cdx2* and *Cdx1* inhibitory RNAs.[22] Molecularly, these changes are associated with decreased levels of posterior *HoxA* cluster genes. Confirming the notion of *Cdx* gene redundancy, ectopic *Cdx4* can compensate not only for itself but also for *Cdx* compound deficiency. Notably, chimera studies performed with $Cdx2^{-/-}$ and wild-type cells show that the *Cdx* effect is cell autonomous, since no rescue is observed in $Cdx2^{-/-}$ cells exposed to a wild-type environment.[22]

To further dissect individual and redundant effects of *Cdx* family members, doxycycline-inducible *Cdx1*, *Cdx2*, and *Cdx4* murine ESCs have been analyzed in parallel. As previously reported, overexpression of *Cdx1* and *Cdx4* during their endogenous expression window strongly enhances the generation of hematopoietic progenitor cells.[4,44,45] Moreover, transient *Cdx4* overexpression enhances lymphoid engraftment from transplanted ESC-derived hematopoietic progenitors, suggesting inductive patterning effects on definitive hematopoiesis and the generation of HSCs.[22] Interestingly, induction of *Cdx2* during embryoid body development shows suppressive effects on hematopoiesis, possibly mediated by induction of additional (anterior) *Hox* genes.[44] Differential impacts of *Cdx1*, *Cdx2*, and *Cdx4* are also detected in preformed CD41$^+$ ckit$^+$ blood progenitors isolated from embryoid bodies. All three *Cdx* genes strongly regulate the hematopoietic potential of CD41$^+$ c-kit$^+$ cells. However, while *Cdx4* expand hematopoietic colonies, *Cdx1* and *Cdx2* have inhibitory effects, possibly by inhibiting differentiation. Notably, ectopic *Cdx* is unable to respecify CD41$^-$ cells to hematopoietic fate.[44]

Together, these data demonstrate important roles for Cdx proteins in the specification of hematopoietic progenitor cells. The coexistence of both redundant and nonredundant Cdx-effects, which has been documented also in the gut,[46] may explain the differential impact of the three *Cdx* genes on adult hematopoiesis and leukemia, which will be discussed below.

Cdx genes in adult hematopoiesis and leukemia

Several genes required during hematopoietic development, such as *Scl/Tal-1* and acute myeloid leukemia (AML)/Runx1, have also been shown to regulate adult blood cell homeostasis and malignant transformation.[47] The effect of the three murine *Cdx* genes on CD41$^+$c-kit$^+$ ESCs derived blood progenitor cells strongly suggests an impact on hematopoietic cell biology beyond early developmental specification. However, in adult whole bone marrow samples of mice and humans, *Cdx1* and *Cdx4* can be detected in only low levels, and *Cdx2* is absent. Lack of *Cdx2* expression in healthy blood cells has been confirmed in analyses of sorted human CD34$^+$ stem and progenitor cells, CD19$^+$ B$^-$ and CD3$^+$ T cells and in murine HSCs, (Lin$^-$ Sca1$^+$ c-kit$^+$), common myeloid progenitors (CMPs), granulocyte-macrophage progenitors (GMPs), and megakaryocyte-erythroid progenitors (MEPs). Interestingly, aberrant activation of *Cdx2* can be observed in most cases of acute myeloid and lymphoid leukemia, suggesting a contribution to oncogenic transformation of blood cells. In support of this hypothesis, retroviral overexpression of *Cdx2* in murine whole bone marrow enhances *in vitro* serial-replating activity and robustly induces acute myeloid leukemia in mice.[48,49] Molecularly, induction of downstream *Hox* genes such as *Hoxa10* and *Hoxb8* is observed. Thus, acquisition of aberrant *Cdx2* expression has been proposed as the mechanism for the deregulated *Hox* expression observed in several leukemias.[48–51] In support of this hypothesis, deletion of the *Cdx2* N-terminal domain abrogated its ability to perturb *Hox* genes as well as its leukemogenic activity in mice.[52] Additionally, expression analysis performed on 115 patients with AML showed a correlation between *Cdx2* and *HOX* gene levels.[52]

The mechanism by which *CDX2* is reactivated in leukemia remains unclear. In AML, expression of *CDX2* has been shown to be predominantly monoallelic.[48] Amplifications at the 13q12.3 locus of the human *CDX2* gene have been observed in only 3 out of 170 AML patients, all three belonging to the complex karyotype group and showing high *CDX2* transcripts. Despite its first description in a t(12;13)(p13;q12)-positive AML,[53] *CDX2* expression levels analyzed in samples from a cohort of 170 patients with AML showed highest expression associated with t(9;11)(p22;p23) translocations, followed by those with normal karyotype, t(15;17)(q22;q11–21), t(8;21)(q22;22), inv(16)(p13q22), or other chromosomal aberrations, and complex karyotype, defined as three or more cytogenetic abnormalities in the absence of t(8;21), inv(16), t(15;17), or t(11q23). *CDX2* expression levels correlate with disease burden and response to therapy, suggesting possible use of *CDX2* as a marker of minimal residual disease in AML. *CDX2* expression has also been detected in subgroups of patients with myelodysplasia and chronic myelogenous leukemia, where, in a low number of analyzed patients, increased *CDX2* expression was associated with transit into secondary AML and respectively with blast and accelerated phase.[48]

As previously mentioned, the downstream pathways regulated by *CDX2* in leukemia are likely to involve *HOX* genes. However, to our knowledge, no data are available showing that modulation of human *CDX2* expression levels affect *HOX* gene expression patterns in human leukemic cells. Analyses available on *CDX2* in human AML are limited to the correlative expression study mentioned above, showing association between *CDX2* and *HOX* expression levels, and data from AML cell lines showing reduced *in vitro* growth and colony-forming capacity after *CDX2* suppression. No functional or molecular data exist in acute lymphoblastic leukemia (ALL), where *CDX2* expression is also a common event.[54,55] Unlike in AML, recent data in pediatric ALL suggest that *HOXA9* expression correlates with better prognosis ($P = 0.03$, $n = 61$ pediatric ALL samples[56]), while *CDX2* is associated with negative prognosis in adult ALL.[55] Intriguingly, *CDX2* expression was not found to correlate with specific *HOX* expression deregulation in pediatric ALL, which may explain why no associations between *CDX2* expression and MLL rearrangements previously associated with *HOX* deregulation have been yet reported in leukemia.[57,58] Moreover, *CDX2* has been directly implicated in the upregulation of Bcl-2 in t(14;18)–positive lymphoid cells,[59] partially by interaction with C/EBP.[60] Furthermore, miR-125b is involved as a *CDX2* target regulating hematopoietic cell differentiation through repression of the core binding factor in hematopoietic malignancies.[61] Finally, as reviewed above, more recent data from developmental studies indicate that major morphogen pathways, which also affect leukemogenesis, are downstream targets of *Cdx*.

The roles of *Cdx1* and *Cdx4* in adult hematopoiesis and leukemia are less well defined. Expression of human *CDX1* and *CDX4* has been noted in subgroups of AML in one report,[62] but could not be confirmed in another.[52] Recent data document low levels of *CDX1* but not of *CDX4* in a subgroup of pediatric ALL samples (Grauer and Lengerke, unpublished). In murine models, *Cdx4* has also been shown to activate *Hox* genes and induce myeloid leukemia.[62,63] However, leukemia occurs with a long latency of almost nine months, suggesting that *Cdx4* by itself is insufficient to drive leukemogenesis.[62] More robust contribution of *Cdx4* to leukemia has been shown in combination with other genetic events such as MLL-AF9[64]and Meis1a.[62] Interestingly, *Cdx4* has also proven to be a downstream target of the leukemogenic *HoxA10*.[65] Confirming these data and the studies on embryonic hematopoiesis, no significant effect on adult hematopoiesis is observed in single knockout *Cdx4* mice.[64]

In summary, these data suggest that *CDX2* is an important coregulator of malignant transformation in blood cells, but its function and molecular regulation in leukemia is poorly understood. The role of *CDX2* may be different in myeloid versus lymphoid neoplasia and its molecular effectors in leukemia likely include non-*HOX* targets. The role of *CDX1* and *CDX4* in adult blood cell homeostasis remains to be defined; however, low or negative expression levels in human leukemia samples suggest modest roles in leukemogenesis when compared to *CDX2*.

Conclusion

Cdx genes are major regulators of embryonic hematopoiesis in zebrafish, and conserved roles have been demonstrated in murine pluripotent stem cells, where ectopic *Cdx* can be used to direct blood specification via activation of *Hox* genes. While functional redundancy among the individual *Cdx* family members and compensation during development make it difficult to discern loss-of-function phenotypes in single gene mutant mice, gene-specific effects have been revealed by extensive *in vitro* analysis of differentiating pluripotent stem cells. These may be due to differential activation of *Hox*-gene combinations, or to other downstream targets such as morphogens or miRNAs.

Non-*Hox*–related effects may explain the activity of *Cdx2* in some leukemia types, especially in MLL-negative ALL and other leukemias in which the role of *Hox* genes is less prominent.

Acknowledgments

C.L. is supported by grants from the Deutsche Krebshilfe (*Max-Eder* program), the Deutsche Forschungsgemeinschaft (*SFB773*), the University of Tübingen (*Fortüne* program), and the Böhringer Ingelheim Foundation (*Exploration Grant* program). G.Q.D. is an investigator of the Howard Hughes Medical Institute and supported by Grants from the NIH (R24DK092760, UO1-HL100001, RC4-DK090913, P50HG005550, and special funds from the ARRA stimulus package-RC2-HL102815), the Roche Foundation for Anemia Research, Alex's Lemonade Stand, the Ellison Medical Foundation, the Doris Duke Charitable Foundation, and the Harvard Stem Cell Institute. G.Q.D. is an affiliated member of the Broad Institute and a senior scientist of the Manton Center for Orphan Disease Research.

Conflicts of interest

The authors declare no conflicts of interest.

References

1. Young, T. & J. Deschamps. 2009. Hox, Cdx, and anteroposterior patterning in the mouse embryo. *Curr. Top. Dev. Biol.* **88:** 235–255.
2. Davidson, A.J. *et al.* 2003. cdx4 mutants fail to specify blood progenitors and can be rescued by multiple hox genes. *Nature* **425:** 300–306.
3. Davidson, A.J. & L.I. Zon. 2006. The caudal-related homeobox genes cdx1a and cdx4 act redundantly to regulate hox gene expression and the formation of putative hematopoietic stem cells during zebrafish embryogenesis. *Dev. Biol.* **292:** 506–518.
4. Lengerke, C. *et al.* 2008. BMP and Wnt specify hematopoietic fate by activation of the Cdx-Hox pathway. *Cell Stem Cell* **2:** 72–82.
5. Wingert, R.A. *et al.* 2007. The cdx genes and retinoic acid control the positioning and segmentation of the zebrafish pronephros. *PLoS Genet.* **3:** 1922–1938.
6. Lengerke, C. *et al.* 2011. Interactions between Cdx genes and retinoic acid modulate early cardiogenesis. *Dev. Biol.* **354:** 134–142.
7. Guo, R.J., E.R. Suh & J.P. Lynch. 2004. The role of Cdx proteins in intestinal development and cancer. *Cancer Biol. Ther.* **3:** 593–601.
8. Gaunt, S.J., D. Drage & A. Cockley. 2003. Vertebrate caudal gene expression gradients investigated by use of chick cdx-A/lacZ and mouse cdx-1/lacZ reporters in transgenic mouse embryos: evidence for an intron enhancer. *Mech. Dev.* **120:** 573–586.
9. Gaunt, S.J., A. Cockley & D. Drage. 2004. Additional enhancer copies, with intact cdx binding sites, anteriorize Hoxa-7/lacZ expression in mouse embryos: evidence in keeping with an instructional cdx gradient. *Int. J. Dev. Biol.* **48:** 613–622.
10. Gaunt, S.J., D. Drage & R.C. Trubshaw. 2005. cdx4/lacZ and cdx2/lacZ protein gradients formed by decay during gastrulation in the mouse. *Int. J. Dev. Biol.* **49:** 901–908.
11. Iimura, T. & O. Pourquie. 2006. Collinear activation of Hoxb genes during gastrulation is linked to mesoderm cell ingression. *Nature* **442:** 568–571.
12. Carapuco, M. *et al.* 2005. Hox genes specify vertebral types in the presomitic mesoderm. *Genes. Dev.* **19:** 2116–2121.
13. Subramanian, V., B.I. Meyer & P. Gruss. 1995. Disruption of the murine homeobox gene Cdx1 affects axial skeletal identities by altering the mesodermal expression domains of Hox genes. *Cell* **83:** 641–653.
14. Young, T. *et al.* 2009. Cdx and Hox genes differentially regulate posterior axial growth in mammalian embryos. *Dev. Cell* **17:** 516–526.
15. Strumpf, D. *et al.* 2005. Cdx2 is required for correct cell fate specification and differentiation of trophectoderm in the mouse blastocyst. *Development* **132:** 2093–2102.
16. van de Ven, C. *et al.* 2011. Concerted involvement of Cdx/Hox genes and Wnt signaling in morphogenesis of the caudal neural tube and cloacal derivatives from the posterior growth zone. *Development* **138:** 3451–3462.
17. Savory, J.G. *et al.* 2009. Cdx2 regulation of posterior development through non-Hox targets. *Development* **136:** 4099–4110.
18. Chawengsaksophak, K. *et al.* 2004. Cdx2 is essential for axial elongation in mouse development. *Proc. Natl. Acad. Sci. USA* **101:** 7641–7645.
19. Chawengsaksophak, K. *et al.* 1997. Homeosis and intestinal tumours in Cdx2 mutant mice. *Nature* **386:** 84–87.
20. van Nes, J. *et al.* 2006. The Cdx4 mutation affects axial development and reveals an essential role of Cdx genes in the ontogenesis of the placental labyrinth in mice. *Development* **133:** 419–428.
21. Lengerke, C. & G.Q. Daley. 2005. Patterning definitive hematopoietic stem cells from embryonic stem cells. *Exp. Hematol.* **33:** 971–979.
22. Wang, Y. *et al.* 2008. Cdx gene deficiency compromises embryonic hematopoiesis in the mouse. *Proc. Natl. Acad. Sci. USA* **105:** 7756–7761.
23. Gering, M. *et al.* 1998. The SCL gene specifies haemangioblast development from early mesoderm. *EMBO J.* **17:** 4029–4045.
24. Charite, J. *et al.* 1998. Transducing positional information to the Hox genes: critical interaction of cdx gene products with position-sensitive regulatory elements. *Development* **125:** 4349–4358.
25. Hunter, C.P. *et al.* 1999. Hox gene expression in a single Caenorhabditis elegans cell is regulated by a caudal homolog and intercellular signals that inhibit wnt signaling. *Development* **126:** 805–814.
26. Savory, J.G. *et al.* 2011. Cdx4 is a Cdx2 target gene. *Mech. Dev.* **128:** 41–48.

27. Beland, M. *et al.* 2004. Cdx1 autoregulation is governed by a novel Cdx1-LEF1 transcription complex. *Mol. Cell Biol.* **24:** 5028–5038.
28. Houle, M. *et al.* 2000. Retinoic acid regulation of Cdx1: an indirect mechanism for retinoids and vertebral specification. *Mol. Cell Biol.* **20:** 6579–6586.
29. Houle, M., J.R. Sylvestre & D. Lohnes. 2003. Retinoic acid regulates a subset of Cdx1 function in vivo. *Development* **130:** 6555–6567.
30. de Jong, J.L. *et al.* 2010. Interaction of retinoic acid and scl controls primitive blood development. *Blood* **116:** 201–209.
31. Pilon, N. *et al.* 2007. Wnt signaling is a key mediator of Cdx1 expression in vivo. *Development* **134:** 2315–2323.
32. Pilon, N. *et al.* 2006. Cdx4 is a direct target of the canonical Wnt pathway. *Dev. Biol.* **289:** 55–63.
33. Hikasa, H. *et al.* 2010. Regulation of TCF3 by Wnt-dependent phosphorylation during vertebrate axis specification. *Dev Cell* **19:** 521–532.
34. Ro, H. & I.B. Dawid. 2011. Modulation of Tcf3 repressor complex composition regulates cdx4 expression in zebrafish. *EMBO J.* **30:** 2894–2907.
35. Keenan, I.D., R.M. Sharrard & H.V. Isaacs. 2006. FGF signal transduction and the regulation of Cdx gene expression. *Dev. Biol.* **299:** 478–488.
36. Yue, R. *et al.* 2009. Beta-arrestin1 regulates zebrafish hematopoiesis through binding to YY1 and relieving polycomb group repression. *Cell* **139:** 535–546.
37. Kovacs, J.J. *et al.* 2009. Arrestin development: emerging roles for beta-arrestins in developmental signaling pathways. *Dev. Cell* **17:** 443–4458.
38. Chen, W. *et al.* 2001. beta-Arrestin1 modulates lymphoid enhancer factor transcriptional activity through interaction with phosphorylated dishevelled proteins. *Proc. Natl. Acad. Sci. USA* **98:** 14889–14894.
39. Hart, D.O. *et al.* 2007. Initiation of zebrafish haematopoiesis by the TATA-box-binding protein-related factor Trf3. *Nature* **450:** 1082–1085.
40. David, R. *et al.* 2008. MesP1 drives vertebrate cardiovascular differentiation through Dkk-1-mediated blockade of Wnt-signalling. *Nat. Cell Biol.* **10:** 338–345.
41. Paik, E.J. *et al.* 2010. A chemical genetic screen in zebrafish for pathways interacting with cdx4 in primitive hematopoiesis. *Zebrafish* **7:** 61–68.
42. Murry, C.E. & G. Keller. 2008. Differentiation of embryonic stem cells to clinically relevant populations: lessons from embryonic development. *Cell* **132:** 661–680.
43. Lengerke, C. *et al.* 2009. Hematopoietic development from human induced pluripotent stem cells. *Ann. N. Y. Acad. Sci.* **1176:** 219–227.
44. McKinney-Freeman, S.L. *et al.* 2008. Modulation of murine embryonic stem cell-derived CD41+c-kit+ hematopoietic progenitors by ectopic expression of Cdx genes. *Blood* **111:** 4944–4953.
45. Lengerke, C. *et al.* 2007. The cdx-hox pathway in hematopoietic stem cell formation from embryonic stem cells. *Ann. N. Y. Acad. Sci.* **1106:** 197–208.
46. Calon, A. *et al.* 2007. Different effects of the Cdx1 and Cdx2 homeobox genes in a murine model of intestinal inflammation. *Gut* **56:** 1688–1695.
47. Izraeli, S. 2004. Leukaemia–a developmental perspective. *Br. J. Haematol.* **126:** 3–10.
48. Scholl, C. *et al.* 2007. The homeobox gene CDX2 is aberrantly expressed in most cases of acute myeloid leukemia and promotes leukemogenesis. *J. Clin. Invest.* **117:** 1037–1048.
49. Rawat, V.P. *et al.* 2004. Ectopic expression of the homeobox gene Cdx2 is the transforming event in a mouse model of t(12;13)(p13;q12) acute myeloid leukemia. *Proc. Natl. Acad. Sci. USA* **101:** 817–822.
50. Rice, K.L. & J.D. Licht. 2007. HOX deregulation in acute myeloid leukemia. *J. Clin. Invest.* **117:** 865–868.
51. Frohling, S. *et al.* 2007. HOX gene regulation in acute myeloid leukemia: CDX marks the spot? *Cell Cycle* **6:** 2241–2245.
52. Rawat, V.P. *et al.* 2008. Overexpression of CDX2 perturbs HOX gene expression in murine progenitors depending on its N-terminal domain and is closely correlated with deregulated HOX gene expression in human acute myeloid leukemia. *Blood* **111:** 309–319.
53. Chase, A. *et al.* 1999. Fusion of ETV6 to the caudal-related homeobox gene CDX2 in acute myeloid leukemia with the t(12;13)(p13;q12). *Blood* **93:** 1025–1031.
54. Riedt, T. *et al.* 2009. Aberrant expression of the homeobox gene CDX2 in pediatric acute lymphoblastic leukemia. *Blood* **113:** 4049–4051.
55. Thoene, S. *et al.* 2009. The homeobox gene CDX2 is aberrantly expressed and associated with an inferior prognosis in patients with acute lymphoblastic leukemia. *Leukemia* **23:** 649–655.
56. Starkova, J. *et al.* 2010. HOX gene expression in phenotypic and genotypic subgroups and low HOXA gene expression as an adverse prognostic factor in pediatric ALL. *Pediatr Blood Cancer* **55:** 1072–1082.
57. Muntean, A.G. & J.L. Hess. 2012. The pathogenesis of mixed-lineage leukemia. *Annu. Rev. Pathol.* **7:** 283–301.
58. Ernst, P. *et al.* 2004. An MLL-dependent Hox program drives hematopoietic progenitor expansion. *Curr. Biol.* **14:** 2063–2069.
59. Heckman, C.A. *et al.* 2000. A-Myb up-regulates Bcl-2 through a Cdx binding site in t(14;18) lymphoma cells. *J. Biol. Chem.* **275:** 6499–6508.
60. Heckman, C.A., M.A. Wheeler & L.M. Boxer. 2003. Regulation of Bcl-2 expression by C/EBP in t(14;18) lymphoma cells. *Oncogene* **22:** 7891–7899.
61. Lin, K.Y. *et al.* 2011. miR-125b, a target of CDX2, regulates cell differentiation through repression of the core binding factor in hematopoietic malignancies. *J. Biol. Chem.* **286:** 38253–38263.
62. Bansal, D. *et al.* 2006. Cdx4 dysregulates Hox gene expression and generates acute myeloid leukemia alone and in cooperation with Meis1a in a murine model. *Proc. Natl. Acad. Sci. USA* **103:** 16924–16929.
63. Yan, J. *et al.* 2006. Cdx4 and menin co-regulate Hoxa9 expression in hematopoietic cells. *PLoS One* **1:** e47.
64. Koo, S. *et al.* 2010. Cdx4 is dispensable for murine adult hematopoietic stem cells but promotes MLL-AF9-mediated leukemogenesis. *Haematologica* **95:** 1642–1650.
65. Bei, L. *et al.* 2011. HoxA10 activates CDX4 transcription and Cdx4 activates HOXA10 transcription in myeloid cells. *J. Biol. Chem.* **286:** 19047–19064.

Ann. N.Y. Acad. Sci. ISSN 0077-8923

ANNALS OF THE NEW YORK ACADEMY OF SCIENCES
Issue: *Hematopoietic Stem Cells VIII*

Hematopoietic stem cells are regulated by Cripto, as an intermediary of HIF-1α in the hypoxic bone marrow niche

Kenichi Miharada, Göran Karlsson, Matilda Rehn, Emma Rörby, Kavitha Siva, Jörg Cammenga, and Stefan Karlsson
Department for Molecular Medicine and Gene Therapy, Lund Strategic Center for Stem Cell Biology, Lund University, Lund, Sweden

Address for correspondence: Kenichi Miharada and Stefan Karlsson, BMC A12, 221 84 Lund, Sweden.
Kenichi.Miharada@med.lu.se and Stefan.Karlsson@med.lu.se

Cripto has been known as an embryonic stem (ES)- or tumor-related soluble/cell membrane protein. In this study, we demonstrated that Cripto has a role as an important regulatory factor for hematopoietic stem cells (HSCs). Recombinant Cripto sustained the reconstitution ability of HSCs *in vitro*. Flow cytometry analysis uncovered that GRP78, one of the candidate receptors for Cripto, was expressed on a subset of HSCs and could distinguish dormant/myeloid-biased HSCs and active/lymphoid-biased HSCs. Cripto is expressed in hypoxic endosteal niche cells where GRP78$^+$ HSCs mainly reside. Proteomics analysis revealed that Cripto-GRP78 binding stimulates glycolytic metabolism-related proteins and results in lower mitochondrial potential in HSCs. Furthermore, conditional knockout mice for HIF-1α, a master regulator of hypoxic responses, showed reduced Cripto expression and decreased GRP78$^+$ HSCs in the endosteal niche area. Thus, Cripto-GRP78 is a novel HSC regulatory signal mainly working in the hypoxic niche.

Keywords: Cripto; GRP78; hypoxia; glycolysis

Introduction

Hematopoietic stem cells (HSCs) can generate all hematopoietic cell types (multilineage potential) and produce new HSC through cell division (self-renewal).[1] HSCs are maintained in the special microenvironment called a *niche*, in which many types of niche component cells and factors keep HSCs in a dormant state.[2–5] The endosteal area in the bone marrow is called the *hypoxic niche* because the oxygen level is relatively low and therefore HSCs have low mitochondrial activity.[6–8] However, key regulators, particularly soluble factors, that sustain dormancy of HSCs within the hypoxic niche and control their metabolic activity remain largely unknown.

Recently we have demonstrated that the soluble Cripto protein (also known as teratocarcinoma derived growth factor-1: TDGF-1), which has been reported to be a pluripotent stem cell– and tumor cell–related factor, plays important roles for HSC regulation in the hypoxic niche as an extracellular factor.[9] Cripto is a member of the EGF-CFC family and has both soluble and cell membrane–associated forms (Fig. 1A), although the difference between the two forms is unclear at present.[10–15] The Cripto gene has been identified as a gene that regulates tumorigenesis, embryogenesis, and embryonic stem (ES) cells,[16–20] and is also a critical regulator of myocardial development.[21,22]

As an ES cell–related gene product, Cripto is known to interact with other ES cells and embryonic developmental gene products, for example, Wnt, transforming growth factor-β (TGF-β) family members, Notch pathway members, and hypoxia inducible factor-1α (HIF-1α).[13,14] While Cripto is one of the primary target genes of these pathways,[13] roles for Cripto as a coreceptor for TGF-β family members have been characterized as well. Classically Cripto has been known to bind to TGF-β and the TGF-β receptor and inhibits phosphorylation of Smad2/3 and the formation of a complex with Smad4 (Fig. 1B).[13–25] Additionally, Cripto functions as an important modulator of other

doi: 10.1111/j.1749-6632.2012.06564.x

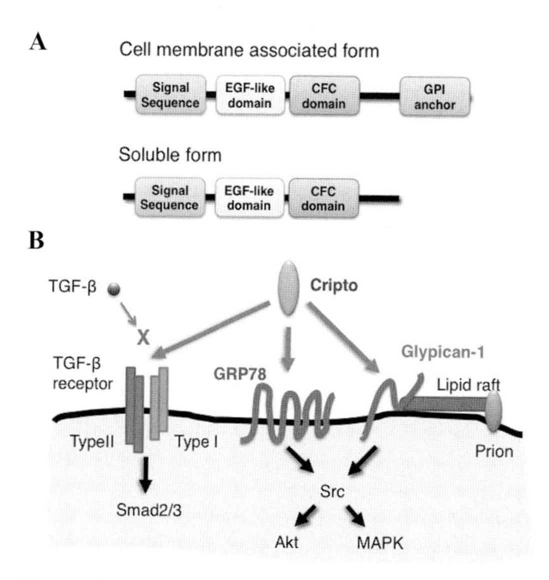

Figure 1. (A) Schematic figure of the protein structure of Cripto. Cripto has both a cell membrane–associated form (above) and a soluble form (below), which is released by cleaving off at a GPI anchor site. (B) Possible candidate receptor proteins for Cripto. Classically, Cripto has been known to inhibit the binding of TGF-β to the TGF-β receptor. Cell surface GRP78, a member of the heat shock protein 70 family, is usually expressed on the ER membrane, whereas some tumor cells have the protein on their cell surface. Cripto is known to bind with cell surface GRP78, activating MAPK/Akt pathways. Cripto also binds to glypican-1, which is a heparan sulfate proteoglycan and associates with cell surface lipid rafts and Prion protein.

TGF-β family members by binding to the TGF-β receptor complex. In the absence of Cripto, Nodal signaling is impaired, whereas activin signaling is high and leads to phosphorylate Smad2/3. In contrast, in the presence of Cripto, activin signaling is reduced due to conformation changes in the receptor complex by Cripto and the TGF-β receptors, while Nodal can bind to the receptors in association with Cripto.[13,14] Furthermore, Cripto can function as a coreceptor for noncanonical Wnt11 in a complex with frizzled-7 and glypican-4.[13] Notch is also known to be a target of Cripto/Nodal signaling. Binding of Cripto to four Notch receptors—mainly in the endoplasmic reticulum (ER) membrane/Golgi complex—has been reported, and Cripto facilitates the posttranslational maturation of Notch receptors.[13,26]

Recently Cripto has been shown to bind to the cell surface protein GRP78 and glypican-1, which subsequently stimulates MAPK-PI3K-Akt pathways (Fig. 1B).[13,14,27–29] GRP78 is a member of the heat shock protein 70 (HSP70) family and is usually expressed on the ER membrane, but some types of tumor cells (e.g., breast cancer cells) express GRP78 on the cell surface as a target receptor for certain ligands, including Cripto.[30] Our results revealed that HSCs are separated into functionally distinct subpopulations based on the expression of GRP78, though both have the capacity of long-term reconstitution. Moreover, proteomics and endosteal cell analyses have clarified that Cripto is expressed on the endosteal niche component cells and governs HSC maintenance by regulating glycolytic metabolism.

Cripto sustains HSC function *in vitro*

We measured the expression level of Cripto mRNA in several hematopoietic cell populations using real-time qPCR. Notably, the quantitative RT-PCR assay showed higher Cripto expression in long-term HSCs than in other more mature populations. Cripto gene expression has been known to correlate well with the immaturity of ES/iPS cells, so that the expression level is reduced immediately after withdrawal of LIF.[16,19] Real-time qPCR results suggested that this was similar in hematopoietic cells (Fig. 2A). To test functions of Cripto as a soluble protein, we used recombinant mouse Cripto (rmCripto) and murine HSCs with the immunophenotype of $CD34^-$ c-kit^+ Sca-I^+ lineage-marker negative cells ($CD34^-$ KSL cells).[31] In a colony-forming unit assay (CFU-assay), $CD34^-$ KSL cells cultured with rmCripto, stem cell factor (SCF), and thrombopoietin (TPO) showed increased colony formation, including increased numbers of mixed lineage colonies, even after 14 days of *in vitro* culture, while cells grown without Cripto generated no mixed colonies. Furthermore, the cells cultured with rmCripto for 14 days could successfully reconstitute long-term in lethally-irradiated mice, while cells grown without rmCripto could not reconstitute recipients. Donor cells from engrafted primary recipient mice could reconstitute secondary recipient mice. These findings suggest that Cripto has a supportive function for maintenance of HSC *in vitro*.

GRP78 distinguishes two types of HSCs

GRP78, reported as one of the receptor proteins for Cripto, is known to be expressed on cells possessing high proliferation potential, for example, tumor cells. In order to analyze whether HSCs express GRP78 on their cell surface, flow cytometry using anti-GRP78 antibody was performed.

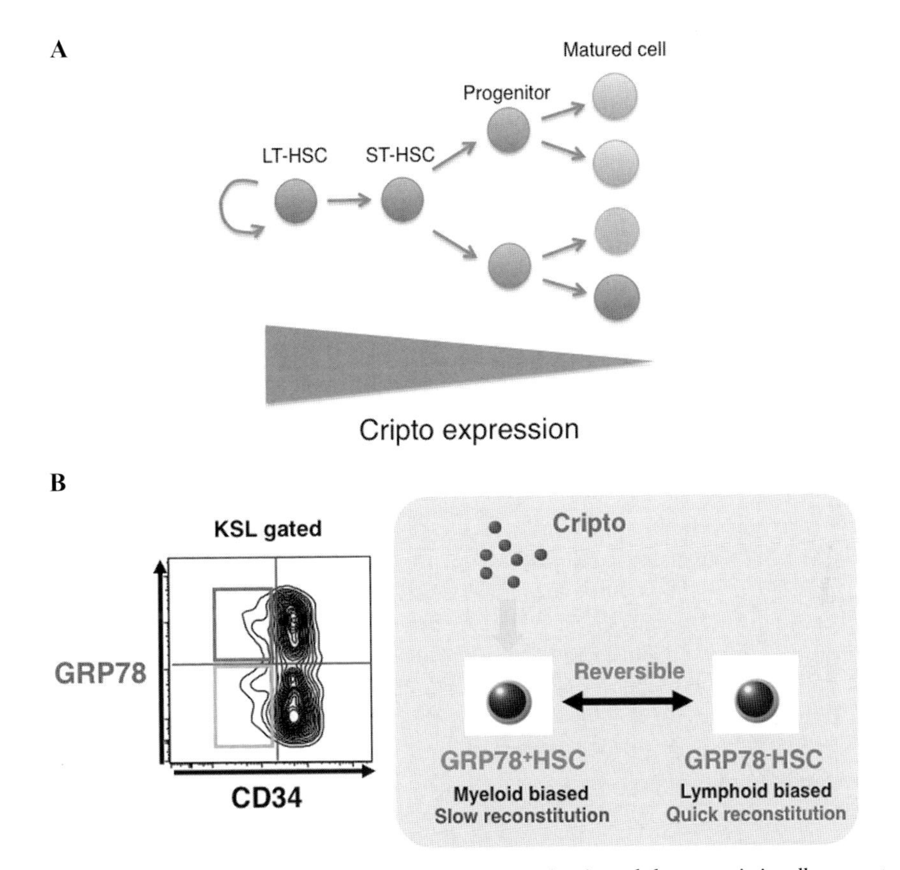

Figure 2. (A) The hematopoietic cell hierarchy and *Cripto* gene expression in each hematopoietic cell compartment. Cripto expression was higher in immature populations, especially the long-term stem cell (LT-HSC) compared to the short-term stem cell (ST-HSC) and other mature populations. (B) Representative flow cytometry analysis for GRP78 staining. CD34$^-$ KSL cells had clearly separated two populations based on the expression of cell surface GRP78. In contrast, glypican-1 was expressed on all CD34$^-$ KSL cells (data not shown). *In vivo* transplantation assays revealed that GRP78$^+$ HSCs were myeloid-biased, slow reconstituting cells, while GRP78$^-$ HSCs were lymphoid-biased, quick-reconstituting cells.

Surprisingly, CD34$^-$ KSL cells were clearly separated into two distinct subpopulations based on the GRP78 expression (Fig. 2B, left). To address which population was more primitive and whether there was a functional difference, GRP78$^+$ CD34$^-$ KSL cells (GRP78$^+$ HSCs) and GRP78$^-$ CD34$^-$ KSL cells (GRP78$^-$ HSCs) were separately transplanted into lethally-irradiated mice. The results showed that GRP78$^-$ HSCs reconstituted the recipient mice quickly, hence peripheral blood (PB) chimerism was higher at early time points and decreased later, while GRP78$^+$ HSCs showed slow reconstitution, and chimerism was increased over time, although both subpopulations achieved long-term reconstitution (Fig. 2B, right). Another difference observed in the lineage distribution was that myeloid-biased reconstitution occurred in the GRP78$^+$ HSCs engrafted mice, while GRP78$^-$ HSCs tended to generate more lymphoid cells (Fig. 2B, right). Four months after transplantation, bone marrow cells from the recipient mice were analyzed. Curiously, both GRP78$^+$ and GRP78$^-$ HSCs could generate each other, suggesting that these subpopulations are not hierarchical. Additionally, combined staining of the side population (SP) and CD150 staining with GRP78 indicated that GRP78$^+$ CD150$^+$ CD34$^-$ KSL cells contained a much higher number of SP-low cells, which was previously reported to contain more dormant cells,[32] while GRP78$^-$ CD150$^-$ CD34$^-$ KSL cells contained almost no SP-low cells.

GRP78$^+$ and GRP78$^-$ HSCs were cultured with or without rmCripto and then transplanted. The results showed that GRP78$^+$ HSCs reacted to rmCripto, but GRP78⁻ HSCs did not. In addition, anti-GRP78 neutralizing antibody (N-20) treatment completely blocked the effect of rmCripto. Interestingly, the *ex vivo* culture of GRP78⁺ HSCs with N-20 alone showed less growth compared with the control condition, which may suggest an autocrine effect of endogenous Cripto, as Cripto expression is higher in LT-HSCs. Taken together, GRP78 has been demonstrated to be a functional receptor for Cripto on HSCs. Additionally, this receptor can distinguish different types of (nonhierarchical) HSCs similar to other markers.[33–38]

Cripto-GRP78 stimulates Akt and controls glycolytic metabolism

It has been reported that functions of Cripto are involved in both TGF-β signaling and MAPK-PI3K-Akt pathways. Intracellular staining for phosphorylated interacting proteins in GRP78⁺ HSCs after Cripto treatment observed significant upregulation of Akt pathway proteins (e.g., pAkt, p4E-BP1), but not Smad pathway components. Neither untreated cells nor GRP78⁻ HSCs showed this activation. Interestingly, even using freshly separated cells, both GRP78⁺ and GRP78⁻ HSCs exhibit different types of activated signaling proteins. This finding confirms that the GRP78 subpopulations are different also with respect to intrinsic signals. TGF-β has recently been reported to be an important regulator for the dormancy of HSCs.[33,39,40] Because the SP-low subpopulation has been reported to have a positive response to cell growth by TGF-β signaling,[32] we tested the reaction of the GRP78 subsets to TGF-β. As expected, GRP78⁺ HSCs indicated less responsiveness to the suppressive effect of TGF-β than GRP78⁻ HSCs, although we did not find a "positive" effect as reported for the low SP cells. Even though Cripto did not stimulate Smad signaling in HSCs, it could still be involved in the regulation of HSCs by interacting to TGF-β and its receptors through non-Smad pathways.

To study more detailed mechanisms of Cripto-GRP78 signaling, 2D-DIGE proteomics assay was performed using the Lhx2 cell line, which was established by overexpressing the homeobox gene *Lhx2* and is known as an "HSC-like" cell line because of its CD34⁻ KSL immunophenotype and ability to engraft irradiated recipients.[41] Interestingly, this cell line produces both GRP78⁺ and GRP78⁻ populations. An assay using proteins extracted from rmCripto treated GRP78⁺ Lhx2 cells demonstrated that Cripto induced upregulation and/or phosphorylation of metabolism-related proteins, especially those involved in glycolysis, e.g., pyruvate kinase, phosphoglycerate kinase, and fructose-bisphosphate aldolase A (Fig. 3A). There was also an increase in several other metabolic enzymes and actin-polymerization–related proteins.

GRP78⁺ HSCs localize to the endosteal area of bone marrow and have low mitochondrial activity

The proteomic analysis revealed that the Cripto-GRP78 interaction leads to stimulation of glycolytic metabolism. Since glycolysis is an important energy production step under hypoxic conditions where HSCs mainly reside, we hypothesized that distribution of each GRP78 subpopulation might be related to the level of hypoxia. Using pimonidazole as a hypoxic probe[41] we found that GRP78⁺ CD34⁻ KSL cells were more hypoxic than GRP78⁻ HSCs and CD34⁺ progenitor cells. The hypoxic environment, referred to as the "hypoxic niche", is located in the endosteal area of bones.[6,7,41] To clarify the correlation between frequency of GRP78⁺/GRP78⁻ HSCs and their localization in bones, central bone marrow cells from the bone cavity and endosteal marrow cells attached to the endosteum were separately collected as described.[42] We found that the endosteal region contained more GRP78⁺ HSCs compared with the central bone marrow (Fig. 3B).

HSCs are also known to exhibit lower mitochondrial activity,[8] which indirectly indicates high glycolysis activity. An assay using MitoTracker® (a probe to indicate mitochondrial mass or membrane activity) revealed that GRP78⁺ HSCs had lower probe intensity than GRP78⁻ HSCs (Fig. 3C). Interestingly, after *in vitro* (normoxic) culture with rmCripto, GRP78⁺ HSCs showed increased MitoTracker intensity compared with nontreated cells, while GRP78⁻ HSCs showed similar increase of intensity in the presence and absence of rmCripto. HSCs derived from fetal liver and spleen that include almost no low-mitochondrial activity population showed no response to rmCripto. These findings suggest that Cripto function and mitochondrial activity (glycolytic activity) in the targeted cells are closely correlated.

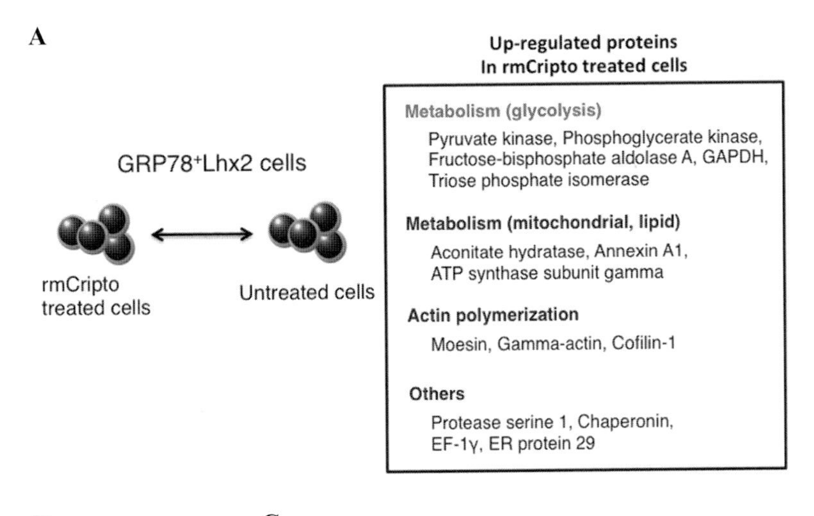

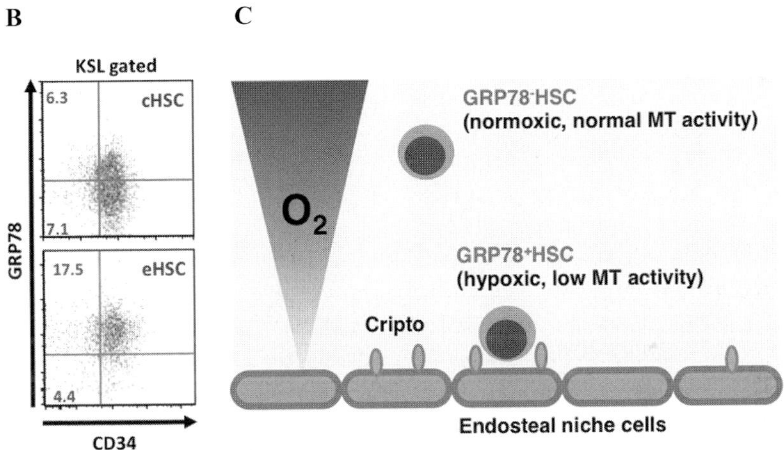

Figure 3. (A) A brief summary of the proteomics analysis using the Lhx2 HSC-like cell line. Presorted GRP78$^+$ Lhx2 cells were cultured with or without rmCripto, and proteins were compared (also with freshly isolated cells). The right panel shows the name of the proteins with increased expression. (B) Representative flow cytometry analysis for GRP78 and CD34 staining on central hematopoietic stem cells (cHSCs) and endosteal hematopoietic stem cells (eHSCs). The cells shown are gated within the KSL population. (C) A schematic image of a relationship between the endosteal niche and Cripto-GRP78. The endosteal niche area is a low-oxygen environment (hypoxia) and contains more GRP78$^+$ HSCs. ALCAM$^-$ Sca-I$^+$ niche cells in this area express cell surface Cripto. GRP78$^+$ HSCs are relatively hypoxic and have a lower mitochondrial (MT) activity, which indirectly indicates higher glycolytic activity.

Endosteal cells express Cripto and maintain GRP78$^+$ HSCs

Because separation analysis for the localization of HSCs showed that GRP78$^+$ HSCs mainly resided in close proximity to the endosteal area, we speculated that the endosteal niche component cells might express Cripto. Soluble Cripto is produced by shedding of the GPI-anchor site of cell membrane–associated Cripto.[13] Therefore, we decided to analyze the expression of cell-surface Cripto protein on the endosteal cells by flow cytometry. Combined staining of ALCAM and Sca-I with hematopoietic markers has been reported to be a method to separate endosteal adherent cells into osteoblastic and mesenchymal populations.[43] The staining protocol demonstrated that cell membrane–associated Cripto was expressed mainly on the ALCAM$^-$ Sca-I$^+$ mesenchymal cell population, whereas ALCAM$^+$ Sca-I$^-$ osteoblastic cells contained less cell-surface Cripto. ALCAM$^-$ Sca-I$^-$ populations had no expression of cell-surface Cripto. To investigate how the cell-surface Cripto in these regions is important in the maintenance of HSCs, we injected GRP78 neutralization antibody (N-20) into normal mice and then analyzed by flow cytometry in order to see if any alteration of HSC localization occurred. N-20–injected mice showed a decreased frequency of GRP78$^+$ HSCs in the endosteal region compared with the control IgG–injected group, whereas GRP78$^-$ HSCs were increased in the central marrow area. These data demonstrated that HSC transition from the endosteal niche to the central marrow area is blocked by binding of Cripto to its receptor on HSCs. These observations strongly suggest that Cripto, as a niche-related factor, may have an important role in maintaining HSCs, specifically GRP78$^+$ HSCs, in the endosteal bone marrow area.

The expression of Cripto is under control of HIF-1α

The master regulator of hypoxic responses is HIF-1α.[41,44–46] HIF-1 is stabilized under hypoxia and dimerizes with HIF-1β to control downstream gene expression. A study using HIF-1α–conditional knockout (cKO) mice has demonstrated that HIF-1α-deleted mice have immunophenotypically normal frequencies of HSCs (e.g., CD34$^-$ KSL, Tie2, SLAM), however, functional HSCs were exhausted as a result of the accelerated cell cycle.[41] Since the promoter region of the *Cripto* gene has hypoxia responsive elements (HRE) to which the HIF-1 complex binds, we asked if lack of HIF-1α had any effect on Cripto-GRP78 signaling by using the same cKO mice.

Analyses of the endosteal niche cells uncovered decreased number of both ALCAM$^+$ Sca-I$^-$ and ALCAM$^-$ Sca-I$^+$ cells in the cKO mice. Moreover, cell surface Cripto–expressing cells were clearly diminished. Lowered expression of *Cripto* mRNA in hematopoietic stem/progenitor cells was also detected. Finally, we analyzed the frequency of GRP78$^+$ HSCs in the HIF-1α cKO mice and found that the percentage of GRP78$^+$ HSCs in the endosteal area was significantly decreased, whereas the central marrow cells showed no difference. Taken together, expression of Cripto is regulated by HIF-1α and HIF-1α protein expression may be critical for maintaining functional HSCs, probably GRP78$^+$ HSCs (Fig. 4).

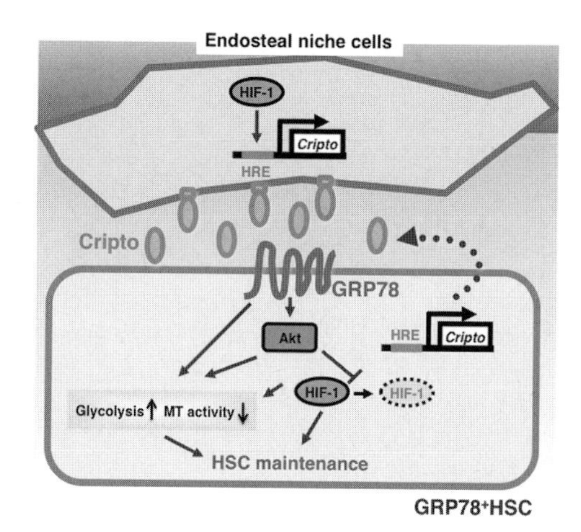

Figure 4. Summary of the role and regulation of Cripto for HSCs. Under hypoxic conditions, Cripto expression is positively regulated by HIF-1α, which is a key regulator of hypoxic responses. Cripto binds to the cell surface GRP78 and subsequently stimulates Akt pathways. The binding also increases glycolytic metabolism that results in inhibition of MT activity. Because Akt activation and phosphorylation have been reported to stabilize HIF-1α protein, the Akt activation caused by Cripto may stimulate HSC maintenance through HIF-1α stabilization.

Conclusion

Hypoxia is a critical environment in which HSCs must be regulated, especially in the endosteal region of the bone marrow. HIF-1α is a key factor to regulate hypoxic responses of many types of cells, including HSCs. The study described in this paper characterized Cripto as an intermediary of HIF-1 regulatory factor that governs HSC in the endosteal niche. In addition to the HIF-1α binding site, the Cripto promoter region also has a Smad complex binding site (SBE) and a TCF–LEF complex binding site (TBE) (which binds factors downstream of Wnt signaling).[13] Thus, many kinds of stem cell–related pathways are involved in Cripto expression. This might be the reason why ES/iPS cells and HSCs have higher expression of Cripto than other mature cell populations. GRP78 is a HSP70 family member, and it is known that its expression and cell membrane transition is regulated by hypoxia.[47,48] Although cell membrane expression of GRP78 has been reported mainly on tumor cells, it can be a useful marker also for tissue-specific stem cells and cancer stem cells. Very recently, GRP78 conditional knockout mice were reported; the paper demonstrated a critical role for GRP78 in the interaction with PTEN signal and leukemia.[49]

In the work presented here we focused on the role of Cripto as an extracellular regulator of HSC. However, the reason why the expression level is higher in immature ES cells and HSCs remains unknown. Because several tumor cells have higher Cripto expression (e.g., teratocarcinoma and breast cancers), leukemia cells and cancer stem cells in stem cell niches may have abnormal Cripto expression/secretion. Thus, Cripto-GRP78 signaling is a novel pathway that may be involved in the regulation of a large variety of stem cells, including cancer stem cells. Since the importance of hypoxia for cancer stem cells has been established, Cripto may be a useful factor to investigate in other stem cell systems than hematopoiesis.

Acknowledgments

This work was supported by the Hemato-Linné grant (Swedish Research Council Linnaeus), the Swedish Cancer Foundation (Cancerfonden), The Swedish Cancer Society (S.K.), the Swedish Children's Cancer Society (S.K.), the Swedish Medical Research Council (S.K.), The Tobias Prize awarded by The Royal Swedish Academy of Sciences financed by The Tobias Foundation (S.K.), and the EU project grant CONSERT, STEMEXPAND, and PERSIST. K.M. was funded by the EU-funded NOVEXPAND project in Marie Curie actions. The Lund Stem Cell Center was supported by a Center of Excellence grant in life sciences from the Swedish Foundation for Strategic Research.

Conflicts of interest

The authors declare no conflicts of interest.

References

1. Orkin, S.H. & L. Li. 2008. Hematopoiesis: an evolving paradigm for stem cell biology. *Cell* **132:** 631–644.
2. Scadden, D.T. 2006. The stem-cell niche as an entity of action. *Nature* **441:** 1075–1079.
3. Arai, F. & T. Suda. 2007. Maintenance of quiescent hematopoietic stem cells in the osteoblastic niche. *Ann. N.Y. Acad. Sci.* **1106:** 41–53.
4. Morrison, S.J. & A.C. Spradling. 2008. Stem cells and niches: mechanisms that promote stem cell maintenance throughout life. *Cell* **132:** 598–611.
5. Li, L. & H. Clevers. 2010. Coexistence of quiescent and active adult stem cells in mammals. *Science* **327:** 542–545.
6. Parmar, K., P. Mauch, J.A. Vergilio, *et al.* 2007. Distribution of hematopoietic stem cells in the bone marrow according to regional hypoxia. *Proc. Natl. Acad. Sci. USA* **104:** 5431–5436.
7. Kubota, Y., K. Takubo & T. Suda. 2008. Bone marrow long label-retaining cells reside in the sinusoidal hypoxic niche. *Biochem. Biophys. Res. Commun.* **366:** 335–339.
8. Simsek, T., F. Kocabas & J. Zheng. 2010. The distinct metabolic profile of hematopoietic stem cells reflects their location in a hypoxic niche. *Cell Stem Cell* **7:** 380–390.
9. Miharada, K., G. Karlsson & M. Rehn. 2010. Cripto regulates hematopoietic stem cells as a hypoxic-niche-related factor through cell surface receptor GRP78. *Cell Stem Cell* **9:** 330–344.
10. Ciccodicola, A., R. Dono, S. Obici, *et al.* 1989. Molecular characterization of a gene of the 'EGF family' expressed in undifferentiated human NTERA2 teratocarcinoma cells. *EMBO J.* **8:** 1987–1991.
11. Salomon, D.S., C. Bianco, A.D. Ebert, *et al.* 2000. The EGF-CFC family: novel epidermal growth factor-related proteins in development and cancer. *Endocr. Relat. Cancer.* **7:** 199–226.
12. Minchiotti, G., S. Parisi, G. Liguori, *et al.* 2002. Role of the EGF-CFC gene Cripto in cell differentiation and embryo development. *Gene* **287:** 33–37.
13. Bianco, C., M.C. Rangel & N.P. Castro. 2010. Role of Cripto-1 in stem cell maintenance and malignant progression. *Am. J. Pathol.* **177:** 532–540.
14. Gray, P.C. & Vale, W. 2012. Cripto/GRP78 modulation of the TGF-β pathway in development and oncogenesis. *FEBS Lett.* doi:10.1016/j.febslet.2012.01.051.
15. Ding, J., L. Yang, Y.T. Yan, *et al.* 1998. Cripto is required for correct orientation of the anterior-posterior axis in the mouse embryo. *Nature* **395:** 702–707.
16. Sato, N., I.M. Sanjuan & M. Heke. 2003. Molecular signature of human embryonic stem cells and its comparison with the mouse. *Dev. Biol.* **260:** 404–413.
17. Minchiotti, G. 2005. Nodal-dependent Cripto signaling in ES cells: from stem cells to tumor biology. *Oncogene* **24:** 5668–5675.
18. Strizzi, L., D. Abbott, D.S. Salomon, *et al.* 2008. Potential for Cripto-1 in defining stem cell-like characteristics in human malignant melanoma. *Cell Cycle* **7:** 1931–1935.
19. Hough, S.R., A.L. Laslett, S.B. Grimmond, *et al.* 2009. A continuum of cell states pluripotency and lineage commitment in human embryonic stem cells. *PLoS One* **4:** e7708.
20. Watanabe, K., M.J. Meyer, L. Strizzi, *et al.* 2010. Cripto-1 is a cell surface marker for a tumorigenic, undifferentiated subpopulation in human embryonal carcinoma cells. *Stem Cells* **28:** 1303–1314.
21. Xu, C., G. Liguori, E.D. Adamson, *et al.* 1998. Specific arrest of cardiogenesis in cultured embryonic stem cells lacking Cripto-1. *Dev. Biol.* **196:** 237–247.
22. Bianco, C., C. Cotten, E. Lonardo, *et al.* 2009. Cripto-1 is required for hypoxia to induce cardiac differentiation of mouse embryonic stem cells. *Am. J. Pathol.* **175:** 2146–2158.
23. Gray, P.C., G. Shani, K. Aung, *et al.* 2006. Cripto binds transforming growth factor β (TGF-β) and inhibits TGF-β signaling. *Mol. Cell Biol.* **26:** 9268–9278.
24. Karlsson, G., U. Blank, J.L. Moody, *et al.* 2007. Smad4 is critical for self-renewal of hematopoietic stem cells. *J. Exp. Med.* **204:** 467–474.
25. Blank, U. & S. Karlsson. 2011. The role of Smad signaling in hematopoiesis and translational hematology. *Leukemia* **25:** 1379–1388.
26. Watanabe, K., T. Nagaoka & J.M. Lee. 2009. Enhancement of Notch receptor maturation and signaling sensitivity by Cripto-1. *J. Cell Biol.* **187:** 343–353.
27. Ebert, A.D., C. Wechselberger, S. Frank, *et al.* 1999. Cripto-1 induces phosphatidylinositol 3′-kinase-dependent

phosphorylation of Akt and glycogen synthase kinase 3β in human cervical carcinoma cells. *Cancer Res.* **59:** 4502–4505.
28. Shani, G., W.H. Fischer, N.J. Justice, *et al.* 2008. GRP78 and Cripto form a complex at the cell surface and collaborate to inhibit transforming growth factor signaling and enhance cell growth. *Mol. Cell Biol.* **28:** 666–677.
29. Kelber, J.A. 2009. Blockade of Cripto binding to cell surface GRP78 inhibits oncogenic Cripto signaling via MAPK/PI3K and Smad2/3 pathways. *Oncogene* **28:** 2324–2336.
30. Gonzalez-Gronow, M., M.A. Selim, J. Papalas, *et al.* 2009. GRP78: a multifunctional receptor on the cell surface. *Antioxid. Redox. Signal.* **11:** 2299–2306.
31. Osawa, M. 1996. Long-term lymphohematopoietic reconstitution by a single CD34-low/negative hematopoietic stem cell. *Science* **273:** 242–245.
32. Challen, G.A., N.C. Boles, S.M. Chambers, *et al.* 2010. Distinct hematopoietic stem cell subtypes are differentially regulated by TGF-β1. *Cell Stem Cell* **6:** 265–278.
33. Dykstra, B., D. Kent, M. Bowie, *et al.* 2007. Long-term propagation of distinct hematopoietic differentiation programs *in vivo*. *Cell Stem Cell* **1:** 218–229.
34. Haug, J.S., X.C. He, J.C. Grindley, *et al.* 2008. N-cadherin expression level distinguishes reserved versus primed states of hematopoietic stem cells. *Cell Stem Cell* **2:** 367–379.
35. Wilson, A., E. Laurenti & G. Oser. 2008. Hematopoietic stem cells reversibly switch from dormancy to self-renewal during homeostasis and repair. *Cell* **135:** 1118–1129.
36. Kent, D.G., M.R. Copley, C. Benz, *et al.* 2009. Prospective isolation and molecular characterization of hematopoietic stem cells with durable self-renewal potential. *Blood* **113:** 6342–6350.
37. Morita, Y., H. Ema & H. Nakauchi. 2010. Heterogeneity and hierarchy within the most primitive hematopoietic stem cell compartment. *J. Exp. Med.* **207:** 1173–1182.
38. Yamazaki, S., A. Iwama, S. Takayanagi, *et al.* 2009. TGF-β as a candidate bone marrow niche signal to induce hematopoietic stem cell hibernation. *Blood* **113:** 1250–1256.
39. Yamazaki, S., H. Ema, G. Karlsson, *et al.* 2011. Nonmyelinating schwann cells maintain hematopoietic stem cells hibernation in the bone marrow niche. *Cell* **147:** 1146–1158.
40. Pinto do, Ó.P., K. Richter & L. Karlsson. 2002. Hematopoietic progenitor/stem cells immortalized by Lhx2 generate functional hematopoietic cells in vivo. *Blood* **99:** 3939–3946.
41. Takubo, K., N. Goda, W. Yamada, *et al.* 2010. Regulation of the HIF-1α level is essential for hematopoietic stem cells. *Cell Stem Cell* **7:** 391–402.
42. Grassinger, J., D.N. Haylock, B. Williams, *et al.* 2010. Phenotypically identical hematopoietic stem cells isolated from different regions of bone marrow have different biologic potential. *Blood* **116:** 3185–3196.
43. Nakamura, Y., F. Arai, H. Iwasaki, *et al.* 2010. Isolation and characterization of endosteal niche cell populations that regulate hematopoietic stem cells. *Blood* **116:** 1422–1432.
44. Danet, G.H., Y. Pan, J.L. Luongo, *et al.* 2003. Expansion of human SCID-repopulating cells under hypoxic conditions. *J. Clin. Invest.* **112:** 126–135.
45. Semenza, G.L. 2007. Hypoxia-inducible factor 1 (HIF-1) pathway. *Sci. STKE* **2007**: cm8.
46. Simon, M.C. & B. Keith. 2008. The role of oxygen availability in embryonic development and stem cell function. *Nat. Rev. Mol. Cell Biol.* **9:** 285–296.
47. Østergaard, L., U. Simonsen, Y. Eskildsen-Helmond, *et al.* 2009. Proteomics reveals lowering oxygen alters cytoskeletal and endoplasmatic stress proteins in human endothelial cells. *Proteomics* **9:** 4457–4467.
48. Hardy, B. & A. Raiter. 2010. Peptide-binding heat shock protein GRP78 protects cardiomyocytes from hypoxia-induced apoptosis. *J. Mol. Med.* **88:** 1157–1167.
49. Wey, S. 2012. Inducible knockout of GRP78/BiP in the hematopoietic system suppresses Pten-null leukemogenesis and AKT oncogenic signaling. *Blood* **119:** 817–825.

Ann. N.Y. Acad. Sci. ISSN 0077-8923

ANNALS OF THE NEW YORK ACADEMY OF SCIENCES
Issue: *Hematopoietic Stem Cells VIII*

G protein–coupled receptor crosstalk and signaling in hematopoietic stem and progenitor cells

Robert Möhle and Adriana C. Drost
Department of Medicine II, University of Tübingen, Tübingen, Germany

Address for correspondence: Robert Möhle, M.D., Department of Medicine II, University of Tübingen, Otfried-Müller-Str. 10, 72076 Tübingen, Germany. robert.moehle@med.uni-tuebingen.de

A variety of G protein–coupled receptors (GPCRs) is expressed in hematopoietic stem and progenitor cells (HPCs), including the chemokine receptor CXCR4, the leukotriene receptor CysLT1, the sphingosine 1-phosphate receptor S1P1, the cannabinoid receptor CB2, and the complement receptor C3aR. While the role of CXCR4 in stem cell homing is largely established, the function of the other GPCRs expressed in HPCs is only partially understood. CXCR4 and CysLT1 inhibit their own activation after ligand binding (homologous desensitization). Stimulation of S1P1 or C3aR has been shown to activate CXCR4 in HPCs that may sensitize CXCR4-dependent stem cell homing. In contrast, activation of CXCR4 results in a loss of CysLT1 function, which is most likely mediated by protein kinase C (PKC) signaling (heterologous desensitization) and could explain the ineffectiveness of CysLT1 antagonists to mobilize HPCs *in vivo*. Further characterization of GPCR crosstalk will allow a better understanding of HPC trafficking.

Keywords: hematopoietic progenitor cells; G protein–coupled receptors; crosstalk; signaling; homing

GPCRs expressed in HPCs

More than a decade ago, the major role of the chemokine receptor CXCR4 and its ligand CXCL12 (SDF-1) in hematopoietic stem cell homing to the bone marrow was identified.[1,2] During the process of hematopoietic stem and progenitor cell (HPC) homing to the bone marrow, interaction of CXCL12 and CXCR4 already occurs at the level of the bone marrow endothelium, which presents the chemokine to attaching cells.[3] A CXCL12 gradient may guide HPC to their niche, where sustained CXCR4 signaling supports stem cell survival. Conversely, blocking of the CXCL12/CXCR4 axis results in stem cell mobilization; indeed, CXCR4 antagonists have successfully been introduced as mobilizing agents in the clinical setting of stem cell transplantation.[4,5]

In addition to CXCR4, several other G protein–coupled receptors (GPCRs) are expressed in HPCs. CysLT1 is a key regulator in inflammatory processes, such as allergic asthma,[6–8] and induces cellular responses in HPCs similar to CXCR4.[9] Furthermore, while CysLT1 contributes to survival and proliferation of HPCs,[10,11] the functional role of CysLT1 in stem cell biology is not fully understood. In the hematopoietic microenvironment, several cell types, including hematopoietic, endothelial, and stromal cells, can produce cysteinyl leukotrienes, the ligands for CysLT1.[11] However, interaction of stem cells with CysLT1 ligands seems not to be a prerequisite for their survival in the bone marrow, as chronic administration of CysLT1 antagonists for treatment and prophylaxis of asthma has not been reported to result in suppression of hematopoiesis.[12] On the other hand, crosstalk exists between different GPCRs, which could lead to redundancy and explain why inhibition of a single receptor expressed on HPCs has only a minor effect. Potential mechanisms of such a GPCR crosstalk are discussed below.

Receptors for sphingosine 1-phosphate (S1P) were initially discovered as endothelial differentiation genes (EDG) receptors[13] due to their expression and functional properties in endothelial cells.[14] Similar to vascular endothelial growth factor (VEGF), S1P is a major angiogenic factor

doi: 10.1111/j.1749-6632.2012.06559.x

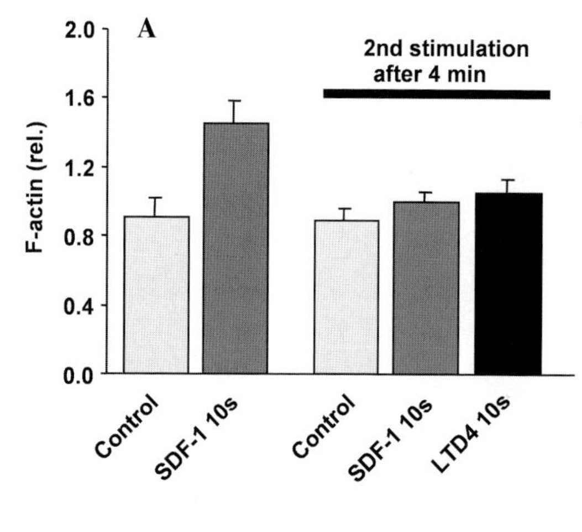

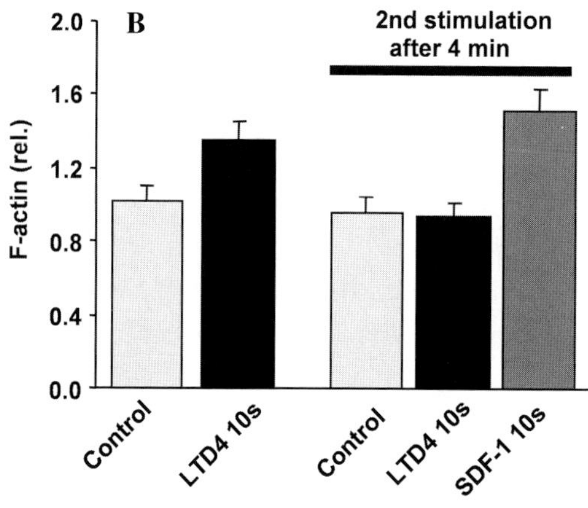

Figure 1. Homologous and heterologous desensitization of CXCR4 and CysLT1 in CD34$^+$ HPCs. For this *in vitro* analysis, actin polymerization (relative amount of filamentous actin ten seconds after exposure to the ligand) was measured by flow cytometry in isolated CD34$^+$ human HPCs as a read-out for GPCR activation as described previously.[11] Values are means ± standard error of the mean (SEM) of at least three experiments. (A) Four min after initial stimulation with the CXCR4 ligand SDF-1 (100 ng/ml), a second challenge with either SDF-1 (homologous desensitization) or with the CysLT1 ligand LTD4 at 100 nM (heterologous desensitization) did not induce actin polymerization. (B) Four min after initial stimulation with the CysLT1 ligand LTD4, the same ligand did not induce actin polymerization for a second time (homologous desensitization). However, the response to the CXCR4 ligand was not affected (lack of heterologous desensitization).

and essentially required during vasculogenesis.[14] Extracellular S1P is also involved in regulation of the immune system and in acute inflammation.[15] We found expression of S1P receptors, particularly S1P1, also in human HPC. The role of S1P as an extracellular mediator and ligand of S1P receptors has to be distinguished from its role as an intracellular second messenger, where the balance between the levels of the sphingolipid metabolites S1P and ceramide control cell survival and apoptosis.[16]

An additional GPCR that recognizes an inflammatory mediator is found on HPCs, the complement receptor C3aR.[17,18] Similar to CXCR4 antagonists, blocking of C3aR augmented G-CSF mobilization of HPCs in a mouse model.[18] Thus far, C3aR antagonists have not been established as stem cell-mobilizing agents in humans. Whether the complement fragment C3a plays a particular role as a paracrine mediator in the hematopoietic microenvironment, in addition to its function during inflammation, remains elusive. However, some studies question functional expression of C3aR in human HPCs,[19] and there is still an ongoing debate about the significance of C3a and its receptor in stem cell trafficking.[19,20]

More than ten years ago the potential role of endogenous cannabinoids, such as anandamide as a growth factor in the bone marrow microenvironment, was discovered.[21] Historically, endogenous cannabinoids and their receptors were identified as mediators in the nervous and immune system.[22] More recently, expression of endocannabinoids by stromal cells and modulation of G-CSF mobilization by CB2 receptor ligands was demonstrated.[23] Interestingly, in contrast to antagonists of CXCR4 and C3aR, which promote circulation of HPCs, CB2 antagonists reduced G-CSF–induced stem cell mobilization in an *in vivo* model; conversely, CB2 agonists augment mobilization. These results underscore a particular role of endogenous cannabinoids in the hematopoietic microenvironment.

Activation of CXCR4 by other GPCRs

Both, S1P1 and C3aR have been shown to augment CXCR4-dependent effects induced by CXCL12 (SDF-1) in HPCs. *In vitro* transendothelial migration in response to CXCL12 and *in vivo* stem cell homing and engraftment were increased when HPCs were primed with an S1P1 agonist.[24,25] Similarly, engagement of C3aR with its ligand C3a supported a variety of CXCL12/CXCR4-mediated cellular responses *in vitro*, including chemotaxis, and supported engraftment of Sca-1$^+$ murine HPCs in a mouse model of stem cell transplantation;[17] however, the underling signaling pathways have not been characterized thus far. It appears that rather than affecting cell migration directly, the main

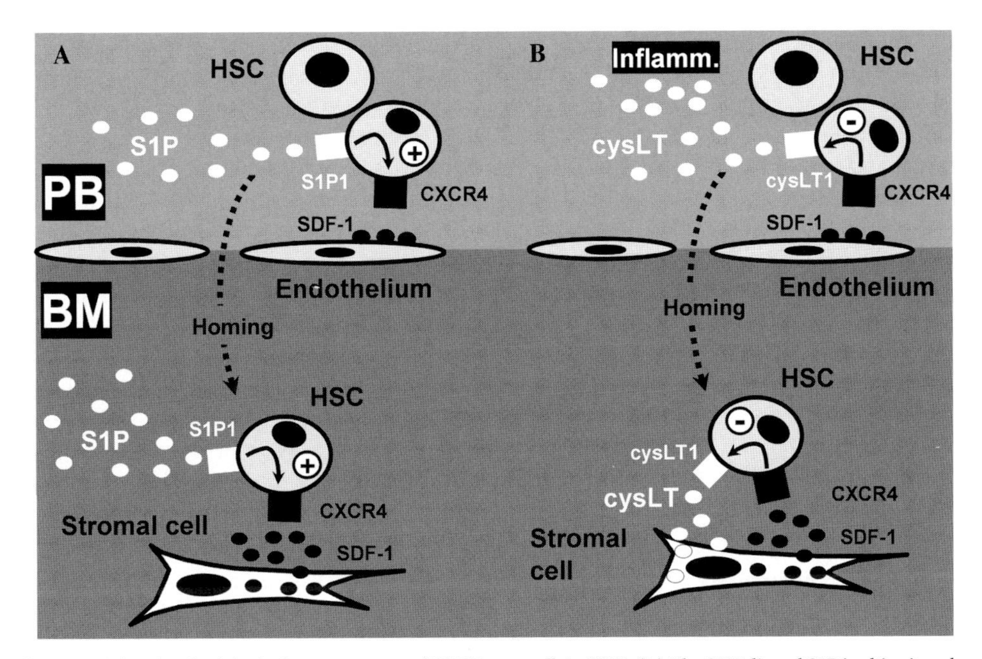

Figure 2. Potential pathophysiological consequences of GPCR crosstalk in HPC. (A) The S1P1 ligand S1P is ubiquitously present, but may be released particularly in the bone marrow microenvironment. Activation of S1P augments CXCR4-mediated effects during stem cell homing at the level of the bone marrow endothelium, where SDF-1 is locally presented to attaching HPC, and in the stem cell niche where SDF-1 is released by stromal and osteoblastic cells. (B) Engagement of CXCR4 during bone marrow homing of HPC desensitizes CysLT1, which prevents redistribution to areas with high cysteinyl leukotriene (cysLT) levels. Stromal cells may also release cysLTs, but desensitization due to CXCR4 engagement may render CysLT1 unresponsive. This could explain the ineffectiveness of CysLT1 antagonists to mobilize HPC *in vivo*.

function of several GPCRs expressed on HPCs is modulation of CXCR4 activity. Regarding C3aR, an additional study, demonstrating that the effect of C3a on CXCR4-mediated migration might be independent from C3aR, questioned the initial findings; however, a more recent study confirmed the inhibition of HPC homing by a C3aR antagonist.[20]

Homologous and heterologous desensitization of GPCRs expressed in HPCs

Homologous desensitization is observed after activation of many GPCRs in different cell types and leads to rapid termination of receptor signaling.[26] The underlying mechanisms comprise internalization of the receptor after ligand binding, phosphorylation by G protein–coupled receptor kinases (GRKs), and formation of complexes with β-arrestins; after a certain time, the receptor is re-expressed on the cell surface. In Figure 1, homologous desensitization of CXCR4 and CysLT1 is shown in human CD34$^+$ HPCs (preliminary data). When the cells were stimulated with the corresponding ligand (SDF-1 for CXCR4 and LTD4 for CysLT1), actin polymerization was observed within ten seconds (a read-out for receptor activation). After four minutes, however, a second challenge with the same ligand did not induce a response in these cells (preliminary data).

Activation of a GPCR can also result in temporary inhibition of another GPCR, as demonstrated for opioid and chemokine receptors.[27] This so-called "heterologous desensitization" does not involve receptor internalization and several signal transduction pathways are employed, particularly protein kinase C (PKC)-dependent signaling. We also analyzed potential heterologous desensitization of CXCR4 and CysLT1 in human CD34$^+$ HPCs. As shown in Figure 1, activation of CXCR4 resulted in heterologous desentization of CysLT1, as demonstrated by a lack of actin polymerization in response to the cysteinyl leukotriene LTD4 four minutes after the initial stimulation with SDF-1 (preliminary data). In contrast, CysLT1 did not cross-desensitize CXCR4, as actin polymerization in response to SDF-1 was still observed four minutes after stimulation with LTD4. In accordance with this finding, we could show that inhibitors of PKC-zeta—the predominat PKC isoform in human HPCs—completely blocked CXCR4-dependent, but not CysLT1-dependent, signaling.[28] Based on these findings it can be concluded that PKC-mediated heterologous GPCR desensitization also occurs in human HPCs. Heterologous desensitization represents a possible mechanism that underlies GPCR redundancy: lack of GPCR activation results in a gain-of-function of other, initially desensitized, GPCRs that eventually replace the function of the first GPCR.

Pathophysiological significance of GPCR crosstalk in HPCs

Bone marrow aplasia due to toxic damage is an emergency present not only after chemotherapy or irradiation but also after accidental ingestion of poisonous plants. Thus, the development of processes that support rapid restoration of hematopoiesis has been of advantage during evolution. Mobilization, circulation, and redistribution of stem cells to areas with more extensively damaged hematopoietic tissue ensure effective onset of hematopoietic recovery. It is conceivable that the presence of inflammation in particular requires rapid restoration of hematopoiesis. Indeed, receptors for mediators released during inflammation (i.e., C3aR and S1P1) are expressed in HPCs and have been shown to support CXCR4-mediated effects *in vitro* and/or homing of HPCs *in vivo*, as mentioned above.[17,24] In Figure 2A, the potential effect of S1P1 activation at the different steps of HPC homing is shown. After cytotoxic therapy, the microvessel network in the bone marrow is disrupted,[29] and activated platelets locally release the S1P1 ligand S1P.[30] However, other sources of S1P may also exist in the bone marrow microenvironment. Similarly, the complement system is activated by inflammation due to infection that is often present during aplasia and in the early neutrophil recovery phase, when HPCs are mobilized. Activation of the C3aR complement receptor supports CXCR4-mediated homing of circulating stem cells,[17] allowing recolonization of the damaged areas of the hematopoietic tissue, which eventually results in rapid hematopoietic regeneration.

We have shown that inflammatory mediators of the cysteinyl leukotriene family induce chemokine-like effects in HPC.[9,11] At optimal doses, the CysLT1 ligand LTD4 induces even stronger cellular effects, such as calcium fluxes and cell migration, than does optimal doses of the CXCR4 ligand CXCL12/SDF-1.[31] One would therefore expect redirection of HPC homing to areas of inflammation during major infection. Indeed, it has been speculated that the pathophysiological role of CysLT1 expression in HPCs is to promote HSC homing to and accumulation in inflamed tissues.[32] However, heterologous desensitization of CysLT1 by CXCR4 will result in attenuation of CysLT1 effects when CXCR4 is engaged, as shown in Figure 2B. Thus, even in the presence of increased concentrations of CysLT1 ligands, CXCR4-mediated homing to the bone marrow may not be impaired. The function of the cannabinoid receptor CB2 in the bone marrow microenvironment, and particularly crosstalk with other GPCRs, remains elusive and needs to be further characterized. Remarkably, in contrast to antagonists of other GPCRs, CB2 antagonists inhibit stem cell mobilization,[23] which implies that activation of CB2 counteracts retention of stem cells in the hematopoietic environment.

Taken together, GPCR crosstalk modulates HPC trafficking and homing to the bone marrow. Further studies are required to characterize these GPCR interactions in detail.

Acknowledgments

This work was supported by the Deutsche José Carreras Leukämie-Stiftung (DJCLS R08/24V).

Conflicts of interest

The authors declare no conflicts of interest.

References

1. Möhle, R. *et al.* 1998. The chemokine receptor CXCR-4 is expressed on CD34+ hematopoietic progenitors and leukemic cells and mediates transendothelial migration induced by stromal cell-derived factor-1. *Blood* **91:** 4523–4530.
2. Aiuti, A. *et al.* 1999. Human CD34(+) cells express CXCR4 and its ligand stromal cell-derived factor-1. Implications for infection by T-cell tropic human immunodeficiency virus. *Blood* **94:** 62–73.
3. Peled, A. *et al.* 2000. The chemokine SDF-1 activates the integrins LFA-1, VLA-4, and VLA-5 on immature human CD34(+) cells: role in transendothelial/stromal migration and engraftment of NOD/SCID mice. *Blood* **95:** 3289–3296.
4. Flomenberg, N. *et al.* 2005. The use of AMD3100 plus G-CSF for autologous hematopoietic progenitor cell mobilization is superior to G-CSF alone. *Blood* **106:** 1867–1874.

5. Liles, W.C. *et al.* 2003. Mobilization of hematopoietic progenitor cells in healthy volunteers by AMD3100, a CXCR4 antagonist. *Blood* **102:** 2728–2730.
6. Lynch, K.R. *et al.* 1999. Characterization of the human cysteinyl leukotriene CysLT1 receptor. *Nature* **399:** 789–793.
7. Sarau, H.M. *et al.* 1999. Identification, molecular cloning, expression, and characterization of a cysteinyl leukotriene receptor. *Mol. Pharmacol.* **56:** 657–663.
8. Denzlinger, C., C. Haberl & W. Wilmanns. 1995. Cysteinyl leukotriene production in anaphylactic reactions. *Int. Arch. Allergy Immunol.* **108:** 158–164.
9. Bautz, F. *et al.* 2001. Chemotaxis and transendothelial migration of CD34(+) hematopoietic progenitor cells induced by the inflammatory mediator leukotriene D4 are mediated by the 7-transmembrane receptor CysLT1. *Blood* **97:** 3433–3440.
10. Boehmler, A.M. *et al.* 2003. Potential role of cysteinyl leukotrienes in trafficking and survival of hematopoietic progenitor cells. *Adv. Exp. Med. Biol.* **525:** 25–28.
11. Boehmler, A.M. *et al.* 2009. The CysLT1 ligand leukotriene D4 supports alpha4beta1- and alpha5beta1-mediated adhesion and proliferation of $CD34^+$ hematopoietic progenitor cells. *J. Immunol.* **182:** 6789–6798.
12. Spector, S.L. 2001. Safety of antileukotriene agents in asthma management. *Ann. Allergy Asthma Immunol.* **86:** 18–23.
13. Hla, T. *et al.* 2001. Lysophospholipids–receptor revelations. *Science* **294:** 1875–1878.
14. Liu, Y. *et al.* 2000. Edg-1, the G protein-coupled receptor for sphingosine-1-phosphate, is essential for vascular maturation. *J. Clin. Invest.* **106:** 951–961.
15. Roviezzo, F. *et al.* 2011. Sphingosine-1-phosphate modulates vascular permeability and cell recruitment in acute inflammation in vivo. *J. Pharmacol. Exp. Ther.* **337:** 830–837.
16. Spiegel, S. & S. Milstien. 2000. Sphingosine-1-phosphate: signaling inside and out. *FEBS Lett.* **476:** 55–57.
17. Reca, R. *et al.* 2003. Functional receptor for C3a anaphylatoxin is expressed by normal hematopoietic stem/progenitor cells, and C3a enhances their homing-related responses to SDF-1. *Blood* **101:** 3784–3793.
18. Ratajczak, J. *et al.* 2004. Mobilization studies in mice deficient in either C3 or C3a receptor (C3aR) reveal a novel role for complement in retention of hematopoietic stem/progenitor cells in bone marrow. *Blood* **103:** 2071–2078.
19. Honczarenko, M. *et al.* 2005. Complement C3a enhances CXCL12 (SDF-1)-mediated chemotaxis of bone marrow hematopoietic cells independently of C3a receptor. *J. Immunol.* **175:** 3698–3706.
20. Wysoczynski, M. *et al.* 2009. Defective engraftment of $C3aR^{-/-}$ hematopoietic stem progenitor cells shows a novel role of the C3a-C3aR axis in bone marrow homing. *Leukemia* **23:** 1455–1461.
21. Valk, P. *et al.* 1997. Anandamide, a natural ligand for the peripheral cannabinoid receptor is a novel synergistic growth factor for hematopoietic cells. *Blood* **90:** 1448–1457.
22. Sugiura, T. & K. Waku. 2002. Cannabinoid receptors and their endogenous ligands. *J. Biochem.* **132:** 7–12.
23. Jiang, S. *et al.* 2011. Cannabinoid receptor 2 and its agonists mediate hematopoiesis and hematopoietic stem and progenitor cell mobilization. *Blood* **117:** 827–838.
24. Kimura, T. *et al.* 2004. The sphingosine 1-phosphate receptor agonist FTY720 supports CXCR4-dependent migration and bone marrow homing of human $CD34^+$ progenitor cells. *Blood* **103:** 4478–4486.
25. Seitz, G. *et al.* 2005. The role of sphingosine 1-phosphate receptors in the trafficking of hematopoietic progenitor cells. *Ann. N.Y. Acad. Sci.* **1044:** 84–89.
26. Luttrell, L.M. & R.J. Lefkowitz. 2002. The role of beta-arrestins in the termination and transduction of G-protein-coupled receptor signals. *J. Cell Sci.* **115:** 455–465.
27. Steele, A.D. *et al.* 2002. Interactions between opioid and chemokine receptors: heterologous desensitization. *Cytokine Growth Factor Rev.* **13:** 209–222.
28. Drost, A.C. *et al.* 2011. Similarities and differences in signal transduction and crosstalk of the G protein-coupled receptors CXCR4 and CysLT1 in $CD34^+$ hematopoietic stem/progenitor cells. *ASH Annu. Meeting Abstracts* **118:** 3392.
29. Kopp, H.G. *et al.* 2005. The bone marrow vascular niche: home of HSC differentiation and mobilization. *Physiology* **20:** 349–356.
30. Yatomi, Y. *et al.* 1995. Sphingosine-1-phosphate: a platelet-activating sphingolipid released from agonist-stimulated human platelets. *Blood* **86:** 193–202.
31. Xue, X. *et al.* 2007. Differential effects of G protein coupled receptors on hematopoietic progenitor cell growth depend on their signaling capacities. *Ann. N.Y. Acad. Sci.* **1106:** 180–189.
32. Kanaoka, Y. & J.A. Boyce. 2004. Cysteinyl leukotrienes and their receptors: cellular distribution and function in immune and inflammatory responses. *J. Immunol.* **173:** 1503–1510.

Ann. N.Y. Acad. Sci. ISSN 0077-8923

ANNALS OF THE NEW YORK ACADEMY OF SCIENCES
Issue: *Hematopoietic Stem Cells VIII*

Molecular and functional characterization of early human hematopoiesis

Elisa Laurenti[1,2] and John E. Dick[1,2]

[1]Campbell Family Institute for Cancer Research/Ontario Cancer Institute, Princess Margaret Hospital, University Health Network, Toronto, Ontario, Canada. [2]Department of Molecular Genetics, University of Toronto, Toronto, Ontario, Canada

Address for correspondence: John E. Dick, Ph.D., Toronto Medical Discovery Tower, Rm 8-301, 101 College Street, Toronto, ON, Canada M5G 1L7. jdick@uhnres.utoronto.ca

Through improvements in xenograft assay methods and in the identification of novel cell surface markers, significant progress has been made in our understanding of the human hematopoietic stem and progenitor hierarchy. The isolation of clonally pure populations of stem cells and early progenitors opens the way to carry out gene expression profiling studies to uncover the molecular regulators of each developmental step and to gain insight into the process of lineage commitment in human hematopoiesis.

Keywords: stem cells; HSC; xenograft assays; lineage determination; gene expression

Introduction

The hematopoietic system is a highly regulated cellular hierarchy, which originates from hematopoietic stem cells (HSCs) and contains two major lineages of mature hematopoietic cells: the lymphoid (comprising B, T, and natural killer cells) and the myeloid (including different classes of granulocytic cell) lineages. Their differentiation pathways are traditionally thought to separate very early in the hierarchy of hematopoiesis. HSCs are unique, as they possess extensive self-renewal, proliferation, and differentiation capacities that enable them to sustain long-term repopulation of all hematopoietic cell types in transplantation assays. In contrast to the mouse, where highly enriched populations of HSCs and other early hematopoietic progenitors can be purified,[1] isolation of homogeneous fractions of HSCs from human umbilical cord blood (CB) or bone marrow (BM)—a prerequisite to understanding their molecular networks—has been hampered by the lack of adequate cell surface markers. The most commonly used cell surface phenotypes for human HSCs are Lin^- $CD34^+$ $CD38^-$ (Ref. 2) and Lin^- $CD90^+$ $CD45RA^-$ $CD71^-$ (Refs. 3 and 4). Even when combined with additional markers, such as the marker associated with the capacity to efflux rhodamine-123, the frequency of repopulating cells in *in vivo* assays is around 1 in 30 (Ref. 5) compared to 1 in ~2–4 for murine HSCs.[6] Similarly, the phenotype of early committed human progenitors remains unclear, with definitions based exclusively on *in vitro* assays that describe both lymphoid-restricted (common lymphoid progenitors (CLPs)[7,8]) as well as myeloid-restricted progenitors (common myeloid progenitors (CMPs)[9]). As a consequence of the poor understanding of the cellular components of the human hematopoietic differentiation scheme, very few molecular regulators have been identified (reviewed in Ref. 10), again lagging behind the progress achieved in the mouse system, which slows down clinical applications in humans.

Isolation of human HSC

A major obstacle to studying HSC biology is that the cells are extremely rare and can only be directly assayed using xenograft assays. Only 1 in 10^6 cells in human BM is a transplantable HSC.[11]

Over 20 years have passed since primary human hematopoietic cells were first engrafted in immune-deficient mice. However, steady improvements in the assay—including use of more immune-deficient strains that have disabled innate and adaptive immunity or impaired macrophages, or use of

doi: 10.1111/j.1749-6632.2012.06577.x

different transplantation routes (neonatal intrahepatic or intrabone) and longer assay times—have greatly enhanced the sensitivity of HSC measurements.[10] As we recently reported, the assay conditions are now robust enough to enable single-cell transplantation of human HSCs.[12] Although it is important to recognize that the xenograft assay is still a surrogate, how the HSCs assayed with this method relate to the HSCs that repopulate humans upon transplantation is still a matter of inference.

Purification of human HSCs requires exclusion of mature lineage (CD2, CD3, CD14, CD16, CD19, CD24, CD56, CD66b, GlyA) markers ("Lin$^-$"), combined with simultaneous detection of several independent cell surface markers, primarily CD34 and CD38. Markers that are applicable in the mouse are not useful in the human. For example, human HSCs express FLT3 receptor,[13] whereas mouse cells do not, and mouse HSCs express CD150, whereas human cells do not.[14] Using the *Scid*-hu model, Baum *et al.* identified CD90 (Thy1) as a stem cell marker[15] that, in combination with CD34, could mediate HSC transplantation in breast cancer patients, thus confirming stem cell identity.[16,17] Further studies introduced CD45RA and CD38 as markers of further differentiated progenitors that are negatively enriched for HSCs.[2,18] Thus, there is substantial evidence that human HSCs are CD34$^+$ CD38$^-$ Thy1$^+$ CD45RA$^-$ (herein referred to as "Thy1$^+$").

Until recently, there had been little understanding of the phenotype of the immediate progeny of HSCs, including short-term HSCs and multipotential progenitors (MPPs). Both of these cell types are defined by transient multilineage repopulation. The first hint of such a progenitor in humans came from transplants of CD34$^+$ CD38^{lo} CB cells into non-obese diabetes (NOD)-*Scid* mice that generated transient myeloerythroid engraftment at two weeks.[19] But while CD38 expression is gradually acquired by differentiating cells, the CD34$^+$ CD38^{lo} fraction is still highly heterogeneous. In a more recent study, loss of Thy1 expression by CD34$^+$ CD38$^-$ CD45RA$^-$ cells (referred to as "Thy1$^-$") was proposed to demarcate Thy1$^+$ HSCs from transiently-engrafting Thy1$^-$ MPP.[4] However, Thy1$^-$ cells could still mediate serial transfer, suggesting that this population was not completely resolved from HSCs. These studies pointed to the need to identify additional markers to separate HSCs from their nearest progeny, which lack stem cell function.

In our efforts to examine potentially different markers between Thy$^+$ and Thy$^-$ cells, we focused on integrins because some integrins had been shown to be effective in the mouse, including integrin α_2 (CD49b), which differentially marks mouse long- and intermediate-term HSCs.[20] In addition, we have found that another integrin, CD49f, is expressed on ~50% of human Thy1$^+$ and ~25% of Thy1$^-$ cells, and when sorted fractions were assayed *in vivo*, HSC activity was restricted to the CD49f$^+$ cells in both fractions.[12] By contrast, Thy1$^-$ CD49f$^-$ cells could mediate a transient multilineage repopulation that peaks at four weeks and becomes undetectable after 16 weeks, which is reflective of MPPs. Thus, these studies enable isolation of HSCs, and an early MPP, and thus set the stage for comparison of gene expression analysis of closely related cell types to gain insight into which changes in gene expression are associated with loss of stem cell function.

Isolation of a human multilymphoid progenitor

The lymphoid lineage consists of T, B, and NK cells, which carry out adaptive and innate immune responses. The myeloid lineage includes a number of distinct, fully differentiated, short-lived cell types, including granulocytes (neutrophils, eosinophils, mast cells, and basophils), monocytes, erythrocytes, and megakaryocytes. Although some older models postulated that these two lineages arise in a single developmental bifurcation where a multipotent progenitor produces a lineage committed progenitor of the myeloid fate (CMP) or lymphoid fate (CLP), more recent studies indicate that the situation is less rigid, and that the earliest progenitors retain more variable fates where, for example, early lymphoid progenitors retain the ability to generate some myeloid cells.[10] Given the uncertainties in establishing precise lineage potential for any given population—particularly for human cells—we have proposed a broader term, *multilymphoid pogenitor* (MLP), to describe any progenitor that gives rise to all lymphoid lineages (B, T, and NK cells), but may or may not have other (myeloid) potentials.[21] Any B, T, and NK progenitor can be referred to as an MLP, whether or not its precise lineage output is ascertained.

Earlier studies indicated that myeloid progenitors, CMPs, granulocyte macrophage progenitors (GMPs), and megakaryocytic erythroid progenitors

(MEPs), could be isolated on the basis of the expression of either the interleukin 3 (IL-3) receptor α chain (CD123) or FLT3 (CD135), and CD45RA.[9,21] Myeloid, but not erythroid, progenitors express CD123 and CD135, and the CMP to GMP transition is marked by acquisition of CD45RA. Single CD135$^+$ CD45RA$^-$ CMPs produced all myeloid, but no lymphoid, lineages *in vitro* and after transplantation. MLP can be isolated using CD34$^+$ CD10$^+$ selection, as these cells give rise to B, T, and NK cells, but not myeloid or erythroid progeny—although the efficiency of this functional readout was found to be low.[7,8] This suggested that human MLPs are largely lymphoid restricted and express the early B cell marker CD10.

We developed a systematic analysis using seven markers and examined the lineage potential of neonate and adult progenitors using improved single cell assays.[21] Multilymphoid (B, T, and NK) potential was restricted to the CD34$^+$ CD38$^-$ Thy1$^{-/lo}$ CD45RA$^+$ ("Thy1$^-$ CD45RA$^+$") compartment composed of just 1% of all CD34$^+$ cells. We found that lymphoid colonies from single Thy1$^-$ CD45RA$^+$ cells almost always contain myeloid cells—predominantly monocytes, macrophages, and dendritic cells. Moreover, Thy1$^-$ CD45RA$^+$ cells transiently engraft NOD-scid-gamma null (NSG) mice and generate both myelo-monocytic and B cells, which indicates that their myeloid potential is not an artifact of *in vitro* culture.[21] Similar findings were reported using Thy1$^-$ CD45RA$^+$ cells from adult BM, although granulocytes were also seen.[22] Thus, human MLPs are not lymphoid restricted and possess myeloid, but not erythroid or megakaryocytic, potential. Furthermore, they coexpress lymphoid-specific and myeloid-shared transcriptional programs, which is consistent with their biological potential (see below). These findings allowed us to propose a model for human hematopoiesis (see Fig. 2B in Ref. 10).

Global molecular programs in human hematopoiesis

Elucidation of the cellular hierarchy in human hematopoiesis provides a valuable resource for mapping the key regulatory networks that control blood differentiation and lineage commitment. Novershtern *et al.* compared transcriptional profiles of 38 human hematopoietic cell types, including precursors and mature cells, thus providing a rich source of candidates for functional validation in primary human cells.[23] Because our group has focused on the more immature progenitor and HSC compartments, we have carried out a comprehensive transcriptional analysis of 10 stem and progenitor populations.[10,12] We applied several statistical and bioinformatic approaches to, first, derive population-specific signatures (gene lists that uniquely define each population). By applying multiple clustering algorithms to our dataset, we showed that gene expression in human HSC/progenitors can be classified into six predominant transcriptional programs (manuscript in preparation). Interestingly, the newly discovered MLP population (see above) participates in three of these: MLPs that express a group of genes also present in the pluripotent HSCs; MLPs that share many expression features with myeloid restricted cells, such as GMPs; and, in agreement with their proliferative status, MLPs that with myeloid and megakaryocytic/erythrocytic precursors jointly use a distinct cluster of genes highly enriched for cell cycle regulators. Thus, their transcriptional profile is consistent with the status of MLP as a poised population that has lost self-renewal capacity, but stands at the crossroads of commitment to either the lymphoid or the myeloid lineage.

Transcription factors (TFs), owing to their capacity to simultaneously activate and repress hundreds of genes, have been shown to be master regulators of lineage commitment fates. The vast majority of TFs that direct cells to acquire lymphoid fates have been identified in the mouse mainly because of the lack of methods to consistently derive B cells from human hematopoietic progenitors. From our bioinformatic analysis we selected 10 candidate TFs predicted to be key regulators of lymphoid or myeloid commitment and generated lentiviral vectors that efficiently knockdown (KD) all of them (manuscript in preparation). The differentiated output of KD MLP was then scored in our novel single cell stromal assay. Five out of 10 TFs altered MLP fate outcome. One TF, BCL11A, was studied extensively. Upon BCL11A KD, there was reduced formation of MLP that are committed to the B cell fate; this effect was traced to a partial block of B cell maturation at the pro-B to pre-B cell transition in an *in vivo* xenograft model. None of the five TFs affected the myeloerythroid switch, as determined by methylcellulose colony assays, indicating that their effect is restricted to MLP. Better staging of these differentiation defects in *in vivo* assays, gene expression profiling following KD, and genetic rescue experiments are currently underway to determine whether the five TFs act independently or in a network and which common pathways they control.

Conclusions

Advances in our ability to investigate human hematopoiesis from a cellular and molecular viewpoint now offer the possibility of complementing the information from murine models of normal hematopoietic development with parallel studies in primary human cells. These studies should aid in the development of more effective preclinical testing of new therapies and in the improvement of our understanding of how malignant processes perturb normal development and initiate a leukemogenic program; for example, we recently demonstrated that leukemia stem cells (LSCs) and HSCs share gene expression programs.[24]

Conflicts of interest

The authors declare no conflicts of interest.

References

1. Orkin, S.H. & L.I. Zon. 2008. Hematopoiesis: an evolving paradigm for stem cell biology. *Cell* **132:** 631–644.
2. Bhatia, M., J.C.Y. Wang, U. Kapp, *et al.* 1997. Purification of primitive human hematopoietic cells capable of repopulating immune-deficient mice. *Proc. Natl. Acad. Sci. U. S. A.* **94:** 5320–5325.
3. Mayani, H. & P.M. Lansdorp. 1994. Thy-1 expression is linked to functional properties of primitive hematopoietic progenitor cells from human umbilical cord blood. *Blood* **83:** 2410–2417.
4. Majeti, R., C.Y. Park & I.L. Weissman. 2007. Identification of a hierarchy of multipotent hematopoietic progenitors in human cord blood. *Cell Stem Cell* **1:** 635–645.
5. McKenzie, J.L., K. Takenaka, O.I. Gan, *et al.* 2007. Low rhodamine 123 retention identifies long-term human hematopoietic stem cells within the Lin-CD34+CD38- population. *Blood* **109:** 543–545.
6. Wilson, A., G.M. Oser, M. Jaworski, *et al.* 2007. Dormant and self-renewing hematopoietic stem cells and their niches. *Ann. N. Y. Acad. Sci.* **1106:** 64–75.
7. Galy, A., M. Travis, D. Cen & B. Chen. 1995. Human t, b, natural killer, and dendritic cells arise from a common bone marrow progenitor cell subset. *Immunity* **3:** 459–473.
8. Hao, Q.L., J. Zhu, M.A. Price, *et al.* 2001. Identification of a novel, human multilymphoid progenitor in cord blood. *Blood* **97:** 3683–3690.
9. Manz, M.G., T. Miyamoto, K. Akashi & I.L. Weissman. 2002. Prospective isolation of human clonogenic common myeloid progenitors. *Proc. Natl. Acad. Sci. U. S. A.* **99:** 11872–11877.
10. Doulatov, S., F. Notta, E. Laurenti & J.E. Dick. 2012. Hematopoiesis: a human perspective. *Cell Stem Cell* **3:** 120–136.
11. Wang, J.C., M. Doedens & J.E. Dick. 1997. Primitive human hematopoietic cells are enriched in cord blood compared with adult bone marrow or mobilized peripheral blood as measured by the quantitative in vivo SCID-repopulating cell assay. *Blood* **89:** 3919–3924.
12. Notta, F., S. Doulatov, E. Laurenti, *et al.* 2011. Isolation of single human hematopoietic stem cells capable of long-term multilineage engraftment. *Science* **333:** 218–221.
13. Sitnicka, E., N. Buza-Vidas, S. Larsson, *et al.* 2003. Human CD34+ hematopoietic stem cells capable of multilineage engrafting NOD/SCID mice express flt3: distinct flt3 and c-kit expression and response patterns on mouse and candidate human hematopoietic stem cells. *Blood* **102:** 881–886.
14. Larochelle, A., M. Savona, M. Wiggins, *et al.* 2011. Human and rhesus macaque hematopoietic stem cells cannot be purified based only on SLAM family markers. *Blood* **117:** 1550–1554.
15. Baum, C.M., I.L. Weissman, A.S. Tsukamoto, *et al.* 1992. Isolation of a candidate human hematopoietic stem-cell population. *Proc. Natl. Acad. Sci. U. S. A.* **89:** 2804–2808.
16. Murray, L., B. Chen, A. Galy, et al. 1995. Enrichment of human hematopoietic stem cell activity in the CD34+Thy-1+Lin- subpopulation from mobilized peripheral blood. *Blood* **85:** 368–378.
17. Negrin, R.S., K. Atkinson, T. Leemhuis, *et al.* 2000. Transplantation of highly purified CD34+Thy-1+ hematopoietic stem cells in patients with metastatic breast cancer. *Biol. Blood Marrow Transplant* **6:** 262–271.
18. Lansdorp, P.M., H.J. Sutherland & C.J. Eaves. 1990. Selective expression of CD45 isoforms on functional subpopulations of CD34+ hemopoietic cells from human bone marrow. *J. Exp. Med.* **172:** 363–366.
19. Mazurier, F., M. Doedens, O.I. Gan, J.E. Dick. 2003. Rapid myeloerythroid repopulation after intrafemoral transplantation of NOD-SCID mice reveals a new class of human stem cells. *Nat. Med.* **9:** 959–963.
20. Benveniste, P., C. Frelin, S. Janmohamed, *et al.* 2010. Intermediate-term hematopoietic stem cells with extended but time-limited reconstitution potential. *Cell Stem Cell* **6:** 48–58.
21. Doulatov, S., F. Notta, K. Eppert, *et al.* 2010. Revised map of the human progenitor hierarchy shows the origin of macrophages and dendritic cells in early lymphoid development. *Nat. Immunol.* **1:** 585–593.
22. Goardon, N., E. Marchi, A. Atzberger, *et al.* 2011. Coexistence of LMPP-like and GMP-like leukemia stem cells in acute myeloid leukemia. *Cancer Cell* **19:** 138–152.
23. Novershtern, N., A. Subramanian, L.N. Lawton, *et al.* 2011. Densely interconnected transcriptional circuits control cell states in human hematopoiesis. *Cell* **144:** 296–309.
24. Eppert, K., K. Takenaka, E.R. Lechman, *et al.* 2011. Stem cell gene expression programs influence clinical outcome in human leukemia. *Nat. Med.* **17:** 1086–1093.

Ann. N.Y. Acad. Sci. ISSN 0077-8923

ANNALS OF THE NEW YORK ACADEMY OF SCIENCES
Issue: *Hematopoietic Stem Cells VIII*

Role of N-cadherin in the regulation of hematopoietic stem cells in the bone marrow niche

Fumio Arai, Kentaro Hosokawa, Hirofumi Toyama, Yoshiko Matsumoto, and Toshio Suda

Department of Cell Differentiation, The Sakaguchi Laboratory of Developmental Biology, School of Medicine, Keio University, Tokyo, Japan

Address for correspondence: Fumio Arai, D.D.S., Ph.D., Department of Cell Differentiation, The Sakaguchi Laboratory of Developmental Biology, School of Medicine, Keio University, 35 Shinano-machi, Shinjuku-ku, Tokyo, 160-8582, Japan. farai@a3.keio.jp

Cell–cell and cell–extracellular matrix interactions between hematopoietic stem cells (HSCs) and their niches are critical for the maintenance of stem cell properties. Here, it is demonstrated that a cell adhesion molecule, N-cadherin, is expressed in hematopoietic stem/progenitor cells (HSPCs) and plays a critical role in the regulation of HSPC engraftment. Furthermore, overexpression of N-cadherin in HSCs promoted quiescence and preserved HSC activity during serial bone marrow (BM) transplantation (BMT). Inhibition of N-cadherin by the transduction of N-cadherin short hairpin (sh) RNA (shN-cad) reduced the lodgment of donor HSCs to the endosteal surface, resulting in a significant reduction in long-term engraftment. shN-cad-transduced cells were maintained in the spleen for six months after BMT, indicating that N-cadherin expression in HSCs is specifically required in the BM. These findings suggest that N-cadherin-mediated cell adhesion is functionally essential for the regulation of HSPC activities in the BM niche.

Keywords: N-cadherin; hematopoietic stem cells; niche; osteoblasts

Introduction

The interaction of stem cells with their supportive microenvironment, the stem cell niche, is critical for sustaining stem cell pools in tissues over long periods of time. The niche is composed of cellular and extracellular matrix components that surround stem cells and facilitate the signaling networks that control the balance between self-renewal and differentiation.[1–4] Hematopoietic stem cells (HSCs) are maintained in a quiescent state in the bone marrow (BM), where they are anchored to specialized niches along the endosteum and into perivascular sites adjacent to the endothelium.

Cell adhesion molecules are crucial for the stem cell–niche interaction. Cadherins are Ca^{2+}-dependent cell adhesion molecules that play pivotal roles in maintaining tissue architecture.[5–7] The function of cadherins in the stem cell niche has been well characterized for *Drosophila* germline stem cells.[8,9]

Recently, we demonstrated that cell adhesion enhanced by the forced overexpression of N-cadherin in HSCs inhibited HSC division, increased HSC lodgment, and preserved the long-term reconstitution (LTR) capacity of HSCs during serial BM transplantation (BMT).[10] Furthermore, *in vitro* culture of HSCs with bone-derived osteoblasts that expressed high levels of N-cadherin enhanced the LTR activity of HSCs.[11] In contrast, the knockdown of N-cadherin expression in HSCs accelerated HSC division *in vitro* and diminished the lodgment of donor HSCs to the endosteal surface *in vivo*, resulting in a significant reduction in long-term (LT) engraftment.[12] These data suggest that N-cadherin-mediated cell adhesion maintains the quiescence and self-renewal activity of HSCs.

The current work adds to these findings by confirming the endogenous expression of N-cadherin proteins in the hematopoietic stem/progenitor cell (HSPC) population by using a newly developed N-cadherin antibody. Furthermore, we showed that

doi: 10.1111/j.1749-6632.2012.06576.x

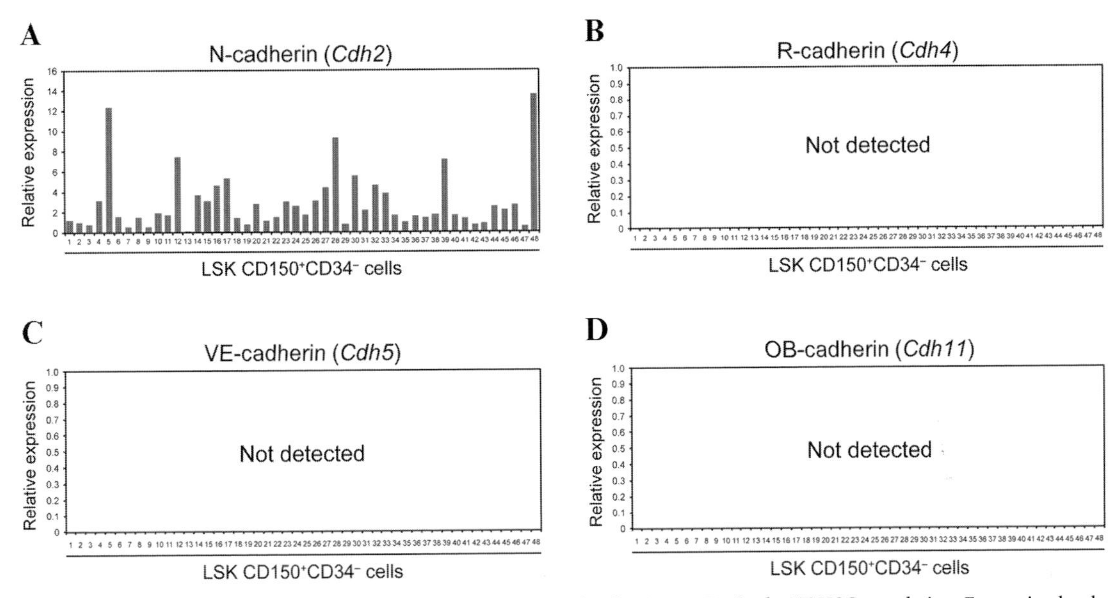

Figure 1. Single-cell gene expression analysis of the expression of cadherin mRNAs in the LT-HSC population. Expression levels of (A) N-cadherin (*Cdh2*), (B) R-cadherin (*Cdh4*), (C) VE-cadherin (*Cdh5*), and (D) OB-cadherin (*Cdh11*) mRNA in single cells ($n = 48$) are shown. LSK CD150$^+$ CD34$^-$ cells expressed varying levels of *Cdh2*. However, *Cdh4, Cdh5*, and *Cdh11* mRNAs were not detected in LSK CD150$^+$ CD34$^-$ cells. LSK CD150$^+$ CD34$^-$ cell #35 was used as the reference cell; the relative expression of *Cdh2* in this cell was set to a value of 1. *Gapdh* was used for the normalization of target gene expression.

N-cadherin functions in HSPCs to increase LTR capacity, which likely occurs via N-cadherin–mediated homophilic and heterophilic interactions with BM niche cells.

Expression of N-cadherin in HSPCs

The expression of N-cadherin in HSPCs is still controversial.[13,14] The major problem is that there have been no reliable anti-N-cadherin antibodies that can be used for the detection of N-cadherin on the surface of living cells.[15] To address this problem, we made use of two different approaches: (1) single cell gene expression analysis and (2) the production of a new N-cadherin antibody.

Recent advances in cDNA amplification and reverse transcription (RT)–quantitative PCR (Q-PCR) allow for the analysis of gene expression in individual cells. Using a microfluidics-based Q-PCR array system (Dynamic Array, Fluidigm, South San Francisco, CA, USA), multiple target genes can be simultaneously measured in multiple single cell samples. By using this method, we analyzed the expression of several cadherin genes, including N-cadherin (*Cdh2*), R-cadherin (*Cdh4*), VE-cadherin (*Cdh5*), and OB-cadherin (*Cdh11*) in individual highly purified LT-HSCs, Lin$^-$ Sca-1$^+$ c-Kit$^+$ (LSK) CD150$^+$ CD34$^-$ cells (Fig. 1). These cadherins seem to be involved in stem cell interactions with the osteoblastic and perivascular BM niches. Single-cell Q-PCR array analysis revealed that LSK CD150$^+$ CD34$^-$ cells express *Cdh2* at various levels (Fig. 1A), but do not express *Cdh4, Cdh5*, or *Cdh11*. (Figs. 1B–1D).

For the production of N-cadherin antibodies that are of appropriate quality for fluorescence-activated cell sorting (FACS), we used the Human Combinatorial Antibody Library (HuCAL) (AbD Serotech, Düsseldorf, Germany), a commercially available phage display library, and isolated recombinant antibodies against mouse N-cadherin. After screening of the phage library, performing an enzyme-linked immunosorbent assay (ELISA), sequencing the clones, and performing quality control ELISA with positive and negative control proteins, we successfully obtained seven clones that reacted with the N-cadherin protein. Because N-cadherin binds to R- and OB-cadherin as well as to N-cadherin, we selected only those antibody clones that reacted with N-cadherin, but not with R- or OB-cadherin. Subsequently, the suitability of the antibodies for flow cytometry was determined using NIH3T3, which express N-cadherin protein[12] (Fig. 2A). We found several clones of the newly developed antibodies were useful for FACS. FACS

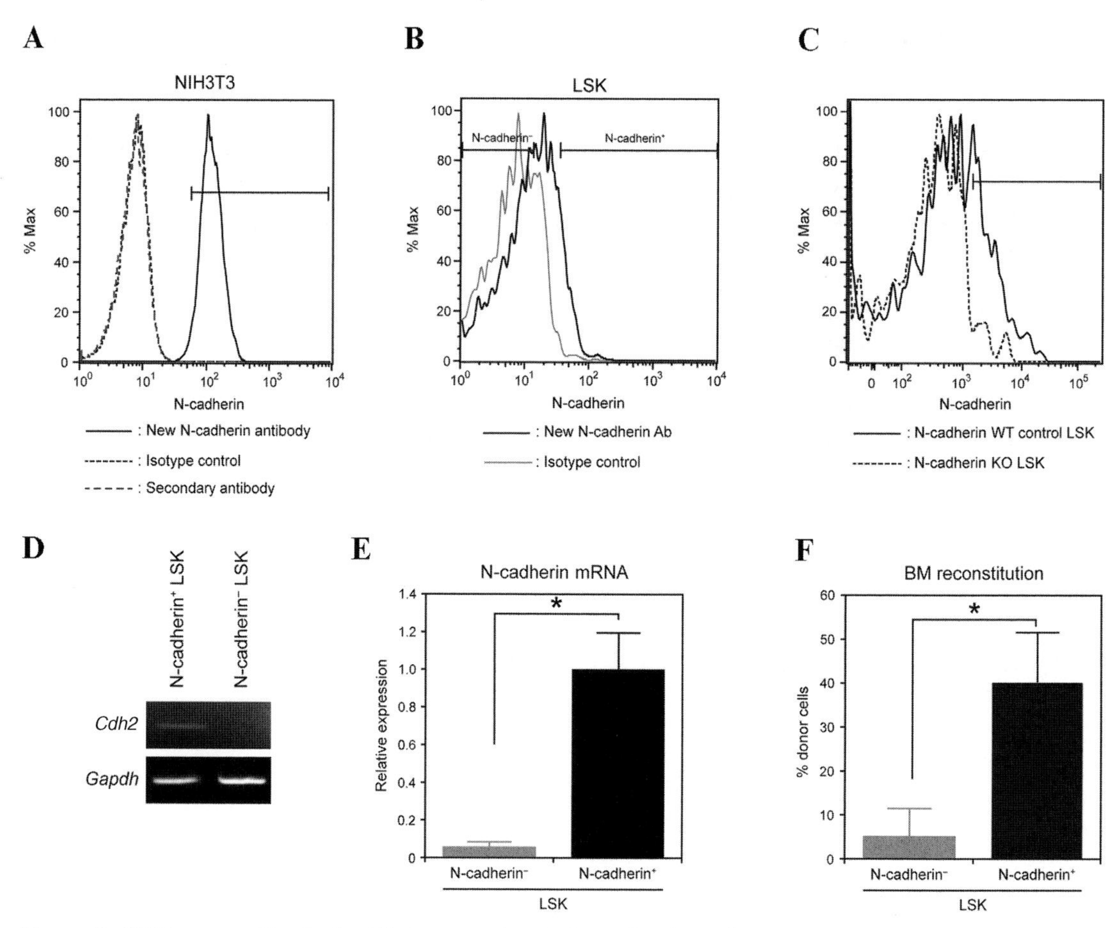

Figure 2. HSPCs express N-cadherin. (A) Representative FACS profile of N-cadherin expression in NIH3T3 cells using a new N-cadherin antibody. (B) Representative FACS profile of N-cadherin expression in LSK cells. For the immunostaining of N-cadherin, BM mononuclear cells were first reacted with primary anti-N-cadherin. The cells were then reacted with a fluorescently labeled secondary antibody and fluorescence-conjugated anti-lineage, Sca-1, and c-Kit antibodies. FACS analysis showed that LSK cells expressed low levels of N-cadherin protein. (C) Representative FACS profile of N-cadherin expression in N-cadherin WT control (solid line) and KO (dashed line) LSK using new N-cadherin antibody. To generate Mx1-Cre(+);N-cadherin$^{flox/flox}$ mice, N-cadherin$^{flox/flox}$ mice were crossed with Mx1-Cre(+) mice. To induce Cre, mice were treated with polyI:C. Mx1-Cre(−);N-cadherin$^{flox/flox}$ mice were used as the control. (D) RT-PCR analysis of the expression of *Cdh2* and *Gapdh* mRNAs in N-cadherin$^+$ and N-cadherin$^-$ LSK cells. The cells were isolated by FACS using the N-cadherin antibody. (E) Q-PCR analysis of the expression of *Cdh2* mRNA in N-cadherin$^+$ and N-cadherin$^-$ LSK cells. The cells were isolated by FACS using the N-cadherin antibody. Data are the means ± SD ($^*P < 0.01$, $n = 4$/each group). (F) BM LTR activity for N-cadherin$^+$ and N-cadherin$^-$ LSK cells. N-cadherin$^+$ and N-cadherin$^-$ LSK cells were isolated from Ly5.1 mice and transplanted into lethally irradiated recipient Ly5.2 mice (10^3 cells/mice). Percentages of donor-derived cells in the peripheral blood at three months after BMT are shown. Data are the means ± SD ($^*P < 0.01$, $n = 5$/group).

analysis with one of the new N-cadherin antibodies showed that LSK cells expressed low levels of N-cadherin protein (Fig. 2B). We confirmed that the reactivity of a new N-cadherin antibody was significantly reduced in N-cadherin knockout mice derived LSK cells compared to the wild-type (WT) control mice LSK cells (Fig. 2C). RT-PCR and Q-PCR analysis revealed significantly higher levels of N-cadherin mRNA in N-cadherin$^+$ LSK cells compared with N-cadherin$^-$ LSK cells (Figs. 2D and 2E). Next, a BMT assay was performed with N-cadherin$^+$ and N-cadherin$^-$ LSK cells. Preliminary data demonstrated that N-cadherin$^+$ cells showed higher BM reconstitution at three months of BMT compared with N-cadherin$^-$ cells (Fig. 2F). These data suggested that HSPCs express N-cadherin.

Expression of N-cadherin in endosteal niche cells

Multiple types of cells, including mesenchymal stem and progenitor cells, osteoblasts, reticular cells, adipocytes, osteoclasts, and other stromal cells have been implicated in the regulation of HSC maintenance in the BM niche.[16–19] Furthermore, the cells in the endosteum are a heterogeneous population in terms of their degree of differentiation and accompanying functions.[20,21] Nonhematopoietic and nonendothelial CD45$^-$ CD31$^-$ Ter119$^-$ endosteal cells were subdivided into three fractions based on their expression of activated leukocyte cell adhesion molecule (ALCAM) (CD166) and Sca-1: ALCAM$^+$ Sca-1$^-$, ALCAM$^-$ Sca-1$^+$, and ALCAM$^-$ Sca-1$^-$ fractions. Gene expression analysis and differentiation assays revealed that mature osteoblasts were enriched in the ALCAM$^+$ Sca-1$^-$ fraction, whereas immature mesenchymal cells were enriched in the ALCAM$^-$ Sca-1$^+$ fraction.[11] These endosteal cell populations expressed cadherin genes to varying degrees.[10,11] In particular, N-cadherin was specifically expressed in the ALCAM$^+$ Sca-1$^-$ fraction.[11] Because N-cadherin mediates both homophilic (N-/N-cadherin) and heterophilic (N-/R-cadherin and N-/OB-cadherin) cell–cell adhesion,[22–25] HSCs may interact with various niche cells through N-cadherin.

Hematopoiesis-supporting activity of endosteal niche cells

In vitro culture of LSK cells with ALCAM$^+$ Sca-1$^-$ bone-derived cells enhanced the LTR activity of HSCs.[11] LSK cells cocultured with ALCAM$^+$ Sca-1$^-$ cells had significantly higher levels of stem cell markers, cell adhesion-related genes, and homing-related genes compared with LSK cells cocultured with ALCAM$^-$ Sca-1$^+$ or ALCAM$^-$ Sca-1$^-$ cells. In particular, the expression of N-cadherin was significantly upregulated in the LSK cells cocultured with ALCAM$^+$ Sca-1$^-$ cells. These data suggest that physiological interactions between HSCs and ALCAM$^+$ Sca-1$^-$ osteoblasts help to maintain the self-renewal activity of HSCs. Because we did not analyze the BM reconstitution of LSK cells cocultured with ALCAM$^+$ Sca-1$^-$ cells after secondary BMT, it is unclear whether the quality of single HSCs was improved by coculture with ALCAM$^+$ Sca-1$^-$ cells.

Role of N-cadherin in the protection of HSCs from stress

We demonstrated that overexpression of N-cadherin is involved in the protection of HSCs under stress conditions.[10] HSCs undergo a great deal of stress during serial BM repopulation, raising the possibility that forced expression of N-cadherin in HSCs would provide the cells with a protective advantage. Control green fluorescent protein (GFP) or WT-N-cadherin-GFP overexpressing HSCs were serially transplanted every three months after BMT. The chimerism of the GFP$^+$ donor cells was then analyzed three months after each transplantation. In recipients transplanted with control GFP-transduced donor cells, the number of GFP$^+$ cells gradually decreased after repeated transplantation, indicating that control HSCs could not overcome the stress of BMT. This also indicated that their reconstitution ability progressively decreased during serial BMT. In contrast, the frequency of GFP$^+$ donor cells was increased relative to gene-untransduced cells (GFP$^-$) after repeated BMT in recipient mice transplanted with WT-N-cadherin-expressing donor cells (GFP$^+$).[10] These data indicate that WT-N-cadherin-overexpressing HSCs overcome the stress of serial BMT, and that enhanced cell adhesion protects HSCs against the loss of self-renewal capacity.

Next, the function of β-catenin–mediated intracellular signaling by N-cadherin in the maintenance of HSCs was analyzed.[10] Although the activation of β-catenin signaling can expand HSCs,[26,27] constitutive activation of β-catenin exhausts the stem cell pool by accelerating the cell cycle and blocking HSC differentiation.[28,29] Therefore, we examined the role of the N-cadherin/β-catenin complex in preserving self-renewal activity following serial BMT. To block the N-cadherin/β-catenin interaction, LSK cells were transduced with a mutant N-cadherin that had been deleted for the catenin-binding region (CBR) (N-cadherin/CBR$^-$). Transduction of N-cadherin/CBR$^-$ into LSK cells enhanced the nuclear accumulation of β-catenin and accelerated cell division.[10] In the BMT assay, the overexpression of the CBR$^-$ mutant led to the expansion of donor cells in recipient mice after the second BMT. However, CBR$^-$ mutant-overexpressing cells were drastically decreased after the third BMT, indicating that N-cadherin/CBR$^-$-overexpressing HSCs were exhausted after serial BMT. These findings suggest that appropriate regulation of the localization of β-catenin is required for the protection of HSCs from the loss of self-renewal capacity during serial BMT.

Knockdown of N-cadherin reduced BM reconstitution in the BM

The function of N-cadherin *in vivo* also remains uncertain. Kiel *et al.*[30] reported that N-cadherin conditional knockout mice do not show defects in HSC number or function. In contrast, we showed that the inhibition of cadherin-mediated homophilic and heterophilic adhesion by the overexpression of dominant negative-N-cadherin reduced the LTR activity of HSPCs.[10]

Furthermore, we recently demonstrated the involvement of N-cadherin in the regulation of HSPC proliferation and engraftment by altering endogenous N-cadherin levels.[12] Transduction of N-cadherin-specific short hairpin (sh) RNA (shN-cad) into LSK cells induced the nuclear accumulation of β-catenin and accelerated the rate of cell division *in vitro.*[12] In contrast, when LSK cells were transduced with WT-N-cadherin and cultured on an N-cadherin-Fc chimeric protein-coated plate, β-catenin was predominantly localized to the juxtamembrane region. In addition, overexpression of WT-N-cadherin in LSK cells decreased the *in vitro* cell division rate.[10] These findings indicate that N-cadherin expression in HSCs affects the subcellular localization of β-catenin, which in turn affects HSC quiescence.

The LTR activity of LSK cells expressing N-cadherin shRNA was next evaluated to analyze the role of N-cadherin in HSC engraftment. Although the transduction of shN-cad did not affect the homing activity of HSCs, the knockdown of N-cadherin reduced the lodgment of donor HSCs after BMT and inhibited LTR of HSCs. On the other hand, co-transduction of sil.mut.N-cad, which has a silent mutation in the shRNA target sequence, rescued the engraftment of shN-cad-expressing cells.[12] More interestingly, when we examined the engraftment of shN-cad-transduced cells in the spleen, which lacks an osteoblastic niche, N-cadherin knockdown donor cells showed LTR at six months after BMT.[12] These findings suggest that N-cadherin-mediated cell adhesion is specifically required for LTR of HSCs in the niche of the BM, but not in the niche of the spleen.

Continuous activation of β-catenin induced the exhaustion of the stem cell pool by accelerating the cell cycle and blocking LT-HSC differentiation.[28,29] However, the phenotype of HSCs with constitutive β-catenin activation only partially overlaps with the phenotype of HSCs in which N-cadherin is knocked down. Although the number of shN-cad-overexpressing donor cells was extremely low, shN-cad-transduced cells differentiated into T, B, and myeloid lineages.[12] These observations suggest that cell adhesion-dependent mechanisms also play a critical role in the regulation of the cell division rate of HSCs.

Conclusions and future directions

Single-cell gene expression analysis demonstrated the endogenous expression of N-cadherin in HSPCs. In addition, our preliminary data using a new anti-N-cadherin antibody indicated that HSPCs express a low level of N-cadherin protein. Our preliminary data also suggested that N-cadherin functions to regulate HSPC LTR. As such, future plans include a more detailed analysis of N-cadherin expression in the LT-HSC fraction, as well as the characterization of N-cadherin$^+$ HSCs.

Finally, we demonstrated that N-cadherin-mediated cell adhesion is essential for the establishment of hematopoiesis after BMT and the maintenance of the HSC compartment, specifically in the BM.[10,12] WT-N-cadherin overexpression in HSCs can inhibit cell cycling, resulting in the LT maintenance of HSCs. These findings suggest a strong correlation between cadherin-mediated cell adhesion and the maintenance of the quiescent state, as well as LT self-renewal of HSCs. Therefore, the control of the levels of adhesive interactions between HSCs and the BM niche could potentially provide the basis for niche-based therapies to protect HSCs.

Acknowledgments

This work was supported by the Funding Program for the Next Generation World-Leading Researchers (NEXT Program) and a Grant-in-Aid from the Japan Society for the Promotion of Science.

Conflicts of interest

The authors declare no conflicts of interest.

References

1. Fuchs, E., T. Tumbar & G. Guasch. 2004. Socializing with the neighbors: stem cells and their niche. *Cell* **116:** 769–778.
2. Li, L. & T. Xie. 2005. Stem cell niche: structure and function. *Annu. Rev. Cell Dev. Biol.* **21:** 605–631.
3. Moore, K.A. & I.R. Lemischka. 2006. Stem cells and their niches. *Science*. **311:** 1880–1885.
4. Wilson, A. & A. Trumpp. 2006. Bone-marrow haematopoietic-stem-cell niches. *Nat. Rev. Immunol.* **6:** 93–106.
5. Gumbiner, B.M. 1996. Cell adhesion: the molecular basis of tissue architecture and morphogenesis. *Cell* **84:** 345–357.
6. Miyatani, S. *et al.* 1989. Neural cadherin: role in selective cell-cell adhesion. *Science* **245:** 631–635.
7. Takeichi, M. 1990. Cadherins: a molecular family important in selective cell-cell adhesion. *Annu. Rev. Biochem.* **59:** 237–252.
8. Song, X. *et al.* 2002. Germline stem cells anchored by adherens junctions in the Drosophila ovary niches. *Science* **296:** 1855–1857.
9. Yamashita, Y.M., D.L. Jones & M.T. Fuller. 2003. Orientation of asymmetric stem cell division by the APC tumor suppressor and centrosome. *Science* **301:** 1547–1550.
10. Hosokawa, K. *et al.* 2010. Cadherin-based adhesion is a potential target for niche manipulation to protect hematopoietic stem cells in adult bone marrow. *Cell Stem Cell* **6:** 194–198.
11. Nakamura, Y. *et al.* 2010. Isolation and characterization of endosteal niche cell populations that regulate hematopoietic stem cells. *Blood* **116:** 1422–1432.
12. Hosokawa, K. *et al.* 2010. Knockdown of N-cadherin suppresses the long-term engraftment of hematopoietic stem cells. *Blood* **116:** 554–563.
13. Kiel, M.J., G.L. Radice & S.J. Morrison. 2007. Lack of evidence that hematopoietic stem cells depend on N-cadherin-mediated adhesion to osteoblasts for their maintenance. *Cell Stem Cell* **1:** 204–217.
14. Haug, J.S. *et al.* 2008. N-cadherin expression level distinguishes reserved versus primed states of hematopoietic stem cells. *Cell Stem Cell* **2:** 367–379.
15. Li, P. & L.I. Zon. 2010. Resolving the controversy about N-cadherin and hematopoietic stem cells. *Cell Stem Cell* **6:** 199–202.
16. Mendez-Ferrer, S. *et al.* 2010. Mesenchymal and haematopoietic stem cells form a unique bone marrow niche. *Nature* **466:** 829–834.
17. Naveiras, O. *et al.* 2009. Bone-marrow adipocytes as negative regulators of the haematopoietic microenvironment. *Nature* **460:** 259–263.
18. Sacchetti, B. *et al.* 2007. Self-renewing osteoprogenitors in bone marrow sinusoids can organize a hematopoietic microenvironment. *Cell* **131:** 324–336.
19. Sugiyama, T. *et al.* 2006. Maintenance of the hematopoietic stem cell pool by CXCL12-CXCR4 chemokine signaling in bone marrow stromal cell niches. *Immunity* **25:** 977–988.
20. Kiel, M.J. & S.J. Morrison. 2008. Uncertainty in the niches that maintain haematopoietic stem cells. *Nat. Rev. Immunol.* **8:** 290–301.
21. Yin, T. & L. Li. 2006. The stem cell niches in bone. *J. Clin. Invest.* **116:** 1195–1201.
22. Straub, B.K. *et al.* 2003. A novel cell-cell junction system: the cortex adhaerens mosaic of lens fiber cells. *J. Cell Sci.* **116:** 4985–4995.
23. Shan, W.S. *et al.* 2000. Functional cis-heterodimers of N- and R-cadherins. *J. Cell Biol.* **148:** 579–590.
24. Matsunami, H. *et al.* 1993. Cell binding specificity of mouse R-cadherin and chromosomal mapping of the gene. *J. Cell Sci.* **106**(Pt 1): 401–409.
25. Prakasam, A.K., V. Maruthamuthu & D.E. Leckband. 2006. Similarities between heterophilic and homophilic cadherin adhesion. *Proc Natl. Acad. Sci. USA*. **103:** 15434–15439.
26. Baba, Y. *et al.* 2006. Constitutively active beta-catenin promotes expansion of multipotent hematopoietic progenitors in culture. *J. Immunol.* **177:** 2294–2303.
27. Reya, T. *et al.* 2003. A role for Wnt signalling in self-renewal of haematopoietic stem cells. *Nature* **423:** 409–414.
28. Kirstetter, P. *et al.* 2006. Activation of the canonical Wnt pathway leads to loss of hematopoietic stem cell repopulation and multilineage differentiation block. *Nat. Immunol.* **7:** 1048–1056.
29. Scheller, M. *et al.* 2006. Hematopoietic stem cell and multilineage defects generated by constitutive beta-catenin activation. *Nat. Immunol.* **7:** 1037–1047.
30. Kiel, M.J. *et al.* 2009. Hematopoietic stem cells do not depend on N-cadherin to regulate their maintenance. *Cell Stem Cell* **4:** 170–179.

Ann. N.Y. Acad. Sci. ISSN 0077-8923

ANNALS OF THE NEW YORK ACADEMY OF SCIENCES
Issue: *Hematopoietic Stem Cells VIII*

Differential requirements for Wnt and Notch signaling in hematopoietic versus thymic niches

Paul P.C. Roozen, Martijn H. Brugman, and Frank J.T. Staal

Department of Immunohematology and Blood Transfusion (IHB), Leiden University Medical Center (LUMC), Leiden, the Netherlands

Address for correspondence: Frank J.T. Staal, Department of Immunohematology and Blood Transfusion (IHB), Leiden University Medical Center, P.O. Box 9600, 2300 RC Leiden, the Netherlands. f.j.t.staal@lumc.nl

All blood cells are derived from multipotent stem cells, the so-called hematopoietic stem cells (HSCs), that in adults reside in the bone marrow. Most types of blood cells also develop there, with the notable exception of T lymphocytes that develop in the thymus. For both HSCs and developing T cells, interactions with the surrounding microenvironment are critical in regulating maintenance, differentiation, apoptosis, and proliferation. Such specialized regulatory microenvironments are referred to as niches and provide both soluble factors as well as cell–cell interactions between niche component cells and blood cells. Two pathways that are critical for early T cell development in the thymic niche are Wnt and Notch signaling. These signals also play important but controversial roles in the HSC niche. Here, we review the differences and similarities between the thymic and hematopoietic niches, with particular focus on Wnt and Notch signals, as well as the latest insights into regulation of these developmentally important pathways.

Keywords: hematopoietic stem cells; thymus; niche; Wnt; Notch; expansion

Introduction

The hematopoietic stem cell (HSC) is, to date, the best understood type of adult stem cell.[1] Knowledge about HSCs provides insight into stem cell homing, proliferation, apoptosis, senescence, maintenance, and differentiation.[1] In mammals, HSCs originate from the mesodermal germ layer, which is responsible for the formation of blood islands, at 7–7.5 days postconception (dpc), resulting in primitive erythropoiesis. Nine days postconception, hematopoietic progenitors appear in the aorta-gonad-mesonephros (AGM) region. Between 9.5 and 11 dpc the fetal liver is colonized by HSCs originating in the AGM region or the yolk sack, as some researchers have proposed. In the fetal liver, these colonizing HSCs expand, and at 12 dpc, definitive hematopoiesis begins, which is indicated by the increase of circulating enucleated erythrocytes and a decrease of nucleated primitive erythrocytes; at 13 dpc, HSCs are no longer detected in the AGM region (reviewed in Ref. 2). The fetal liver serves as the hub for the colonization of the thymus, spleen, and the bone marrow (BM). The spleen is colonized as early as 12 dpc, but immature lymphocytes and erythroid cells are not found until 16 dpc. It is unclear when the colonization of the BM by HSCs precisely occurs, but hematopoiesis in the BM starts 18 dpc and remains the main source of hematopoietic cells throughout adult life.[3,4] Around 12 dpc, the HSCs colonize the thymus and start to form T cells. The thymus, with its own unique signals that overlap with signals provided in the HSC niche, serves as the site for T cell development. In the thymus, these signals obviously have a different function, namely, to support T cell commitment and further differentiation and selection processes.

HSC maintenance and differentiation is regulated by an intricate interplay between signaling pathways initiated by cytokines, chemokines, or extracellular matrix components, as well as by physiological gradients of other molecules, including calcium[5] and oxygen (e.g., hypoxia[6] and reactive oxygen species[7–9]). These signaling pathways include, but are not limited to, activation of signaling through c-Kit via stem cell factor (Scf),[10,11] c-Mpl through

doi: 10.1111/j.1749-6632.2012.06626.x

its ligand thrombopoietin,[12–14] cannabinoid receptors CB 1 and 2,[15] CXCR4,[16,17] bone morphogenetic protein (BMP),[18] canonical Wnt,[18–24] IL-3,[20] and Notch.[21,25–27]

As can be deduced from the pathways mentioned above, maintenance and differentiation of HSCs are influenced by a complex interplay between different signaling pathways. After a short introduction to the canonical Wnt- and Notch-signaling pathways, we will focus on the interplay between both pathways, with regard to HSC maintenance and differentiation in the fetal liver, BM, and T cell differentiation in the thymus niches.

Canonical Wnt signaling

Wnt proteins make up a large family of cysteinrich, glycosylated,[28] and lipid-modified[29,30] secreted molecules that are involved in a wide range of developmental and physiological processes.[31] To date, in mammals 19 Wnt family members[32] and 10 Wnt receptors—so called frizzled (Fz) receptors—have been identified.[33] Besides the seven transmembrane-spanning Fz receptors, there are also several single-pass transmembrane coreceptors. This group of coreceptors consists of the low-density lipoprotein receptor related proteins (Lrp) 5 and 6 (see Ref. 34), the receptor tyrosine kinase-like orphan receptor (Ror) 1 and 2 (see Ref. 35), and the receptor-like tyrosine kinase (Ryk) (see Ref. 36). Wnt signaling can be separated into two distinct pathways, namely, the *canonical* β-catenin–dependent pathway and the *noncanonical* β-catenin–independent pathway.[37] Both the Ror and Ryk receptors are thought to participate in noncanonical Wnt signaling (excellently reviewed in Refs. 38–40). Although, beyond the scope of the present review, the general view is that induction of noncanonical Wnt signaling pathways induces quiescence of HSCs and enhances their engraftment.

The canonical Wnt pathway is conserved across species[31,41] and can be activated by the following Wnt proteins: Wnt1,[42,43] Wnt3, Wnt3a,[30,44,45] Wnt8a, Wnt8b,[46] Wnt10a,[46] and Wnt10b.[47,48] When the canonical Wnt pathway is induced (Fig. 1), cytoplasmic β-catenin is targeted for proteosomal degradation, which is brought about by phosphorylation and ubiquitination.[49] This process is orchestrated by a protein complex commonly called "the destruction complex," which is formed by the scaffolding proteins axis inhibition protein 1 and 2 (Axin1 and Axin2), the tumor suppressor adenomatous polyposis coli (APC) protein, the serine/threonine kinase glycogen synthase kinase 3 beta (GSK3β), and casein kinase 1 alpha (Ck1α) phosphorylating β-catenin.[49–53] The phosphorylated form of β-catenin interacts with FWD1, a member of the Skp1/Cullin/F box protein FWD1 ubiquitin ligase (SCF) complex, linking the destruction complex to the ubiquitin-proteasome pathway.[54] The proteosomal degradation of β-catenin ensures that Wnt target gene expression is low, which occurs through the subsequent recruitment of transcriptional repressors, such as Groucho, histone deacetylases (HDAC), and by the transcription factors T cell factor 1 (HUGO classification *Tcf7*), lymphoid enhancer binding factor 1 (Lef1), and related factors Tcf3 and Tcf4, which are thought to be less important in the hematopoietic system.[55]

Upon interaction of Wnt ligands with Fz receptors and the essential coreceptors Lrp5 or 6[41] at the cell surface, the destruction complex is inhibited, resulting in inhibition of β-catenin phosphorylation and its subsequent stabilization. This stabilization is mediated through signalosome formation and recruitment of dishevelled (Dvl), which in turn disrupts the destruction complex via the recruitment of Axin1, Ck1α, and GSK3β to the cell membrane.[56–60] In the signalosome, Lrp phosphorylation increases its affinity for Axin1,[60–62] as a result, unphosphorylated β-catenin accumulates in the cytoplasm and subsequently translocates to the nucleus.[63–65] In the nucleus, β-catenin binding to Tcf1/Lef1 results in the transcription of Wnt target genes.[55,66] Although many Wnt target genes remain unknown and several are controversial, the genes for c-Myc, cyclin D1, Tcf1, Lef1, and Axin2[55] are well-known targets.

Wnt signaling is regulated at different levels by prevention of Wnt receptor interaction or by intracellular signal modification. One way of modulating the interaction between Wnt and its receptors is by expression of soluble decoy Wnt receptors, such as secreted frizzled related protein (sFrp) and Wnt inhibitory factor 1 (Wif1).[67] The interaction between Wnt and its receptor can also be prevented by the Wnt antagonists Dickkopf (Dkk) 1–4, which bind and block Lrp coreceptors.[68–70] Besides inhibition at the cell membrane, Wnt signaling can also be inhibited in the nucleus by the cell-autonomous inhibitor of β-catenin and Tcf, Icat, which prevents β-catenin from interacting with Tcf1 and Lef1.[71]

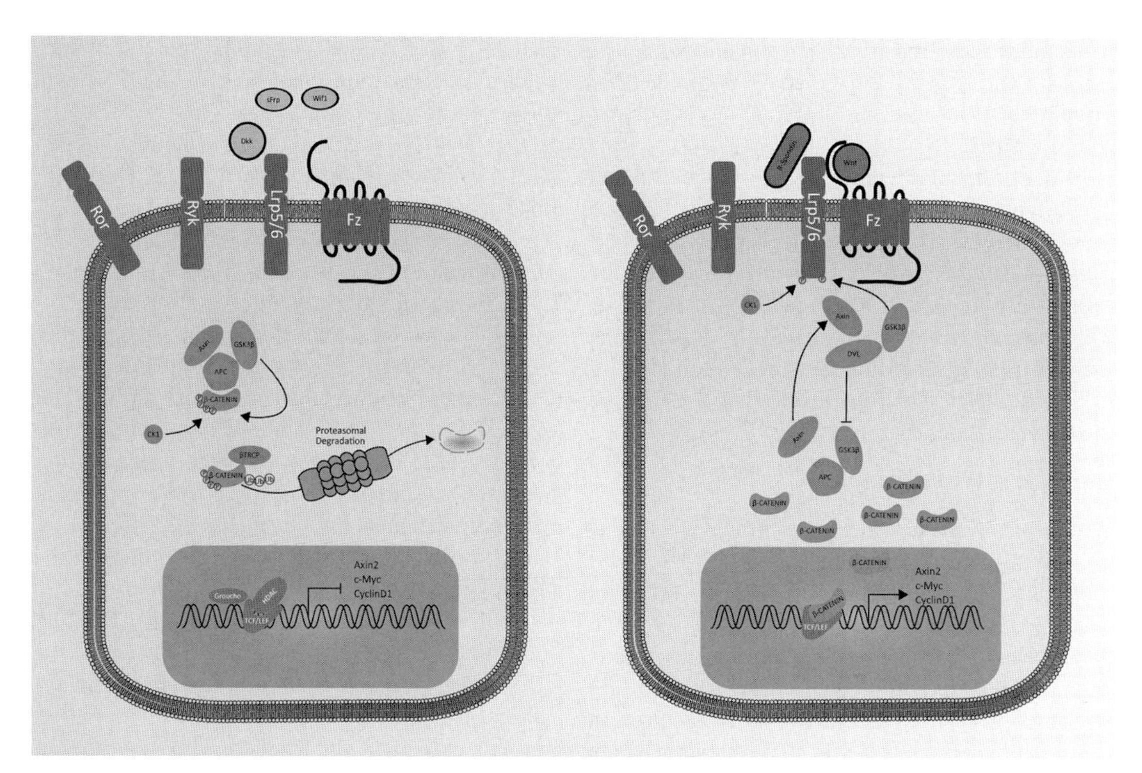

Figure 1. Schematic of the Wnt-signaling pathway. Shown is the state of the pathway upon binding of a Wnt ligand to a Fz/LRP receptor complex. Only the canonical pathway is depicted. For more details see text and Ref. 56.

In addition, several different isoforms of Tcf1 exist, each with a different affinity for β-catenin and, thus, with different effects on Wnt signaling. The shorter Tcf1 isoforms function as negative regulators of the pathway because they cannot bind β-catenin, whereas the longer Tcf1 isoforms initiate gene transcription by binding β-catenin.[72,73] Besides different Tcf1 isoforms, both long and short forms of Lef1 exist. The long Lef1 isoform binds β-catenin and leads to target gene transcription, whereas the short Lef1 isoform does not bind β-catenin and therefore functions as a transcription repressor.[74]

Besides negative regulation of canonical Wnt signaling, the Wnt pathway can also be enhanced. For example, tetraspanins (Tspan), a four-pass transmembrane protein family,[75] and R-spondin, a secreted factor that binds to LGR receptors, are capable of initiating canonical Wnt signaling. Tspan12 has been shown to interact with Fz4 promoting complex formation with Norrin, which results in β-catenin activation. However, it is likely that β-catenin activation is caused by complex formation, not by Norrin.[60,76] The R-spondin family in mammals is made up by R-spondins1–4, and there are orthologues in *Drosophila*, *C. elegans*, and *Xenopus*. The expression of R-spondin1 was found to be coordinated with the expression of Wnt1 and Wnt3a during development, suggesting a functional link between these two protein families.[77,78] Furthermore, R-spondin1 has been shown to stabilize β-catenin and stimulate proliferation of intestinal crypt cells.[79] Besides the previously mentioned regulatory proteins, the extracellular matrix (ECM) also plays a role in regulation of Wnt signaling. In the ECM the concentration of Wnt near the cell surface of secreting cells is maintained at high levels by binding of Wnt to glycosaminoglycans (GAG), including heparan sulfate proteoglycan (HSPC).[80,81] Some proteoglycans directly influence HSC biology; for example, the proteoglycan decorin has been proposed to play a role in maintenance of HSCs.[82]

Canonical Notch signaling

Similar to the Wnt pathway, the Notch pathway can be divided into canonical and noncanonical pathways. Furthermore, both Wnt and Notch signaling are conserved between species and function in a variety of developmental stages and cell types.[83,84]

In mammals, four Notch (Notch 1–4) receptors and five canonical ligands of the Delta-Serrate-Lag (DSL) type can be identified. In mammals DSL-type ligands Jagged (Jag) 1 and 2 and Delta-like (Dll) 1–4 are responsible for Notch signaling. As both Notch receptors and ligands are cell membrane bound, Notch signaling only occurs through direct cell–cell interactions (*trans*-interactions). There is little evidence that different receptor ligand combinations result in different signaling output, with the exemption of Dll3.[85] Ladi *et al.* showed that Dll3 is capable of inhibiting *trans*-interactions,[86] although it is unclear if Dll3 is expressed on the cell surface.[87,88]

When a DLS-type ligand binds to a Notch receptor via *trans*-interaction this results in the proteolytic cleavage by ADAM proteases and γ-secretase (Fig. 2). ADAM proteases cleave the extracellular domain, which is lysosomaly degraded by the ligand-expressing cell.[89] This proteolytic cleavage leaves Notch extracellular truncated (NEXT) on the receptor-expressing cell, which in turn undergoes endocytosis. During endocytosis, γ-secretase cleaves the intracellular Notch (ICN) domain,[90–92] which then subsequently migrates to the nucleus. In the nucleus ICN interacts with a member of the CBF1/suppressor of hairless/LAG1 (CSL) family (i.e., the RBP-Jκ in mammals, Su(H) in *Drosophila*, and LAG1 in *C. elegans*), which is stabilized by the essential coactivator Mastermind-like (Maml) 1 or 2[93,94] and results in displacement of corepressors (CoR), recruitment of other coactivators (CoA), and polymerase II responsible for transcription of Notch target genes, including c-Myc,[95–98] cyclins D1[96,99,100] and D3,[101] cyclin dependent kinase 5,[98,100] HERP family members (including Hes1, Hes7, and Hey1; reviewed in Ref. 102), and Notch1 and Notch3.[97] Often, the expression of Hes1 is taken as an important read-out for active canonical Notch signaling.

As can be deduced from the preceding section, canonical Notch signaling does not contain a signal amplification step, and one ligand–receptor interaction is thus equal to one ICN–CSL interaction, making the Notch-signaling pathway extremely dosage sensitive. This dosage sensitivity was shown in several studies in which

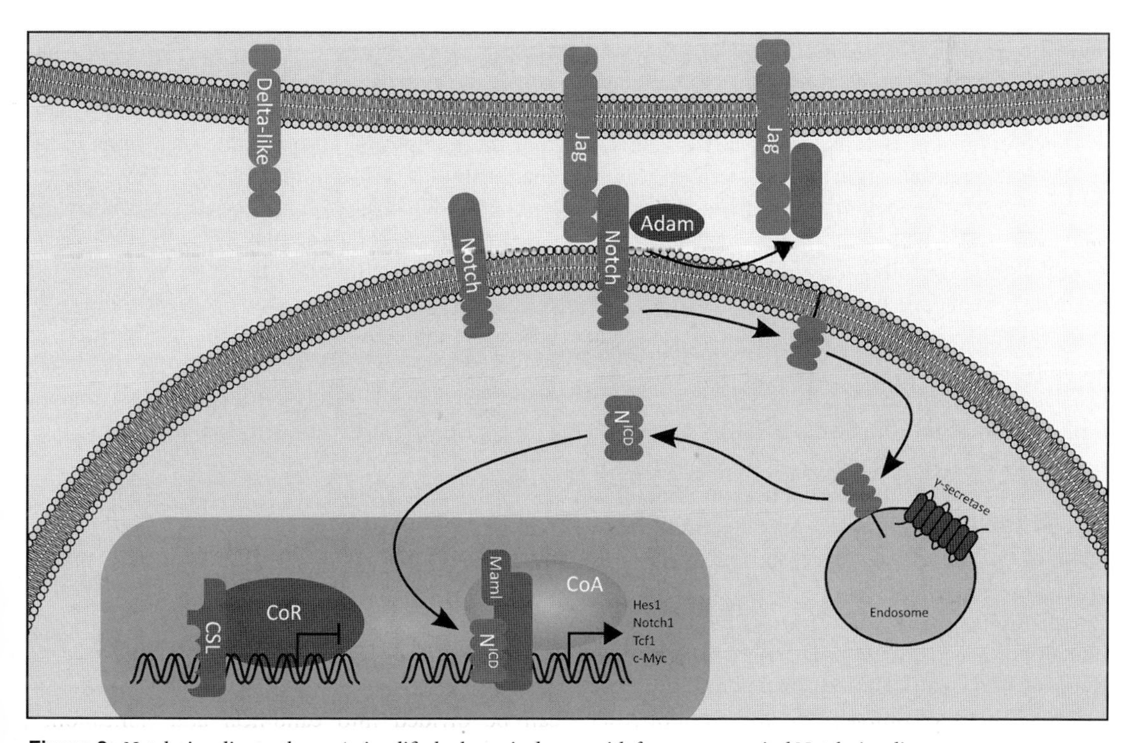

Figure 2. Notch signaling pathway. A simplified scheme is shown, with focus on canonical Notch signaling.

haploinsufficiency models were used. Zhang *et al.* showed that haploinsufficiency of Notch1 resulted in supernumerary hair cell development in the inner ear of mice,[103] and that haploinsufficiency of both Notch2 and Jag1 was causally related to the Alagille syndrome.[104] Another study showed that Dll4$^{+/-}$ was embryonically lethal.[105]

Because deregulation of Notch signaling results in aberrant development, death, or disease, tight regulation of Notch is required. One way of regulation is via signaling specificity obtained through spatial and temporal restriction of DSL-type ligands and Notch receptors.[85,99,106,107] An example of this temporal and spatial regulation is the expression of the ligands Jagged1 and Delta 1–4 and Notch1, 2, and 4 (which are not expressed in the yolk sac blood islands, the para-aortic splanchnopleura, the hematopoietic aortic clusters, and the early stages of embryonic liver hematopoiesis in humans). Expression of Notch1 and 2 and the ligand Delta 4 was only detected in the embryonic liver between days 34 and 38 postconception. Another method of regulation is by signal directionality via *trans*-activation and *cis*-inhibition (for example via expression of Dll1 or Dll3 and a Notch receptor on the same cell).[108,109] Regulation of Notch signaling can also occur by means of expression of the auxiliary proteins; for example, the ubiquitin ligases Mind bomb and Neuralized regulate Notch ligand expression.[110,111] Other auxiliary proteins expressed in Notch-expressing cells include Deltex 1–4, which control ICN internalization and ubiquitination, and thereby regulate Notch signaling both positively and negatively.[112–115]

Besides the fact that deregulation of canonical Notch signaling has been implicated in disease, Notch dosages also influence embryonic and adult hematopoiesis. For example, Wilson and colleagues reported that the endogenous Notch inhibitors Numb and Numb-like are dispensable during adult hematopoiesis, because deletion of them in the BM of postnatal and adult mice did not influence HSC self-renewal or T cell differentiation.[116] More recent reports, however, showed that Numb and Numb-like play a crucial roles in erythroid differentiation during primitive mouse and zebrafish hematopoiesis. Bresciani *et al.* showed, for example, that knockdown of both Numb and Numb-like leads to a defect in primitive erythroid differentiation due to enhanced Notch activation,[117] while the study by Cheng *et al.* showed that active Wnt signaling and inhibition of Notch signaling by Numb are required for primitive erythroid specification.[118] Kwon *et al.* showed that membrane bound Notch1 physically interacts with β-catenin in embryonic stem cells (ESCs), mesenchymal stromal cells (MSCs), and neural stem cells (NSCs), and thereby negatively influences accumulation of β-catenin—which points to one way in which Wnt and Notch signals interact, showing the way a cell responds to signals is dependent on additional signals that occur in these cells.[119]

Furthermore, Notch dosage has also been reported to influence T cell development. For example, using an Rbpjκ knockout model, different levels of Notch signaling were shown to steer differentiation of naive CD4 to either Th1 or Th2 cells. This was linked to signaling through either Jagged or Delta ligands in antigen presenting cells.[120] During thymic organ cultures, T cell development was altered or completely abolished, resulting in a shift into non-T cell differentiation, depending upon the level of Notch signaling inhibition.[121] Although Notch is essential for the generation of HSCs, and Notch dosage clearly influences both embryonic and adult hematopoiesis, its role in adult HSCs is not clear. Because several gain-of-function and loss-of-function studies (discussed later) currently do not provide conclusive evidence that Notch signaling is essential for adult HSC function, it is tempting to speculate that there might be an optimal Notch signal dose that promotes HSC maintenance.

The canonical Notch pathway also interacts with other pathways known to play a role in HSC maintenance and development. It has been shown, for example, that ICN is capable of blocking NF-κB target gene transcription through binding of p50/cRel.[122] Notch has also been linked to SMAD signaling, where Notch fine-tunes SMAD signaling, and SMAD signaling, in turn, enhances Notch signaling.[123–126]

T cell development and the thymic niche

T cell development is a highly regulated, multistep process aimed at generating mature, functional T cells bearing T cell receptors (TCRs) that are capable of recognizing a broad range of antigens in the context of self-MHC.[127] Using cell-surface markers several T cell developmental stages can be distinguished (Fig. 3). Thymocytes are primarily

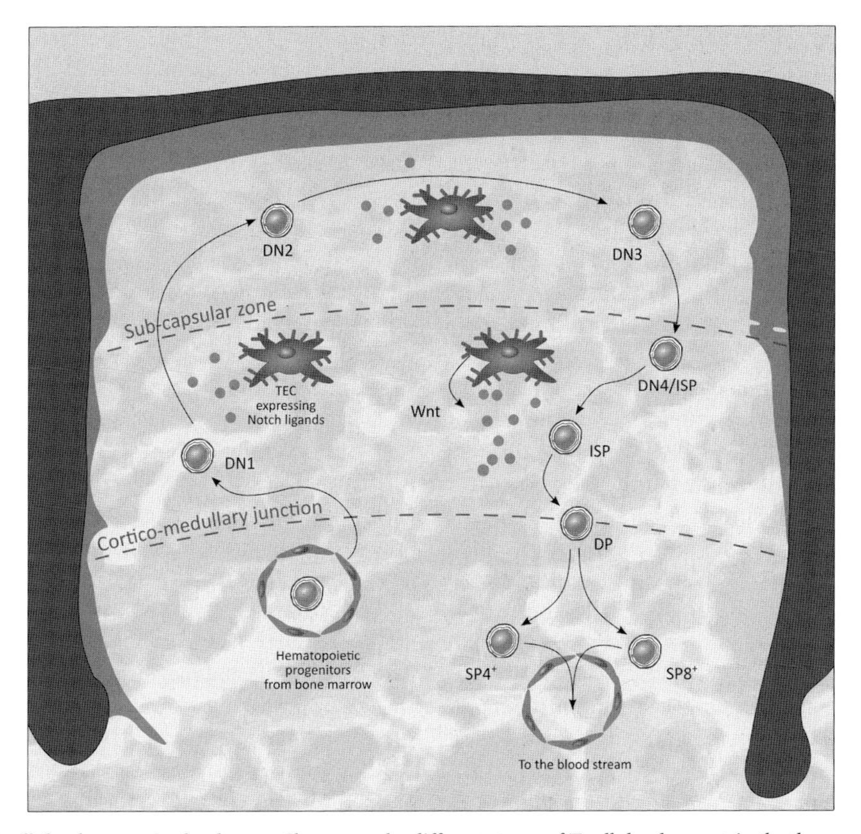

Figure 3. T cell development in the thymus. Shown are the different stages of T cell development in the thymus as commonly referred to. Both mouse and human T cell development are depicted, with an emphasis on similarities rather than differences.

subdivided into double negative (DN), double positive (DP), and single positive (SP) populations, referring to the expression of the coreceptors CD4 and CD8. The most immature thymocytes lack expression of both CD4 and CD8 and are therefore called DN.[128] In mice and humans, additional but different surface markers are used to further subdivide the DN stage. In the mouse, the markers CD25 and CD44 are used: $CD44^+CD25^-$ cells are DN1 cells, $CD44^+CD25^+$ cells are DN2 cells, DN3 cells express CD25 but no CD44, and DN4 cells express neither CD25 nor CD44 (for a comparison of human and mouse T cell development see Ref. 128, and references therein). The most immature human thymocyte population is characterized by the expression of CD34 but lacks CD1a and CD38 expression, and resembles the murine DN1 population.[129] The next stage of differentiation is marked by the expression of both CD34 and CD38 and resembles the murine DN2 stage. The most mature human DN stage that can be discerned are cells expressing CD34 and CD38, as well as CD1a. Thymocytes undergo a substantial number of cell divisions, 6 to 10, in the first DN stages. This expansion is limited by the availability of stromal niches in the thymus, and competition for them by DN cells. The earliest DN thymocytes are capable of differentiation into a large variety of blood cells, including monocytes, B cells, dendritic cells, and NK cells,[130,131] and in humans even into erythrocytes at very low frequencies.[132] Once in the thymus, DN cells migrate through different anatomical zones that likely provide different microenvironmental signals that help to establish a T cell–development program and to direct proliferation of developing thymocytes.[127,133] These cells become more and more restricted to the T cell lineage and begin expressing genes critical for TCR rearrangement, assembly, and signaling, such as recombinase activating gene 1 (Rag1) and Rag2, CD3 chains, and Lck, among others.[134] Thus, the thymic niche provides signals that direct T cell specification and commitment, as well as factors that induce proliferation, such as Scf, IL-7, Flt3L, and Wnt proteins.[135]

Notch signaling in the thymus

The best characterized function of Notch in hematopoiesis is in the commitment of hematopoietic progenitors to the T cell lineage.[136,137] The role of Notch signaling in T cell commitment was initially observed using both gain- and loss-of-function assays. Constitutive expression of the Notch activated form (ICN) in mouse hematopoietic progenitor cells transplanted into recipient mice resulted in thymic-independent T cell development in the BM. Activation of Notch signaling allowed robust T cell development until the DP stage, concomitantly with a failure to generate B cell progenitors.[136,137] Later, these findings were also extended to the human system, with ICN expression in cord blood CD34$^+$ cells promoting T cell development at the expense of B cell development, both *in vitro* and in transplanted NOD-SCID mice.[121] An essential role of Notch signaling in T cell commitment was also confirmed in complementary loss-of-function studies. Conditional deletion of Notch1 in BM progenitors resulted in blocked T cell development in the thymus at the most early intrathymic stage. Importantly Notch signaling inhibition resulted in ectopic development of Notch1–deficient immature B cells in the thymus.[129] In agreement with those studies, similar results were observed with conditional deletion of the *Rbpj* gene and with overexpression of the Notch antagonist Deltex in BM progenitor cells.[138,139] The similar phenotype observed with deletion of *Notch1* and *Rbpj* indicates that Notch1 signals through CSL in a nonredundant way to specify the T cell lineage.[140] This idea is consistent with the absence of a T cell development defect in mice deficient for Notch2[141] and Notch3.[142] Of interest, somatic activating mutations in Notch1 are causative in the development of human and murine T cell lineage leukemias.[143–146] This indicates that regulation of Notch signaling strength is critically important for normal T cell development.

Wnt signaling in the thymus

The role of Wnt signaling in the thymus has recently been reviewed extensively.[135,147] For roles of Wnt signaling in later stages of T cell development, where the comparison with the HSC niche becomes less obvious, we refer the reader to those reviews. Among the Wnt proteins, specifically Wnt1 and Wnt4 have been shown to serve as growth factors for DN cells, and therefore are essential for thymocyte proliferation;[42] mice deficient in Wnt1 and Wnt4 display low thymic cellularity. In addition, overexpression of Wnt4 was shown to selectively expand thymic output from transduced HSCs.[42] Recently, we showed that another Wnt protein, Wnt3a, plays a crucial role in fetal thymopoiesis, with Wnt3a$^{-/-}$ thymi showing reduced numbers of DP cells and a block of the preceding CD8$^+$ ISP stage.[23]

Studies on mice deficient for Wnt-responsive transcription factors revealed crucial roles for Tcf1 in T cell development and Lef1 in B cell development.[148–150] Tcf1$^{-/-}$ mice have a severe reduction of thymic cellularity and a partial block in thymocyte differentiation at the transition from the CD8$^+$ ISP stage to the CD4$^+$CD8$^+$ double positive stage. ISP and DN thymocytes of Tcf1$^{-/-}$ mice do not proliferate as strongly as their wild-type counterparts.[151] These data indicated that lack of Tcf1 mainly results in lack of proliferation and therefore expansion of the thymocytes. Although Lef1$^{-/-}$ mice have normal T cell development, mice deficient in both Lef1 and Tcf1 have a complete block in T cell differentiation at the ISP stage, which indicates redundancy between these two factors.[148] Recent work from our laboratory indicates that Tcf1 has another essential function in the thymus besides acting as the nuclear effector of Wnt signaling in thymocytes, namely a function as a critical tumor suppressor gene for the development of thymic lymphomas, the murine counterpart of human T cell acute lymphoblastic leukemia (T-ALL). Mice deficient for Tcf1 develop thymic lymphomas with high frequency due to ectopic upregulation of Lef1 and, paradoxically, extremely high Wnt signaling levels that form the initial step of leukemia development,[152] which is often followed by additional oncogenic hits such as Notch1 mutations.

The HSC niche

There is ample evidence that indicates cross-talk between HSCs and the niche cells in their close vicinity, leading to the definition of one adhesion and signaling unit termed the *stem-cell-niche synapse*—in analogy to the neuronal and immunological synapses (Fig. 4).[153] A wide variety of factors are involved in this synapse, and they mediate mainly two types of interactions. They promote cell–cell

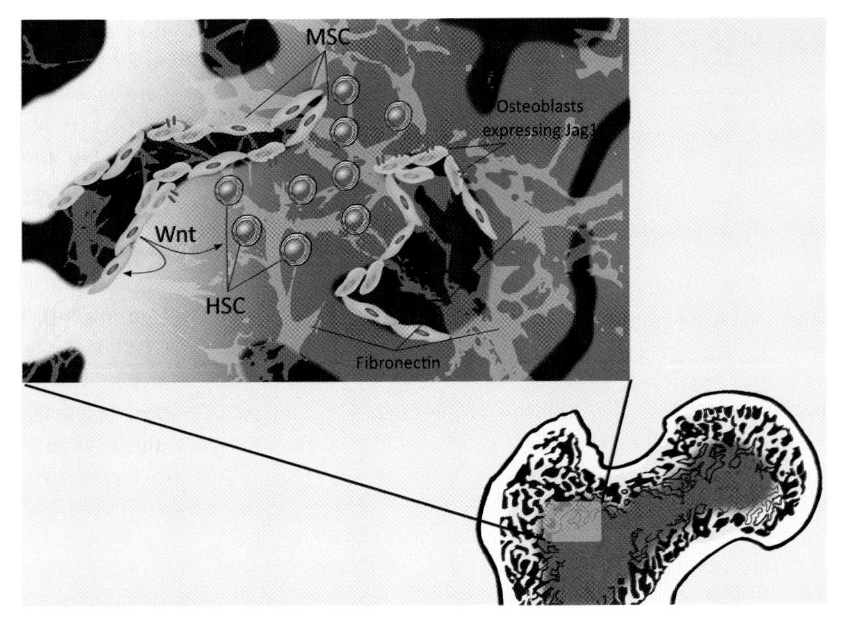

Figure 4. Composition of the hematopoietic niche. Many different cell types make up the niche components that provide the necessary supportive signals for HSC maintenance, survival, self-renewal, and differentiation. Both Wnt and Notch signals are abundantly expressed in the hematopoietic niche.

and cell–ECM adhesion to maintain cells in close proximity, and they promote activation of specific signaling pathways that influence HSC fate decisions, survival, and proliferation. The activation of signaling pathways can be mediated by direct cell–cell interactions through binding of membrane-associated ligands and receptors, such as Notch signaling, or by binding of soluble factors to specific receptors located both on the HSC and on the niche cell, such as Wnt, SMAD/TGF-β/BMP/activin and hedgehog signaling, and also hematopoietic cytokines like Scf, which binds to the Kit receptor, and thrombopoietin, which binds to the Mpl receptor.[153,154] The SMAD/TGF-β/BMP and hedgehog pathways have recently been reviewed in a series of excellent reviews.[155,156]

Notch in the HSC niche

Notch signaling is essential for generating definitive HSCs during embryonic development; however, a potential role for Notch in adult HSC function is more controversial.[157]

Germline mutant mice for Notch1 and for the downstream effector of all Notch receptors RBP-Jκ showed impaired generation of hematopoietic progenitors in the AGM region,[21] accompanied by the lack of expression of the hematopoietic transcription factors Aml/Runx1, Gata2, and Scl/Tal1 by the aortic endothelium.[158] In contrast, Notch2 germline mutation did not affect the generation of HSCs.[21] In addition, Jag1-, but not Jag2-, null embryos also failed to generate HSCs in the AGM, indicating that Notch signaling mediated by Notch1–Jag1 interaction is required for the onset of definitive hematopoiesis during mouse embryonic development.[159]

A role for Notch signaling in the regulation of adult HSC function was initially proposed from *in vitro* studies in which both human and mouse hematopoietic progenitors, cultured in the presence of Notch ligands, could be expanded without losing *in vivo* reconstitution capacity.[26,160–162] In line with these results, overexpression of ICN, as well as the Notch target gene Hes1, in hematopoietic progenitor cells enhanced self-renewal capacity of long-term HSCs.[162] Osteoblasts expressing Jag1 were identified as being part of the HSC niche.[163] An *in vivo* approach to activate osteoblasts by constitutive activation of the parathyroid hormone (PTH) receptor resulted in increased numbers of these cells in BM that showed increased expression of Jag1. Such PTH-activated osteoblasts supported an increase in the number of HSCs but no other hematopoietic progenitors, consistent with expansion of HSCs through enhanced self-renewal. Importantly, HSCs in the BM of mice with an enlarged osteoblast niche showed increased Notch1 activation *in vivo*, and Notch inhibition with γ-secretase inhibitor reduced the supportive capacity of transgenic stroma to wild-type levels in *in vitro* long-term cocultures.[163] In agreement with these observations transgenic Notch-reporter mice showed Notch signaling activity in prospectively defined HSCs that was downregulated upon HSC differentiation. Furthermore, inhibition of Notch signaling by over-expression of dominant-negative forms of XSu(H) (the *Xenopus* homologue of SCL/RBP-Jκ) and Mastermind resulted in increased differentiation *in vitro* and HSC depletion *in vivo*.[27] In contrast, loss-of-function assays resulted in controversial results showing that both Notch1 and Jag1 are dispensable for HSC self-renewal and differentiation.[164] Conditional deletion of Notch1 or Jag1 in HSCs and BM stromal cells did not affect HSC repopulation capacity, even when Notch1-deficient stem cells were transplanted into Jag1-deficient recipient mice. This unexpected lack of phenotype may be explained by functional redundancy with other Notch receptors and ligands also expressed in HSCs and in BM. However, recent work also indicates that despite use of several loss-of-function approaches, loss of Notch signals led to normal ongoing hematopoiesis, while in the same experiments T cell development was affected. Thus, deletion of Notch signaling had no effect on adult HSCs in the mouse.[165,166]

Wnt signaling in the HSC niche

The issue of Wnt signaling in the HSC niche has, until recently, been highly controversial.[23,24,30,167–170] Although many papers have shown evidence that Wnt signaling plays a role in HSC biology, several others have indicated the opposite. Recent work from our laboratory has provided a unifying concept that explains these contradictory results, namely a differential effect of Wnt dosage on the functional outcome for HSCs.[22] Interestingly, this also is the case for myeloid precursor cells and thymocytes, which have different optimal levels of Wnt signaling from low to high for HSCs, myeloid cells, and thymocytes.[22] In short, several loss-of-function approaches have shown that this pathway is necessary for normal HSC function; these studies have used either Wnt3a-deficient mice[23] over-expression of the Wnt negative regulator Dkk1 in the osteoblastic stem cell niche,[24] or Vav-Cre–mediated conditional deletion of β-catenin.[171] Besides these loss-of-function experiments, gain-of-function approaches to activate the pathway in HSCs were performed with disparate results, using stabilized forms of β-catenin either with enhancement of HSC function and maintenance of an immature phenotype,[8,30,172] or with exhaustion of the HSC pool followed by failure in repopulation capacity in transplantation assays.[167,173] As recently reviewed extensively elsewhere,[174] all of these discrepancies can be explained by the differential dosage model in which each lineage has an "appropriate" level of Wnt signaling.[22] This model can even explain the results of loss-of-function experiments that reported lack of effect of deletion of β-catenin,[175,176] as the mutants were not Wnt-null mutants, as significant Wnt activity remained that apparently was sufficient to support normal HSC function.

Interactions between the Wnt and Notch pathways

The canonical Wnt and Notch pathways are interconnected through a surprising number of proteins involving in both Wnt and Notch signaling. Both the APC protein and Axin have been shown to regulate NEXT trafficking.[177,178] β-catenin activity has also been shown to be regulated through binding of Notch, working with Axin in a synergistic fashion.[177–180] In turn, β-catenin, together with CLS and ICN, can regulate transcriptional activity.[181,182] Dvl, another protein functioning in the Wnt pathway, inhibits Notch signals by sequestering ICN.[183] GSK3β, responsible for phosphorylation of β-catenin, thereby marking it for destruction, is also capable of phosphorylating ICN, resulting in inhibition of Notch signaling inhibition.[184,185] The essential coactivatior Maml interacts with both GSK3β and β-catenin; and while the interaction between Maml and GSK3β results in a reduced Notch target gene transcription,[186] β-catenin and Maml have a synergistic effect and increase cyclin D1 and c-Myc expression,[187] which function as shared target genes between the pathways. Besides direct interactions between proteins involved in both Wnt and Notch signaling, activation of the Wnt signaling pathway has been shown to induce Notch2[188] and upregulate expression of both Jag1[189] and Dll4.[190] Because both the Wnt and Notch pathways interact in numerous ways, resulting in both positive and negative regulation, it is tempting to speculate that both interact during HSC maintenance and differentiation to provide regulatory cues.

Interaction among Notch, Wnt, and hematopoiesis stimulatory cytokines

HSCs have been cultured *ex vivo* with the aim to expand the number of repopulating cells. Initially, these *ex vivo* cultures were performed using recombinant cytokines such as Scf, Tpo, and Flt3L.[191] After a gene expression study of fetal liver, Zhang and Lodish identified a series of proteins (Scf, Tpo, Igf2, and Fgf-1)[192] that enhanced hematopoiesis in the mouse. However, fetal liver cells also function as stromal support cells for HSC,[193] which is in line with Wagner *et al.*,[194] who demonstrated that MSC-produced matrix components that result in an increase in asymmetric divisions, and that cytokines were mainly required for expansion of cell numbers. The combination of matrix and cytokine signals thus seems to provide the combination of signals necessary for efficient HSC expansion.

As precisely tuned low levels of Wnt signals are required to enhance transplantation,[22] understanding the interations among Wnt, Notch, and other HSC-promoting cytokines is of interest. In cell culture, several compounds have been tested for their ability to enhance HSCs, such as StemRegenin™[195] and PGE2 (Cellagen Technology, San Diego, CA, USA).[196,197] Although StemRegenin mainly functions as a repressor of HSC differentiation, dmPGE2 interacts with Wnt through cAMP. In addition, it was shown that PGE2 modulates HSCs by upregulating CXCR4, thereby improving HSC homing.[198] PGE2 also influences trabecular bone morphology and upregulates LSK cell numbers in mice.[199]

Modeling approaches, focused on the interaction of FGF, Wnt, and Notch signals,[200–202] show that during vertebrate development interactions among FGF, Wnt, and Notch pathways can result in an "oscillating pattern" of signaling. These patterns are thought to be due to a protein in downstream FGF signaling that inhibits transcription of Axin; similarly, GSK3β can interfere with ICN signals. In cancer, FGF and Wnt interactions have been shown, primarily through a system where Akt is phosphorylated due to FGF signals, to downregulate GSK3β activity by phosphorylating Ser 9.[203,204]

Wnt and Notch signals, however, are not only important on the HSC side of niche interaction. In the niche, osteoblasts are suggested to be most important cells for HSC maintenance.[163,205] Recently, it was shown that expressing Wnt3a in OP9 stromal cells more strongly leads to an osteoblastic phenotype, which subsequently leads to improved hematopoiesis.[82] The fact that bone progenitors are important is evidence of the fact that bone progenitor dysfunction causes myelodysplasia.[206] The importance of Wnt signals in the HSC niche was shown in multiple myeloma, where MSC secreting the Wnt inhibitor Dkk1, resulting in osteolytic lesions, were found.[207] Treatment with a Wnt signal activator, BIO, resolved the osteolytic activity,[208] demonstrating the importance of osteoblast-dependent differentiation of MSC in the BM niche for HSC function. The role of Notch for the MSC niche seems to be a negative one, as the Notch target gene Hey1 has been shown to inhibit mineralization and Runx2 transcriptional activity.[209]

Thymic versus hematopoietic niche

As discussed earlier, in both the thymic DN niche and the HSC niche Wnt and Notch components are expressed and functional. There is controversy regarding the physiological role of Notch signaling for self-renewal of HSCs, but most evidence indicates that Wnt plays a crucial role. In the thymus, Wnt signaling provides crucial proliferative signals to the most immature thymocytes, which have limited self-renewal, although there is some capacity for the DN pool to maintain itself. Notch signaling in the thymus is very different from Notch signaling in the HSC niche. In the thymus, Notch clearly provides a differentiation rather than a self-renewal signal. One obvious explanation for this difference is the abundance of Delta ligands (Delta1, but mainly Delta4, the main Notch ligand) in the thymus. However, this cannot be the sole explanation, as other Notch ligands also are capable of inducing T cell development, albeit often with lower efficacy. How exactly this is regulated is unclear, as all four Notch ligands use the same, or highly similar, signaling pathways. Nevertheless, it seems likely that Notch signaling strength may also provide clues as to what kind of functional response is induced.

In summary, Wnt signaling seems to provide self-renewal and stemness maintenance signals in both niches, whereas Notch signaling in the thymus is a differentiation and lineage commitment signal. Interestingly, recent evidence indicates that Tcf1, the nuclear effector of canonical Wnt signaling in T cells, is induced by Notch1 signaling.[210] The large number of shared target genes and interaction points in the pathways, and the interplay of both pathways with other signaling pathways (TCR signaling, NF-κB, hedgehog,[156] and TGF-β–SMAD[155]) indicate a high level of regulation to fine-tune functional outcomes.

Acknowledgments

This work was supported in part by grants provided by the Netherlands Organization for Health Research and Development (ZonMW; TOP grant to F.J.T.S.), the Netherlands Institute for Regenerative Medicine (NIRM), Children Cancer Free (KiKa 09–36), and the Association of International Cancer Research (AICR).

Conflicts of interest

The authors declare no conflicts of interest.

References

1. Staal, F.J. *et al.* 2011. Stem cell self-renewal: lessons from bone marrow, gut and iPS toward clinical applications. *Leukemia* **25:** 1095–1102.
2. Luis, T.C., N.M. Killmann & F.J. Staal. 2011. Signal transduction pathways regulating hematopoietic stem cell biology: introduction to a series of Spotlight Reviews. *Leukemia* **26:** 86–90.
3. Galloway, J.L. & L.I. Zon. 2003. Ontogeny of hematopoiesis: examining the emergence of hematopoietic cells in the vertebrate embryo. *Curr. Top Dev. Biol.* **53:** 139–158.
4. Orkin, S.H. & L.I. Zon. 2008. Hematopoiesis: an evolving paradigm for stem cell biology. *Cell* **132:** 631–644.
5. Adams, G.B. *et al.* 2006. Stem cell engraftment at the endosteal niche is specified by the calcium-sensing receptor. *Nature* **439:** 599–603.
6. Parmar, K. *et al.* 2007. Distribution of hematopoietic stem cells in the bone marrow according to regional hypoxia. *Proc. Natl. Acad. Sci. U. S. A.* **104:** 5431–5436.
7. Cobas, M. *et al.* 2004. Beta-catenin is dispensable for hematopoiesis and lymphopoiesis. *J. Exp. Med.* **199:** 221–229.
8. Baba, Y., K.P. Garrett & P.W. Kincade. 2005. Constitutively active beta-catenin confers multilineage differentiation potential on lymphoid and myeloid progenitors. *Immunity* **23:** 599–609.
9. Tothova, Z. *et al.* 2007. FoxOs are critical mediators of hematopoietic stem cell resistance to physiologic oxidative stress. *Cell* **128:** 325–339.
10. Matsuoka, Y. *et al.* 2011. Low level of c-kit expression marks deeply quiescent murine hematopoietic stem cells. *Stem Cells* **29:** 1783–1791.
11. Zhu, H.H. *et al.* 2011. Kit-Shp2-Kit signaling acts to maintain a functional hematopoietic stem and progenitor cell pool. *Blood* **117:** 5350–5361.
12. Buza-Vidas, N. *et al.* 2006. Cytokines regulate postnatal hematopoietic stem cell expansion: opposing roles of thrombopoietin and LNK. *Genes Dev.* **20:** 2018–2023.
13. Bersenev, A. *et al.* 2008. Lnk controls mouse hematopoietic stem cell self-renewal and quiescence through direct interactions with JAK2. *J. Clin. Invest.* **118:** 2832–2844.
14. Ninos, J.M. *et al.* 2006. The thrombopoietin receptor, c-Mpl, is a selective surface marker for human hematopoietic stem cells. *J. Transl. Med.* **4:** 9.
15. Jiang, S., Y. Fu & H.K. Avraham. 2011. Regulation of hematopoietic stem cell trafficking and mobilization by the endocannabinoid system. *Transfusion* **51**(Suppl. 4): 65S–71S.
16. Mendt, M. & J.E. Cardier. 2012. Stromal-derived factor-1 and its receptor, CXCR4, are constitutively expressed by mouse liver sinusoidal endothelial cells: implications for the regulation of hematopoietic cell migration to the liver during extramedullary hematopoiesis. *Stem Cells Dev*, Jan 26 Epub ahead of print. doi:10.1089/scd.2011.0565.
17. Juarez, J.G. *et al.* 2012. Sphingosine-1-phosphate facilitates trafficking of hematopoietic stem cells and their mobilization by CXCR4 antagonists in mice. *Blood* **119:** 707–716.
18. Trompouki, E. *et al.* 2011. Lineage regulators direct BMP and Wnt pathways to cell-specific programs during differentiation and regeneration. *Cell* **147:** 577–589.
19. Cook, B.D., S. Liu & T. Evans. 2011. Smad1 signaling restricts hematopoietic potential after promoting hemangioblast commitment. *Blood* **117:** 6489–6497.
20. Boisset, J.C. *et al.* 2010. In vivo imaging of haematopoietic cells emerging from the mouse aortic endothelium. *Nature* **464:** 116–120.
21. Kumano, K. *et al.* 2003. Notch1 but not Notch2 is essential for generating hematopoietic stem cells from endothelial cells. *Immunity* **18:** 699–711.
22. Luis, T.C. *et al.* 2011. Canonical wnt signaling regulates hematopoiesis in a dosage-dependent fashion. *Cell Stem Cell* **9:** 345–356.
23. Luis, T.C. *et al.* 2009. Wnt3a deficiency irreversibly impairs hematopoietic stem cell self-renewal and leads to defects in progenitor cell differentiation. *Blood* **113:** 546–554.
24. Fleming, H.E. *et al.* 2008. Wnt signaling in the niche enforces hematopoietic stem cell quiescence and is necessary to preserve self-renewal in vivo. *Cell Stem Cell* **2:** 274–283.
25. Clements, W.K. *et al.* 2011. A somitic Wnt16/Notch pathway specifies haematopoietic stem cells. *Nature* **474:** 220–224.
26. Varnum-Finney, B. *et al.* 2011. Notch2 governs the rate of generation of mouse long- and short-term repopulating stem cells. *J. Clin. Invest.* **121:** 1207–1216.
27. Duncan, A.W. *et al.* 2005. Integration of Notch and Wnt signaling in hematopoietic stem cell maintenance. *Nat. Immunol.* **6:** 314–322.
28. Smolich, B.D. *et al.* 1993. Wnt family proteins are secreted and associated with the cell surface. *Mol. Biol. Cell* **4:** 1267–1275.

29. Takada, R. *et al.* 2006. Monounsaturated fatty acid modification of Wnt protein: its role in Wnt secretion. *Dev. Cell* **11:** 791–801.
30. Reya, T. *et al.* 2003. A role for Wnt signalling in self-renewal of haematopoietic stem cells. *Nature* **423:** 409–414.
31. Logan, C.Y. & R. Nusse. 2004. The Wnt signaling pathway in development and disease. *Annu. Rev. Cell Dev. Biol.* **20:** 781–810.
32. Wong, G.T., B.J. Gavin & A.P. McMahon. 1994. Differential transformation of mammary epithelial cells by Wnt genes. *Mol. Cell Biol.* **14:** 6278–6286.
33. Wang, H.Y., T. Liu & C.C. Malbon. 2006. Structure-function analysis of Frizzleds. *Cell Signal* **18:** 934–941.
34. Li, X. *et al.* 2005. Sclerostin binds to LRP5/6 and antagonizes canonical Wnt signaling. *J. Biol. Chem.* **280:** 19883–19887.
35. Green, J.L., S.G. Kuntz & P.W. Sternberg. 2008. Ror receptor tyrosine kinases: orphans no more. *Trends Cell Biol.* **18:** 536–544.
36. Fradkin, L.G., J.M. Dura & J.N. Noordermeer. 2010. Ryks: new partners for Wnts in the developing and regenerating nervous system. *Trends Neurosci.* **33:** 84–92.
37. Korswagen, H.C. 2002. Canonical and non-canonical Wnt signaling pathways in Caenorhabditis elegans: variations on a common signaling theme. *Bioessays* **24:** 801–810.
38. Semenov, M.V. *et al.* 2007. SnapShot: noncanonical Wnt signaling pathways. *Cell* **131:** 1378.
39. Sugimura, R. & L. Li. 2010. Noncanonical Wnt signaling in vertebrate development, stem cells, and diseases. *Birth Defects Res. C Embryo Today* **90:** 243–256.
40. Kokolus, K. & M.J. Nemeth. 2010. Non-canonical Wnt signaling pathways in hematopoiesis. *Immunol. Res.* **46:** 155–164.
41. MacDonald, B.T., K. Tamai & X. He. 2009. Wnt/beta-catenin signaling: components, mechanisms, and diseases. *Dev. Cell* **17:** 9–26.
42. Staal, F.J. *et al.* 2001. Wnt signaling is required for thymocyte development and activates Tcf-1 mediated transcription. *Eur. J. Immunol.* **31:** 285–293.
43. Lobov, I.B. *et al.* 2005. WNT7b mediates macrophage-induced programmed cell death in patterning of the vasculature. *Nature* **437:** 417–421.
44. Willert, K. *et al.* 2003. Wnt proteins are lipid-modified and can act as stem cell growth factors. *Nature* **423:** 448–452.
45. Reya, T. *et al.* 2000. Wnt signaling regulates B lymphocyte proliferation through a LEF-1 dependent mechanism. *Immunity* **13:** 15–24.
46. Dosen, G. *et al.* 2006. Wnt expression and canonical Wnt signaling in human bone marrow B lymphopoiesis. *BMC Immunol.* **7:** 13.
47. Van Den Berg, D.J. *et al.* 1998. Role of members of the Wnt gene family in human hematopoiesis. *Blood* **92:** 3189–3202.
48. Austin, T.W. *et al.* 1997. A role for the Wnt gene family in hematopoiesis: expansion of multilineage progenitor cells. *Blood* **89:** 3624–3635.
49. Aberle, H. *et al.* 1997. Beta-catenin is a target for the ubiquitin-proteasome pathway. *EMBO J.* **16:** 3797–3804.
50. Hino, S. *et al.* 2003. Casein kinase I epsilon enhances the binding of Dvl-1 to Frat-1 and is essential for Wnt-3a-induced accumulation of beta-catenin. *J. Biol. Chem.* **278:** 14066–14073.
51. Ikeda, S. *et al.* 1998. Axin, a negative regulator of the Wnt signaling pathway, forms a complex with GSK-3beta and beta-catenin and promotes GSK-3beta-dependent phosphorylation of beta-catenin. *EMBO J.* **17:** 1371–1384.
52. Kishida, S. *et al.* 1998. Axin, a negative regulator of the wnt signaling pathway, directly interacts with adenomatous polyposis coli and regulates the stabilization of beta-catenin. *J. Biol. Chem.* **273:** 10823–10826.
53. Liu, C. *et al.* 2002. Control of beta-catenin phosphorylation/degradation by a dual-kinase mechanism. *Cell* **108:** 837–847.
54. Kitagawa, M. *et al.* 1999. An F-box protein, FWD1, mediates ubiquitin-dependent proteolysis of beta-catenin. *EMBO J.* **18:** 2401–2410.
55. Hurlstone, A. & H. Clevers. 2002. T-cell factors: turn-ons and turn-offs. *EMBO J.* **21:** 2303–2311.
56. Staal, F.J., T.C. Luis & M.M. Tiemessen. 2008. WNT signalling in the immune system: WNT is spreading its wings. *Nat. Rev. Immunol.* **8:** 581–593.
57. Schwarz-Romond, T. *et al.* 2007. The DIX domain of Dishevelled confers Wnt signaling by dynamic polymerization. *Nat. Struct. Mol. Biol.* **14:** 484–492.
58. Schwarz-Romond, T., C. Metcalfe & M. Bienz. 2007. Dynamic recruitment of axin by Dishevelled protein assemblies. *J. Cell Sci.* **120:** 2402–2412.
59. Kishida, S. *et al.* 1999. DIX domains of Dvl and axin are necessary for protein interactions and their ability to regulate beta-catenin stability. *Mol. Cell Biol.* **19:** 4414–4422.
60. Bilic, J. *et al.* 2007. Wnt induces LRP6 signalosomes and promotes dishevelled-dependent LRP6 phosphorylation. *Science* **316:** 1619–1622.
61. Davidson, G. *et al.* 2005. Casein kinase 1 gamma couples Wnt receptor activation to cytoplasmic signal transduction. *Nature* **438:** 867–872.
62. MacDonald, B.T. *et al.* 2008. Wnt signal amplification via activity, cooperativity, and regulation of multiple intracellular PPPSP motifs in the Wnt co-receptor LRP6. *J. Biol. Chem.* **283:** 16115–16123.
63. Staal, F.J. *et al.* 2002. Wnt signals are transmitted through N-terminally dephosphorylated beta-catenin. *EMBO Rep.* **3:** 63–68.
64. Mi, K., P.J. Dolan & G.V. Johnson. 2006. The low density lipoprotein receptor-related protein 6 interacts with glycogen synthase kinase 3 and attenuates activity. *J. Biol. Chem.* **281:** 4787–4794.
65. Wu, G. *et al.* 2009. Inhibition of GSK3 phosphorylation of beta-catenin via phosphorylated PPPSPXS motifs of Wnt coreceptor LRP6. *PLoS One* **4:** e4926.
66. Roose, J. *et al.* 1998. The Xenopus Wnt effector XTcf-3 interacts with Groucho-related transcriptional repressors. *Nature* **395:** 608–612.
67. Hsieh, J.C. *et al.* 1999. A new secreted protein that binds to Wnt proteins and inhibits their activities. *Nature* **398:** 431–436.
68. Niehrs, C. 2006. Function and biological roles of the Dickkopf family of Wnt modulators. *Oncogene* **25:** 7469–7481.

69. Bafico, A. *et al.* 2001. Novel mechanism of Wnt signalling inhibition mediated by Dickkopf-1 interaction with LRP6/Arrow. *Nat. Cell Biol.* **3:** 683–686.
70. Mao, B. *et al.* 2001. LDL-receptor-related protein 6 is a receptor for Dickkopf proteins. *Nature* **411:** 321–325.
71. Daniels, D.L. & W.I. Weis. 2002. ICAT inhibits beta-catenin binding to Tcf/Lef-family transcription factors and the general coactivator p300 using independent structural modules. *Mol. Cell* **10:** 573–584.
72. van de Wetering, M. *et al.* 1992. The human T cell transcription factor-1 gene. Structure, localization, and promoter characterization. *J. Biol. Chem.* **267:** 8530–8536.
73. Van de Wetering, M. *et al.* 1996. Extensive alternative splicing and dual promoter usage generate Tcf-1 protein isoforms with differential transcription control properties. *Mol. Cell Biol.* **16:** 745–752.
74. Hovanes, K. *et al.* 2001. Beta-catenin-sensitive isoforms of lymphoid enhancer factor-1 are selectively expressed in colon cancer. *Nat. Genet.* **28:** 53–57.
75. Serru, V. *et al.* 2000. Sequence and expression of seven new tetraspans. *Biochim. Biophys. Acta* **1478:** 159–163.
76. Junge, H.J. *et al.* 2009. TSPAN12 regulates retinal vascular development by promoting Norrin- but not Wnt-induced FZD4/beta-catenin signaling. *Cell* **139:** 299–311.
77. Kamata, T. *et al.* 2004. R-spondin, a novel gene with thrombospondin type 1 domain, was expressed in the dorsal neural tube and affected in Wnts mutants. *Biochim. Biophys. Acta* **1676:** 51–62.
78. Kazanskaya, O. *et al.* 2004. R-Spondin2 is a secreted activator of Wnt/beta-catenin signaling and is required for Xenopus myogenesis. *Dev. Cell* **7:** 525–534.
79. Kim, K.A. *et al.* 2005. Mitogenic influence of human R-spondin1 on the intestinal epithelium. *Science* **309:** 1256–1259.
80. Bradley, R.S. & A.M. Brown. 1990. The proto-oncogene int-1 encodes a secreted protein associated with the extracellular matrix. *EMBO J.* **9:** 1569–1575.
81. Reichsman, F., L. Smith & S. Cumberledge. 1996. Glycosaminoglycans can modulate extracellular localization of the wingless protein and promote signal transduction. *J. Cell Biol.* **135:** 819–827.
82. Ichii, M. *et al.* 2012. The canonical Wnt pathway shapes niches supportive of hematopoietic stem/progenitor cells. *Blood* **119:** 1683–1692.
83. Gazave, E. *et al.* 2009. Origin and evolution of the Notch signalling pathway: an overview from eukaryotic genomes. *BMC Evol. Biol.* **9:** 249.
84. Barolo, S. & J.W. Posakony. 2002. Three habits of highly effective signaling pathways: principles of transcriptional control by developmental cell signaling. *Genes Dev.* **16:** 1167–1181.
85. D'Souza, B., L. Meloty-Kapella & G. Weinmaster. 2010. Canonical and non-canonical Notch ligands. *Curr. Top Dev. Biol.* **92:** 73–129.
86. Ladi, E. *et al.* 2005. The divergent DSL ligand Dll3 does not activate Notch signaling but cell autonomously attenuates signaling induced by other DSL ligands. *J. Cell Biol.* **170:** 983–992.
87. Geffers, I. *et al.* 2007. Divergent functions and distinct localization of the Notch ligands DLL1 and DLL3 in vivo. *J. Cell Biol.* **178:** 465–476.
88. Chapman, G. *et al.* 2011. Notch inhibition by the ligand DELTA-LIKE 3 defines the mechanism of abnormal vertebral segmentation in spondylocostal dysostosis. *Hum Mol. Genet.* **20:** 905–916.
89. Bozkulak, E.C. & G. Weinmaster. 2009. Selective use of ADAM10 and ADAM17 in activation of Notch1 signaling. *Mol. Cell Biol.* **29:** 5679–5695.
90. Vaccari, T. *et al.* 2008. Endosomal entry regulates Notch receptor activation in Drosophila melanogaster. *J. Cell Biol.* **180:** 755–762.
91. Hansson, E.M. *et al.* 2010. Control of Notch-ligand endocytosis by ligand-receptor interaction. *J. Cell Sci.* **123:** 2931–2942.
92. Jorissen, E. & B. De Strooper. 2010. Gamma-secretase and the intramembrane proteolysis of Notch. *Curr. Top Dev. Biol.* **92:** 201–230.
93. Kovall, R.A. & S.C. Blacklow. 2010. Mechanistic insights into Notch receptor signaling from structural and biochemical studies. *Curr. Top Dev. Biol.* **92:** 31–71.
94. McElhinny, A.S., J.L. Li & L. Wu. 2008. Mastermind-like transcriptional co-activators: emerging roles in regulating cross talk among multiple signaling pathways. *Oncogene* **27:** 5138–5147.
95. Rao, P. & T. Kadesch. 2003. The intracellular form of notch blocks transforming growth factor beta-mediated growth arrest in Mv1Lu epithelial cells. *Mol. Cell Biol.* **23:** 6694–6701.
96. Satoh, Y. *et al.* 2004. Roles for c-Myc in self-renewal of hematopoietic stem cells. *J. Biol. Chem.* **279:** 24986–24993.
97. Weng, A.P. *et al.* 2006. c-Myc is an important direct target of Notch1 in T-cell acute lymphoblastic leukemia/lymphoma. *Genes Dev.* **20:** 2096–2109.
98. Palomero, T. *et al.* 2006. NOTCH1 directly regulates c-MYC and activates a feed-forward-loop transcriptional network promoting leukemic cell growth. *Proc. Natl. Acad. Sci. U. S. A.* **103:** 18261–18266.
99. Cohen, B. *et al.* 2010. Cyclin D1 is a direct target of JAG1-mediated Notch signaling in breast cancer. *Breast Cancer Res. Treat* **123:** 113–124.
100. Ronchini, C. & A.J. Capobianco. 2001. Induction of cyclin D1 transcription and CDK2 activity by Notch(ic): implication for cell cycle disruption in transformation by Notch(ic). *Mol. Cell Biol.* **21:** 5925–5934.
101. Joshi, I. *et al.* 2009. Notch signaling mediates G1/S cell-cycle progression in T cells via cyclin D3 and its dependent kinases. *Blood* **113:** 1689–1698.
102. Iso, T., L. Kedes & Y. Hamamori. 2003. HES and HERP families: multiple effectors of the Notch signaling pathway. *J. Cell Physiol.* **194:** 237–255.
103. Zhang, N. *et al.* 2000. A mutation in the Lunatic fringe gene suppresses the effects of a Jagged2 mutation on inner hair cell development in the cochlea. *Curr. Biol.* **10:** 659–662.
104. McDaniell, R. *et al.* 2006. NOTCH2 mutations cause Alagille syndrome, a heterogeneous disorder of the notch signaling pathway. *Am. J. Hum Genet.* **79:** 169–173.

105. Krebs, L.T. *et al.* 2004. Haploinsufficient lethality and formation of arteriovenous malformations in Notch pathway mutants. *Genes Dev.* **18:** 2469–2473.
106. Wang, H. & W. Chia. 2005. Drosophila neural progenitor polarity and asymmetric division. *Biol. Cell* **97:** 63–74.
107. Schroder, N. & A. Gossler. 2002. Expression of Notch pathway components in fetal and adult mouse small intestine. *Gene Expr. Patterns* **2:** 247–250.
108. Miller, A.C., E.L. Lyons & T.G. Herman. 2009. cis-Inhibition of Notch by endogenous Delta biases the outcome of lateral inhibition. *Curr. Biol.* **19:** 1378–1383.
109. del Alamo, D., H. Rouault & F. Schweisguth. 2011. Mechanism and significance of cis-inhibition in Notch signalling. *Curr. Biol.* **21:** R40–R47.
110. Itoh, M. *et al.* 2003. Mind bomb is a ubiquitin ligase that is essential for efficient activation of Notch signaling by Delta. *Dev. Cell* **4:** 67–82.
111. Yeh, E. *et al.* 2001. Neuralized functions as an E3 ubiquitin ligase during Drosophila development. *Curr. Biol.* **11:** 1675–1679.
112. Wilkin, M. *et al.* 2008. Drosophila HOPS and AP-3 complex genes are required for a Deltex-regulated activation of notch in the endosomal trafficking pathway. *Dev. Cell* **15:** 762–772.
113. Yamada, K. *et al.* 2011. Roles of Drosophila deltex in Notch receptor endocytic trafficking and activation. *Genes Cells* **16:** 261–272.
114. Diederich, R.J. *et al.* 1994. Cytosolic interaction between deltex and Notch ankyrin repeats implicates deltex in the Notch signaling pathway. *Development* **120:** 473–481.
115. Mukherjee, A. *et al.* 2005. Regulation of Notch signalling by non-visual beta-arrestin. *Nat. Cell Biol.* **7:** 1191–1201.
116. Wilson, A. *et al.* 2007. Normal hemopoiesis and lymphopoiesis in the combined absence of numb and numblike. *J. Immunol.* **178:** 6746–6751.
117. Bresciani, E. *et al.* 2010. Zebrafish numb and numblike are involved in primitive erythrocyte differentiation. *PLoS One* **5:** e14296.
118. Cheng, X. *et al.* 2008. Numb mediates the interaction between Wnt and Notch to modulate primitive erythropoietic specification from the hemangioblast. *Development* **135:** 3447–3458.
119. Kwon, C. *et al.* 2011. Notch post-translationally regulates beta-catenin protein in stem and progenitor cells. *Nat. Cell Biol.* **13:** 1244–1251.
120. Amsen, D. *et al.* 2004. Instruction of distinct CD4 T helper cell fates by different notch ligands on antigen-presenting cells. *Cell* **117:** 515–526.
121. De Smedt, M. *et al.* 2002. Active form of Notch imposes T cell fate in human progenitor cells. *J. Immunol.* **169:** 3021–3029.
122. Wang, J. *et al.* 2001. Human Notch-1 inhibits NF-kappa B activity in the nucleus through a direct interaction involving a novel domain. *J. Immunol.* **167:** 289–295.
123. Blokzijl, A. *et al.* 2003. Cross-talk between the Notch and TGF-beta signaling pathways mediated by interaction of the Notch intracellular domain with Smad3. *J. Cell Biol.* **163:** 723–728.
124. Dahlqvist, C. *et al.* 2003. Functional Notch signaling is required for BMP4-induced inhibition of myogenic differentiation. *Development* **130:** 6089–6099.
125. Fu, Y. *et al.* 2009. Differential regulation of transforming growth factor beta signaling pathways by Notch in human endothelial cells. *J. Biol. Chem.* **284:** 19452–19462.
126. Itoh, F. *et al.* 2004. Synergy and antagonism between Notch and BMP receptor signaling pathways in endothelial cells. *EMBO J.* **23:** 541–551.
127. Rothenberg, E.V., J.E. Moore & M.A. Yui. 2008. Launching the T-cell-lineage developmental programme. *Nat. Rev. Immunol.* **8:** 9–21.
128. Weerkamp, F., K. Pike-Overzet & F.J. Staal. 2006. T-sing progenitors to commit. *Trends Immunol.* **27:** 125–131.
129. Weerkamp, F. *et al.* 2005. Age-related changes in the cellular composition of the thymus in children. *J. Allergy Clin. Immunol.* **115:** 834–840.
130. Bhandoola, A. *et al.* 2007. Commitment and developmental potential of extrathymic and intrathymic T cell precursors: plenty to choose from. *Immunity* **26:** 678–689.
131. Ciofani, M. & J.C. Zuniga-Pflucker. 2007. The thymus as an inductive site for T lymphopoiesis. *Annu. Rev. Cell Dev. Biol.* **23:** 463–493.
132. Weerkamp, F. *et al.* 2006. Human thymus contains multipotent progenitors with T/B lymphoid, myeloid, and erythroid lineage potential. *Blood* **107:** 3131–3137.
133. Petrie, H.T. & J.C. Zuniga-Pflucker. 2007. Zoned out: functional mapping of stromal signaling microenvironments in the thymus. *Annu. Rev. Immunol.* **25:** 649–679.
134. Dik, W.A. *et al.* 2005. New insights on human T cell development by quantitative T cell receptor gene rearrangement studies and gene expression profiling. *J. Exp. Med.* **201:** 1715–1723.
135. Staal, F.J. & H.C. Clevers. 2003. Wnt signaling in the thymus. *Curr. Opin. Immunol.* **15:** 204–208.
136. Pui, J.C. *et al.* 1999. Notch1 expression in early lymphopoiesis influences B versus T lineage determination. *Immunity* **11:** 299–308.
137. Radtke, F. *et al.* 1999. Deficient T cell fate specification in mice with an induced inactivation of Notch1. *Immunity* **10:** 547–558.
138. Izon, D.J., J.A. Punt & W.S. Pear. 2002. Deciphering the role of Notch signaling in lymphopoiesis. *Curr. Opin. Immunol.* **14:** 192–199.
139. Izon, D.J. *et al.* 2002. Deltex1 redirects lymphoid progenitors to the B cell lineage by antagonizing Notch1. *Immunity* **16:** 231–243.
140. Radtke, F. *et al.* 2004. Notch regulation of lymphocyte development and function. *Nat. Immunol.* **5:** 247–253.
141. Saito, T. *et al.* 2003. Notch2 is preferentially expressed in mature B cells and indispensable for marginal zone B lineage development. *Immunity* **18:** 675–685.
142. Krebs, L.T. *et al.* 2003. Characterization of Notch3-deficient mice: normal embryonic development and absence of genetic interactions with a Notch1 mutation. *Genesis* **37:** 139–143.
143. Mansour, M.R. *et al.* 2009. Prognostic implications of NOTCH1 and FBXW7 mutations in adults with T-cell acute lymphoblastic leukemia treated on the MRC

UKALLXII/ECOG E2993 protocol. *J. Clin. Oncol.* **27:** 4352–4356.

144. Vilimas, T. *et al.* 2007. Targeting the NF-kappaB signaling pathway in Notch1-induced T-cell leukemia. *Nat. Med.* **13:** 70–77.
145. Lee, S.Y. *et al.* 2005. Mutations of the Notch1 gene in T-cell acute lymphoblastic leukemia: analysis in adults and children. *Leukemia* **19:** 1841–1843.
146. Weng, A.P. *et al.* 2004. Activating mutations of NOTCH1 in human T cell acute lymphoblastic leukemia. *Science* **306:** 269–271.
147. Staal, F.J. & J.M. Sen. 2008. The canonical Wnt signaling pathway plays an important role in lymphopoiesis and hematopoiesis. *Eur. J. Immunol.* **38:** 1788–1794.
148. Okamura, R. *et al.* 1998. Overlapping functions of Tcf-1 and Lef-1 in T lymphocyte development. *Immunity* **8:** 11–20.
149. Okamura, R.M. *et al.* 1998. Redundant regulation of T cell differentiation and TCRalpha gene expression by the transcription factors LEF-1 and TCF-1. *Immunity* **8:** 11–20.
150. Verbeek, S. *et al.* 1995. An HMG-box-containing T-cell factor required for thymocyte differentiation. *Nature* **374:** 70–74.
151. Schilham, M.W. *et al.* 1998. Critical involvement of Tcf-1 in expansion of thymocytes. *J. Immunol.* **161:** 3984–3991.
152. Tiemessen, M.M. The nuclear effector of Wnt signlaing, TCF1, functions as a T cell specific tumor suppressor for the development of lymphomas. Submitted.
153. Wilson, A. & A. Trumpp. 2006. Bone-marrow haematopoietic-stem-cell niches. *Nat. Rev. Immunol.* **6:** 93–106.
154. Blank, U., G. Karlsson & S. Karlsson. 2008. Signaling pathways governing stem-cell fate. *Blood* **111:** 492–503.
155. Blank, U. & S. Karlsson. 2011. The role of Smad signaling in hematopoiesis and translational hematology. *Leukemia* **25:** 1379–1388.
156. Mar, B.G. *et al.* 2011. The controversial role of the Hedgehog pathway in normal and malignant hematopoiesis. *Leukemia* **25:** 1665–1673.
157. Pajcini, K. V., N.A. Speck & W.S. Pear. 2011. Notch signaling in mammalian hematopoietic stem cells. *Leukemia* **25:** 1525–1532.
158. Robert-Moreno, A. *et al.* 2005. RBPjkappa-dependent Notch function regulates Gata2 and is essential for the formation of intra-embryonic hematopoietic cells. *Development* **132:** 1117–1126.
159. Robert-Moreno, A. *et al.* 2008. Impaired embryonic haematopoiesis yet normal arterial development in the absence of the Notch ligand Jagged1. *EMBO J.* **27:** 1886–1895.
160. Karanu, F.N. *et al.* 2000. The notch ligand jagged-1 represents a novel growth factor of human hematopoietic stem cells. *J. Exp. Med.* **192:** 1365–1372.
161. Varnum-Finney, B. *et al.* 2000. Pluripotent, cytokine-dependent, hematopoietic stem cells are immortalized by constitutive Notch1 signaling. *Nat. Med.* **6:** 1278–1281.
162. Delaney, C. *et al.* 2010. Notch-mediated expansion of human cord blood progenitor cells capable of rapid myeloid reconstitution. *Nat. Med.* **16:** 232–236.
163. Calvi, L.M. *et al.* 2003. Osteoblastic cells regulate the haematopoietic stem cell niche. *Nature* **425:** 841–846.
164. Mancini, S.J. *et al.* 2005. Jagged1-dependent Notch signaling is dispensable for hematopoietic stem cell self-renewal and differentiation. *Blood* **105:** 2340–2342.
165. Gering, M. & R. Patient. 2008. Notch in the niche. *Cell Stem Cell* **2:** 293–294.
166. Maillard, I. *et al.* 2008. Canonical notch signaling is dispensable for the maintenance of adult hematopoietic stem cells. *Cell Stem Cell* **2:** 356–366.
167. Kirstetter, P. *et al.* 2006. Activation of the canonical Wnt pathway leads to loss of hematopoietic stem cell repopulation and multilineage differentiation block. *Nat. Immunol.* **7:** 1048–1056.
168. Luis, T.C. *et al.* 2010. Wnt3a nonredundantly controls hematopoietic stem cell function and its deficiency results in complete absence of canonical Wnt signaling. *Blood* **116:** 496–497.
169. Nemeth, M.J. *et al.* 2007. Wnt5a inhibits canonical Wnt signaling in hematopoietic stem cells and enhances repopulation. *Proc. Natl. Acad. Sci. U. S. A.* **104:** 15436–15441.
170. Suda, T. & F. Arai. 2008. Wnt signaling in the niche. *Cell* **132:** 729–730.
171. Zhao, C. *et al.* 2007. Loss of beta-catenin impairs the renewal of normal and CML stem cells in vivo. *Cancer Cell* **12:** 528–541.
172. Malhotra, S. *et al.* 2008. Contrasting responses of lymphoid progenitors to canonical and noncanonical Wnt signals. *J. Immunol.* **181:** 3955–3964.
173. Scheller, M. *et al.* 2006. Hematopoietic stem cell and multilineage defects generated by constitutive beta-catenin activation. *Nat. Immunol.* **7:** 1037–1047.
174. Luis, T.C. *et al.* 2012. Wnt signaling strength regulates normal hematopoiesis and its deregulation is involved in leukemia development. *Leukemia* **26:** 414–421.
175. Jeannet, G. *et al.* 2008. Long-term, multilineage hematopoiesis occurs in the combined absence of beta-catenin and gamma-catenin. *Blood* **111:** 142–149.
176. Koch, U. *et al.* 2008. Simultaneous loss of – and {gamma}-catenin does not perturb hematopoiesis or lymphopoiesis. *Blood* **111:** 160–164.
177. Munoz-Descalzo, S. *et al.* 2011. Modulation of the ligand-independent traffic of Notch by Axin and Apc contributes to the activation of Armadillo in Drosophila. *Development* **138:** 1501–1506.
178. Hayward, P., T. Balayo & A. Martinez Arias. 2006. Notch synergizes with axin to regulate the activity of armadillo in Drosophila. *Dev. Dyn.* **235:** 2656–2666.
179. Hayward, P. *et al.* 2005. Notch modulates Wnt signalling by associating with Armadillo/beta-catenin and regulating its transcriptional activity. *Development* **132:** 1819–1830.
180. Sanders, P.G. *et al.* 2009. Ligand-independent traffic of Notch buffers activated Armadillo in Drosophila. *PLoS Biol.* **7:** e1000169.
181. Shimizu, T. *et al.* 2008. Stabilized beta-catenin functions through TCF/LEF proteins and the Notch/RBP-Jkappa

complex to promote proliferation and suppress differentiation of neural precursor cells. *Mol. Cell Biol.* **28:** 7427–7441.
182. Yamamizu, K. *et al.* 2010. Convergence of Notch and beta-catenin signaling induces arterial fate in vascular progenitors. *J. Cell Biol.* **189:** 325–338.
183. Axelrod, J.D. *et al.* 1996. Interaction between Wingless and Notch signaling pathways mediated by dishevelled. *Science* **271:** 1826–1832.
184. Espinosa, L. *et al.* 2003. Phosphorylation by glycogen synthase kinase-3 beta down-regulates Notch activity, a link for Notch and Wnt pathways. *J. Biol Chem.* **278:** 32227–32235.
185. Foltz, D.R. *et al.* 2002. Glycogen synthase kinase-3beta modulates notch signaling and stability. *Curr. Biol.* **12:** 1006–1011.
186. Saint Just Ribeiro, M. *et al.* 2009. GSK3beta is a negative regulator of the transcriptional coactivator MAML1. *Nucleic Acids Res.* **37:** 6691–6700.
187. Alves-Guerra, M.C., C. Ronchini & A.J. Capobianco. 2007. Mastermind-like 1 is a specific coactivator of beta-catenin transcription activation and is essential for colon carcinoma cell survival. *Cancer Res.* **67:** 8690–8698.
188. Ungerback, J. *et al.* 2011. The Notch-2 gene is regulated by Wnt signaling in cultured colorectal cancer cells. *PLoS One* **6:** e17957.
189. Estrach, S. *et al.* 2006. Jagged 1 is a beta-catenin target gene required for ectopic hair follicle formation in adult epidermis. *Development* **133:** 4427–4438.
190. Corada, M. *et al.* 2010. The Wnt/beta-catenin pathway modulates vascular remodeling and specification by up-regulating Dll4/Notch signaling. *Dev. Cell* **18:** 938–949.
191. Jacobsen, S.E. *et al.* 1996. Ability of flt3 ligand to stimulate the in vitro growth of primitive murine hematopoietic progenitors is potently and directly inhibited by transforming growth factor-beta and tumor necrosis factor-alpha. *Blood* **87:** 5016–5026.
192. Zhang, C.C. & H.F. Lodish. 2004. Insulin-like growth factor 2 expressed in a novel fetal liver cell population is a growth factor for hematopoietic stem cells. *Blood* **103:** 2513–2521.
193. Chou, S. & H.F. Lodish. Fetal liver hepatic progenitors are supportive stromal cells for hematopoietic stem cells. *Proc. Natl. Acad. Sci. U. S. A.* **107:** 7799–7804.
194. Wagner, W. *et al.* 2007. Molecular and secretory profiles of human mesenchymal stromal cells and their abilities to maintain primitive hematopoietic progenitors. *Stem Cells* **25:** 2638–2647.
195. Boitano, A.E. *et al.* Aryl hydrocarbon receptor antagonists promote the expansion of human hematopoietic stem cells. *Science* **329:** 1345–1348.
196. Goessling, W. *et al.* 2011. Prostaglandin E2 enhances human cord blood stem cell xenotransplants and shows long-term safety in preclinical nonhuman primate transplant models. *Cell Stem Cell* **8:** 445–458.
197. Goessling, W. *et al.* 2009. Genetic interaction of PGE2 and Wnt signaling regulates developmental specification of stem cells and regeneration. *Cell* **136:** 1136–1147.
198. Hoggatt, J. *et al.* 2009. Prostaglandin E2 enhances hematopoietic stem cell homing, survival, and proliferation. *Blood* **113:** 5444–5455.
199. Frisch, B.J. *et al.* 2009. In vivo prostaglandin E2 treatment alters the bone marrow microenvironment and preferentially expands short-term hematopoietic stem cells. *Blood* **114:** 4054–4063.
200. Dequéant, M.-L. *et al.* 2006. A complex oscillating network of signaling genes underlies the mouse segmentation clock. *Science* **314:** 1595–1598.
201. Goldbeter, A. & O. Pourquié. 2008. Modeling the segmentation clock as a network of coupled oscillations in the Notch, Wnt and FGF signaling pathways. *J. Theor. Biol.* **252:** 574–585.
202. Lee, E. *et al.* 2003. The roles of APC and Axin derived from experimental and theoretical analysis of the Wnt pathway. *PLoS Biol.* **1:** E10.
203. Frame, S. & P. Cohen. 2001. GSK3 takes centre stage more than 20 years after its discovery. *Biochem. J.* **359:** 1–16.
204. Shaw, M. & P. Cohen. 1999. Role of protein kinase B and the MAP kinase cascade in mediating the EGF-dependent inhibition of glycogen synthase kinase 3 in Swiss 3T3 cells. *FEBS Lett.* **461:** 120–124.
205. Mendez-Ferrer, S. *et al.* 2011. Mesenchymal and haematopoietic stem cells form a unique bone marrow niche. *Nature* **466:** 829–834.
206. Raaijmakers, M.H. *et al.* 2011. Bone progenitor dysfunction induces myelodysplasia and secondary leukaemia. *Nature* **464:** 852–857.
207. Tian, E. *et al.* 2003. The role of the Wnt-signaling antagonist DKK1 in the development of osteolytic lesions in multiple myeloma. *N. Engl. J. Med.* **349:** 2483–2494.
208. Gunn, W.G. *et al.* 2006. A crosstalk between myeloma cells and marrow stromal cells stimulates production of DKK1 and interleukin-6: a potential role in the development of lytic bone disease and tumor progression in multiple myeloma. *Stem Cells* **24:** 986–991.
209. Zamurovic, N. *et al.* 2004. Coordinated activation of notch, Wnt, and transforming growth factor-beta signaling pathways in bone morphogenic protein 2-induced osteogenesis. Notch target gene Hey1 inhibits mineralization and Runx2 transcriptional activity. *J. Biol. Chem.* **279:** 37704–37715.
210. Weber, B.N. *et al.* 2011. A critical role for TCF-1 in T-lineage specification and differentiation. *Nature* **476:** 63–68.

Ann. N.Y. Acad. Sci. ISSN 0077-8923

ANNALS OF THE NEW YORK ACADEMY OF SCIENCES
Issue: *Hematopoietic Stem Cells VIII*

Phenotypic and functional heterogeneity of human bone marrow– and amnion-derived MSC subsets

Kavitha Sivasubramaniyan,[1,*] Daniela Lehnen,[1,*] Roshanak Ghazanfari,[1] Malgorzata Sobiesiak,[1] Abhishek Harichandan,[1] Elisabeth Mortha,[1] Neli Petkova,[1] Sabrina Grimm,[1] Flavianna Cerabona,[1] Peter de Zwart,[2] Harald Abele,[3] Wilhelm K. Aicher,[4] Christoph Faul,[1] Lothar Kanz,[1] and Hans-Jörg Bühring[1]

[1]Department of Internal Medicine II, Division of Hematology, Immunology, Oncology, Rheumatology and Pulmonology, University Clinic of Tübingen, Tübingen, Germany. [2]Department of Arthroplasty, BG-Trauma-Center, University of Tübingen, Tübingen, Germany. [3]Department of Gynecology and Obstetrics, University Clinic of Tübingen, Tübingen, Germany. [4]Department of Orthopedic Surgery, University Clinic of Tübingen, Tübingen, Germany

Address for correspondence: Hans-Jörg Bühring, Ph.D., Department of Internal Medicine II, Division of Hematology, Oncology, Immunology, Rheumatology and Pulmonology, University Clinic of Tübingen, Laboratory for Stem Cell Research, Otfried-Müller-Str. 10, 72076 Tübingen, Germany. hans-joerg.buehring@uni-tuebingen.de

Bone marrow–derived mesenchymal stromal/stem cells (MSCs) are nonhematopoietic cells that are able to differentiate into osteoblasts, adipocytes, and chondrocytes. In addition, they are known to participate in niche formation for hematopoietic stem cells and to display immunomodulatory properties. Conventionally, these cells are functionally isolated from tissue based on their capacity to adhere to the surface of culture flasks. This isolation procedure is hampered by the unpredictable influence of secreted molecules, the interactions between cocultured hematopoietic and other unrelated cells, and by the arbitrarily selected removal time of nonadherent cells before the expansion of MSCs. Finally, functionally isolated cells do not provide biological information about the starting population. To circumvent these limitations, several strategies have been developed to facilitate the prospective isolation of MSCs based on the selective expression, or absence, of surface markers. In this report, we summarize the most frequently used markers and introduce new targets for antibody-based isolation procedures of primary bone marrow- and amnion-derived MSCs.

Keywords: mesenchymal stem cells; prospective isolation; MSC subsets; bone marrow; placenta

Introduction

Mesenchymal stromal/stem cells (MSCs) are multipotent cells, derived from various tissues, that are able to form fibroblast-like colonies or *colony-forming unit-fibroblasts* (CFU-F).[1,2] Under appropriate conditions MSC can be induced to differentiate into defined cell types of mesodermal lineages, including osteoblasts, adipocytes, and chondrocytes.[3] In addition, they contribute to the formation of protective niches for hematopoietic stem cells (HSCs) by interacting with HSCs and secreting regulatory molecules and cytokines to control the process of hematopoiesis.[4,5] MSC are also known to inhibit the proliferation of T and B cells and the differentiation of dendritic cells.[6] In therapy, MSCs are used for replacement therapy of damaged tissues in patients with osteoarthritis, spinal disk injury, cardiovascular, neurological, and immunological diseases.[7–10] Because of their microenvironment forming ability and multilineage differentiation capacity, they present an attractive cell source for cotransplantation with HSCs.

After expansion in culture, bone marrow–derived MSCs express the surface markers CD29, CD73, CD90, CD105, CD106, CD140b, and CD166, but not CD31, CD45, CD34, CD133, or MHC class

*Both the authors contributed equally to this work.

doi: 10.1111/j.1749-6632.2012.06551.x

II.[11–13] Cultured MSCs from other sources, such as chorion- and amnion-derived cells of placenta, adipose tissue, peripheral blood, umbilical cord blood, amniotic fluid, fetal hepatic and pulmonary tissue, skin, and prostate,[11,14–23] also seem to be negative for CD31, CD45, CD80, but uniformly express CD9, CD10, CD13, CD29, CD73, CD90, CD105, and CD106, and additional tissue-specific expression of other surface antigens has also been reported. For example, only adipose tissue–derived MSCs express high levels of CD34, and only amnion-derived MSCs are positive for stage-specific embryonic antigen (SSEA)-4 and tumor rejection antigen (TRA)-1–81.[14,24] In contrast, bone marrow–derived MSCs, but not placenta-derived MSCs, express CD271 as well as tissue-nonspecific alkaline phosphatase (TNAP).[3,25–27] MSCs from different sources not only display differential expression patterns of surface antigens but also variable differentiation capacity. In a recent publication it was demonstrated that bone marrow–derived MSCs display an increased chondrogenic differentiation potential compared with MSCs of other sources.[28] As MSCs represent an attractive tool for cartilage tissue repair strategies, bone marrow is considered to be the preferred MSC source for such therapeutic approaches.[28]

Isolation procedures of MSCs

Functional isolation of MSCs

Conventional procedures for preparing MSCs for research and clinical purposes rely on the *in vitro* expansion of unselected bone marrow cells, based on their capacity to adhere to the plastic surface in culture dishes. These functionally isolated MSCs are expanded in media of defined compositions.[12,29–31] This isolation procedure is, however, accompanied by several limitations, including the undesired interactions of MSCs with hematopoietic cells and hematopoietic cell–derived growth factors in the first culture period, by removal of nonadherent cells and replacement with fresh media at defined times (which may not be ideal), and by the coexpansion of other adherent cells, mainly macrophages and endothelial cells. In addition, evaluating functionally isolated MSCs does not provide information about the antigenic composition of the starting population. As a consequence, most publications describe retrospective antigen expression profiles of MSC progeny, but not of the initiating cells. Not surprisingly, a variety of surface markers, such as CD109 and CD318, are expressed on cultured MSCs but not on their primary counterparts.[3,32] Other markers like CD271, CD56, and SSEA-3, which are known to be highly expressed on primary MSCs or MSC subsets, are rapidly downregulated in culture.[3] Despite the limitations of functional isolation protocols, these procedures are still prevalent for large-scale MSC preparations in clinical settings because no expensive good manufacturing practice (GMP) antibodies for immunoselection are required.

Prospective isolation of bone marrow–derived MSCs

In contrast to functional isolation procedures, the prospective isolation of MSCs allows for precise definition and provides biological information on the starting population. Several markers were identified that are suitable to isolate MSCs from primary bone marrow or other tissues, including antibodies specific for a variety of cell surface molecules, CD49a, CD63, CD73 (SH3/SH4), CD105 (SH2), CD106, CD140b, CD146, CD271, CD349, TNAP, Hsp90-β, as well as orphan antigens defined by antibodies STRO-1, W5C5, 2B1H4, and others (Table 1A and 1B). In other approaches, bone marrow–derived MSCs were enriched by negative selection, employing markers such as CD14, CD34, CD45, and/or CD235 (glycophorin A) and other lineage-negative markers (Table 1C).

Distinct markers are required for the selection of MSCs from other sources than bone marrow because of some unique phenotypic peculiarities. For example, placenta-derived MSCs are preferentially isolated using antibodies against CD349, SSEA-4, and TRA-1–81, which contrasts with antibodies used for bone marrow–derived MSC, preferably isolated by CD271 or TNAP selection.[3,14] Other markers such as CD34 and CD117 are more suitable to select adipose- and amniotic fluid–derived MSCs,[34–36,58] respectively (Table 1A).

Prospectively isolated MSCs are generally analyzed for their clonogenic potential by the CFU-F assay.[1,2] Candidate antibodies selective for MSC can be evaluated by screening their reactivity with cell populations that express established key MSC markers, such as CD271 or STRO-1. This procedure is illustrated in Figure 1A. An example showing bone marrow cells double stained with antibodies against CD271 and CD140b (platelet-derived growth factor ((PDGF) receptor-β), is illustrated in

Table1. Published surface markers suitable for the prospective isolation of MSC from various tissues

Markers used	Tissue	References
(A) Known antigens for positive selection		
CD9 (MRP-1; MIC3)	Synovial membrane	33
CD10 (Neprilysin; CALLA)	Placenta	14
CD26 (DPP4)	Placenta	14
CD34 (Hematopoietic progenitor cell antigen)	Adipose tissue	17, 20, 21, 34–36
CD44 (PGP-1; ECMR-3)	Bone marrow	37
CD49a (Integrin α1)	Bone marrow	38–42
CD49e (Integrin α5)	Bone marrow	43
CD56 (NCAM)	Bone marrow	3, 25, 26
CD63 (MLA1; TSPAN30)	Bone marrow	42
CD73 (NT5E)	Bone marrow	42, 44–46
CD90 (Thy-1)	Adipose tissue	17, 21
	Bone marrow	47
	Synovial membrane	33
	Endometrium	48
CD105 (Endoglin)	Synovial membrane	49
	Bone marrow	37, 45, 46, 50–52
	Cartilage	53
	Endometrium	54
	Wharton's jelly	55
CD106 (VCAM-1)	Bone marrow	56, 57
	Umbilical cord	18
CD117 (c-kit)	Amnionic fluid	58
CD130 (gp130)	Bone marrow	44
CD140b (PDGFRB)	Endometrium	59
CD146 (MCAM)	Bone marrow	44, 60
	Adipose tissue	22, 61
	Endometrium	48, 54, 59
CD166 (ALCAM)	Synovial membrane	33
	Cartilage	53
	Bone marrow	42
	Fetal membranes	11
CD200 (MRC; OX2)	Bone marrow	44
CD271 (LNGFR)	Amnion	11
	Bone marrow	3, 25, 26, 31, 51, 62–65
	Chorion	11
	Adipose tissue	35
CD309 (Flk-1; VEGFR-2)	Bone marrow	66
CD349 (Frizzled-9)	Placenta	14
ALDH	Bone marrow	67
GD2 (Neural ganglioside)	Bone marrow	68
	Umbilical cord	16
HSP90β	Bone marrow	69
Integrin $\alpha_V\beta_5$	Bone marrow	44
TNAP	Bone marrow	3, 25–27

Continued

Table1. *Continued*

Markers used	Tissue	References
(B) Unknown (antibody-defined) antigens for positive selection		
3G5	Adipose tissue	22
D7-FIB	Bone marrow	63
STRO-1	Bone marrow	27, 40, 42, 56, 57
	Adipose tissue	22
W5C5	Bone marrow	70
	Endometrium	71
(C) Known antigens for negative selection		
CD3 (T cell surface glycoprotein)	Peripheral blood	23
CD14 (LPS receptor)	Peripheral blood	23
CD31 (PECAM-1)	Bone marrow	66
	Adipose tissue	17, 20, 21
CD34 (Hematopoietic progenitor cell antigen)	Bone marrow	27, 66
	Peripheral blood	23
CD45 (Leukocyte common antigen)	Bone marrow	40, 43, 52, 63
	Lung	19
	Adipose tissue	21
CD105 (Endoglin)	Adipose tissue	17, 21
CD144 (Cadherin-5)	Adipose tissue	20
CD146 (MCAM)	Adipose tissue	17, 21
CD235a (GlycophorinA)	Bone marrow	40
Lin^- (various antigens)	Bone marrow	30, 51

Figure 1B. The dual fluorescence plot demonstrates that $CD271^{bright}$, but not $CD271^{dim}$, cells give rise to clonogenic MSCs. As revealed by Giemsa staining, these populations differ considerably in their morphological appearance. Thus, $CD271^{bright}$ cells are characterized by a relatively bright nuclear staining and a high cytoplasmic content, whereas $CD271^{dim}$ cells show a lymphoblastoid appearance with darker nuclear staining. Figure 1B additionally shows that CD140b is a more selective marker for MSC isolation than is CD271, as CD140b is expressed on $CD271^{bright}$, but not $CD271^{dim}$, cells. As anticipated, clonogenic cells (CFU-F) were exclusively found in the $CD140b^+$ population. Using this screening approach, additional antibodies with specificity for MSCs have been identified,[3,14,70] some of which are presented in this report.

Prospective isolation of bone marrow–derived MSC subsets

Several groups have reported that MSCs are heterogeneous with respect to their growth and differentiation potential.[32,72,73] However, little information exists about markers that discriminate between developmentally, functionally, and morphologically distinct MSC subsets. Recently, we introduced a monoclonal antibody against CD56 that recognizes a distinct MSC subset with high selectivity.[3] This antibody (39D5) detects a CD56 epitope that is highly expressed on about 0.5–15% of $CD271^{bright}$ cells, but not on NK cells (Fig. 1C). Giemsa staining revealed that $CD56^-$ cells contain a large bright cytoplasm with vacuoles, whereas cells of the $CD56^+$ subset contain a smaller cytoplasm with basophilic granules. Interestingly, $CD56^+$ cells were about two times more clonogenic than $CD56^-$ cells (Fig. 1C). Further analysis has shown that this increased number of clonogenic cells is correlated with an increased proliferation rate.[3] Surface marker expression analysis of sorted $CD56^+$ and $CD56^-$ cells revealed that $CD56^+$ cells, but not $CD56^-$ cells, coexpress CD166, and a subset of $CD56^-$ cells, but not $CD56^+$ cells, express CD349.[3] However, when these cells were placed in culture, both cell types expressed

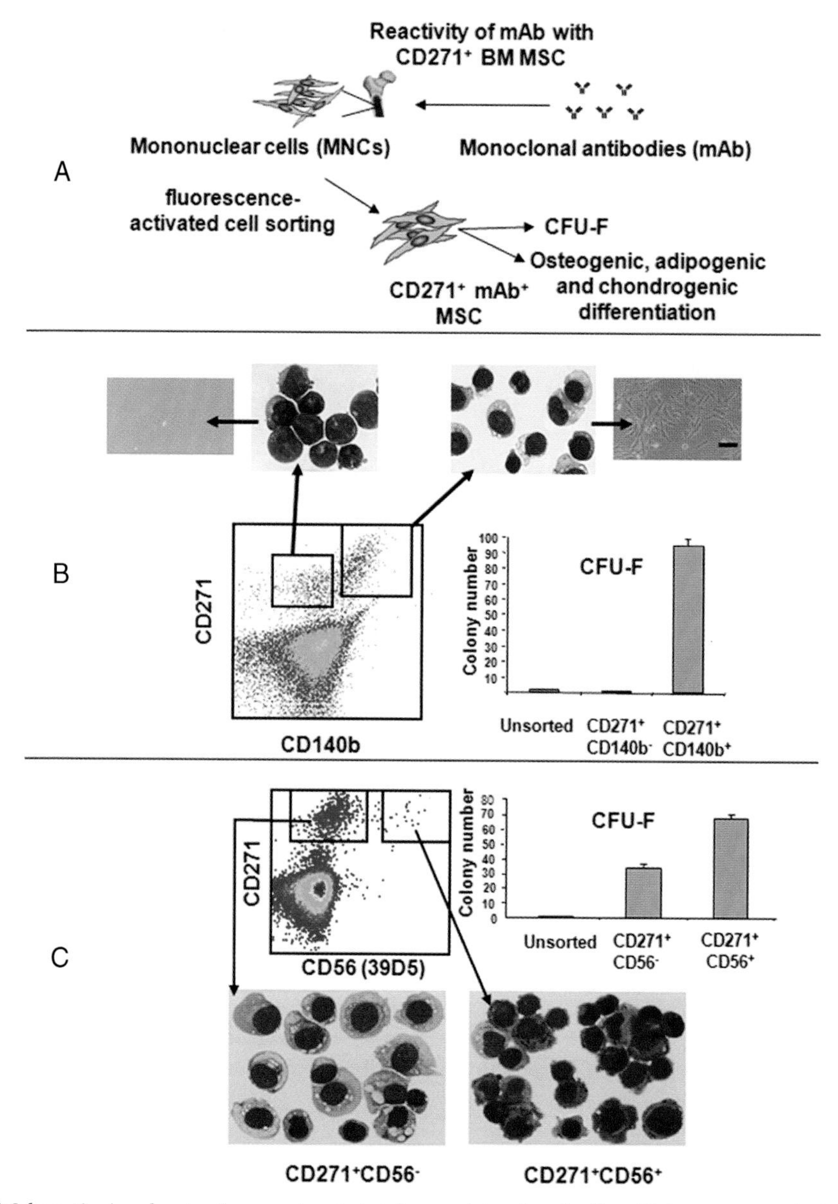

Figure 1. (A) Schematic view showing the procedure to test the reactivity of antibodies with bone marrow–derived MSCs. Only antibodies that react with CD271$^+$ cells are considered to be potential MSC markers. The second prerequisite for an MSC marker is that the cell population recognized by the antibodies gives rise to CFU-F. (B) Example of a suitable MSC marker. CD140b-reactive antibodies selectively recognize CD271$^+$ cells that give rise to CFU-F. CD271$^-$ cells and CD271$^+$ cells not coexpressing CD140b do not give rise to CFU-F. The *y*-axis shows the CFU-F number per 2,000 plated cells. (C) CD56 is a suitable marker to distinguish between two MSC subsets. Both, CD56$^+$ and CD56$^-$ cells give rise to MSCs. CD56$^+$ MSCs have a smaller average size and contain basophilic granules, whereas CD56$^-$ MSCs contain a large cytoplasm with many vacuoles.

high levels of CD166 and moderate levels of CD349 (frizzled-9). Other surface molecules upregulated during culture include the tumor antigen CDCP1 (CD318), which is not expressed on primary MSCs. In contrast, CD271 and CD56 were rapidly downregulated after cells were placed in culture.[74] When cultured cells of either population were induced to differentiate into defined cell lineages, only MSCs derived from the CD56$^-$ population were able to differentiate into adipocytes.[3] In contrast, effective

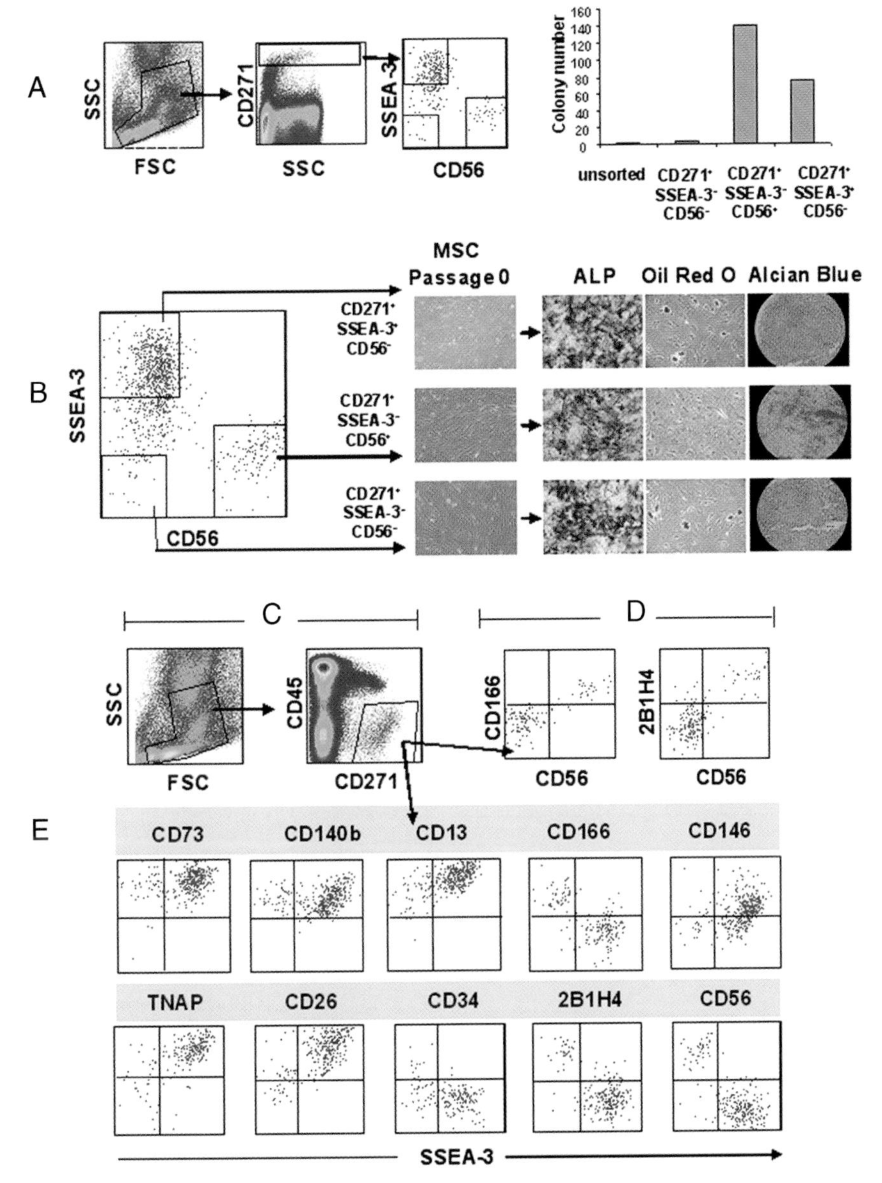

Figure 2. (A) CD56 and SSEA-3 define two distinct MSC subsets. Bone marrow cells stained with antibodies against CD271, CD56, and SSEA-3 were gated on $CD271^{bright}$ cells and sort windows were set as shown in the plot of SSEA-3 versus CD56. After sorting, the ability of cells to give rise to CFU-F was analyzed. The *y*-axis shows the CFU-F number per 2,000 plated cells. MSCs express either CD56 or SSEA-3. A small population is negative for both markers. The frequency of CFU-F in the $CD56^+$ population is about two times higher compared with the SSEA-3^+ population. (B) SSEA-3^+, but not $CD56^+$, MSCs are able to differentiate into Oil Red O-positive adipocytes. In contrast, $CD56^+$, but not SSEA-3^+, MSCs give rise to Alcian Blue-positive chondrocytes. (C) Four-color staining of bone marrow cells with antibodies against CD45 and CD271 and selected markers. Cells are gated on $CD271^+CD45^-$ cells. (D) Display of bone marrow cells gated on the $CD271^+CD45^-$ population and stained with CD56 and selected markers. CD166 and 2B1H4 antigens are selectively expressed on cells of the $CD56^+$ subset. (E) Coexpression analysis of markers on SSEA-3^+ and SSEA-3^- MSCs gated on the $CD271^+CD45^-$ population. Note that the ecto-enzymes TNAP and CD26, as well as the adhesion molecule CD146, are preferentially expressed on SSEA-3^+ cells. In contrast, SSEA-3^+ cells are negative for CD56, CD166, and the 2B1H4 antigen.

Table 2. Differential expression of surface markers on primary and cultured bone marrow–derived MSCs

Primary MSC		
$CD56^+$	$CD56^-$	Cultured MSC (passage 2)
$CD10^-$	$CD10^+$	$CD10^+$
$CD13^+$	$CD13^+$	$CD13^+$
$CD26^-$	$CD26^+$	$CD26^+$
$CD34^-$ (subp.$^+$)	$CD34^-$	$CD34^-$ (some clones$^+$)
$CD49a^+$	$CD49a^+$	$CD49a^+$
$CD49b^-$	$CD49b^-$	$CD49b^-$
$CD56^+$	$CD56^-$	$CD56^-$
$CD90^+$	$CD90^+$	$CD90^+$
$CD105^{dim}$	$CD105^{bright}$	$CD105^+$
$CD133^-$	$CD133^-$	$CD133^-$
$CD140b^+$	$CD140b^+$	$CD140b^+$
$CD146^{-/dim}$	$CD146^+$	$CD146^+$
$CD166^+$	$CD166^-$	$CD166^+$
$CD318^-$	$CD318^-$	$CD318^+$
$CD271^+$	$CD271^+$	$CD271^{-/dim}$
$TNAP^{-/dim}$	$TNAP^{bright}$	$TNAP^{+/-}$
$SSEA3^-$	$SSEA3^+$	$SSEA3^-$
$W5C5^+$	$W5C5^+$	$W5C5^+$
$2B1H4^+$	$2B1H4^-$	$2B1H4^+$

chondrocyte differentiation was induced only in MSCs derived from the $CD56^+$ subset, suggesting that this subpopulation is the preferred source for therapeutic approaches in the field of cartilage repair and spinal disk injuries.[3]

We have previously shown that $CD56^+$ MSCs express low levels of TNAP, whose expression is upregulated during osteogenic differentiation.[3] In a model proposed by Gronthos *et al.*, cell surface expression of TNAP, absent on early STRO-1^+ stem cells, is upregulated during osteogenic differentiation.[75] The STRO-$1^+$$TNAP^-$ population may correspond to the recently described $CD56^+TNAP^{-/dim}$ subset identified by our group. In agreement with this hypothesis, cells of the $CD56^+$ subset mature at a later time into osteoblasts, compared with $CD56^-$ cells. We therefore propose an extended model in which STRO-$1^+CD56^+TNAP^{-/dim}$ MSCs represent an immature precursor with multilineage differentiation capacity. Cells committed to the chondrocyte lineage diverge at very early ($CD56^+$) stages of MSC differentiation. This chondrogenic potential, which is rapidly lost upon differentiation into $TNAP^+CD56^-$ cells, is accompanied by the induction of the adipogenic differentiation potential.

Several reports underline the important role of CD56 expression on fibroblasts to support the growth of HSCs.[76–78] The contribution of CD56 was initially described from interactions between mouse and monkey cells,[76,77] but a more recent report showed that CD56 expressed on a mouse stromal line plays a crucial role in supporting human hematopoiesis.[78] The authors showed that coculture of $CD34^+CD38^-$ cord blood cells with a $CD56^+$ stromal cell line resulted in a significantly greater expansion rate of $CD34^+$ hematopoietic cells compared with stromal cells, which did not express CD56. This enhancing effect could be blocked by the addition of an inhibitory CD56 antibody, suggesting that direct interactions between CD56 molecules from different cells are essential. It remains open whether CD56 on human MSCs plays a similar hematopoiesis-supporting activity. As human HSCs do not express CD56, it is unlikely that a potential supporting effect is caused by homotypic interactions between CD56 molecules on stromal and hematopoietic cells. Rather, $CD56^+$ stromal cells may interact with extracellular matrix components such as heparin sulphate or chondroitin sulphate proteoglycans.

We have recently shown that SSEA-3, but not SSEA-4, TRA-1–60, or TRA-1–81, is a candidate marker for MSCs derived from primary femur bone marrow.[25] In Figure 2A, bone marrow cells were stained with antibodies against CD271, SSEA-3, and CD56 and gated on $CD271^{bright}$SSEA-3^-CD56^-, $CD271^{bright}$SSEA-3^+CD56^-, and $CD271^{bright}$SSEA-3^-CD56^+ cells. The clonogenic potential and the differentiation capacity of the sorted populations were then determined by CFU-F assay and appropriate differentiation protocols. In this setting, clonogenic cells were about 44-fold enriched for CFU-F in the $CD271^{bright}$SSEA-3^+CD56^- population, 83-fold in the $CD271^{bright}$SSEA-3^-CD56^+ population, but only about twofold in the $CD271^{bright}$SSEA-3^-CD56^- fraction. Not surprisingly, $CD271^{bright}$SSEA-3^+CD56^- MSCs gave rise to osteoblasts and adipocytes but not to chondrocytes (Fig. 2B). In contrast, $CD271^{bright}$SSEA-3^-CD56^+ cells were able to differentiate into chondrocytes but not adipocytes. Collectively, SSEA-3 and CD56 are expressed on distinct MSC subsets that differ in their proliferation and differentiation capacity.

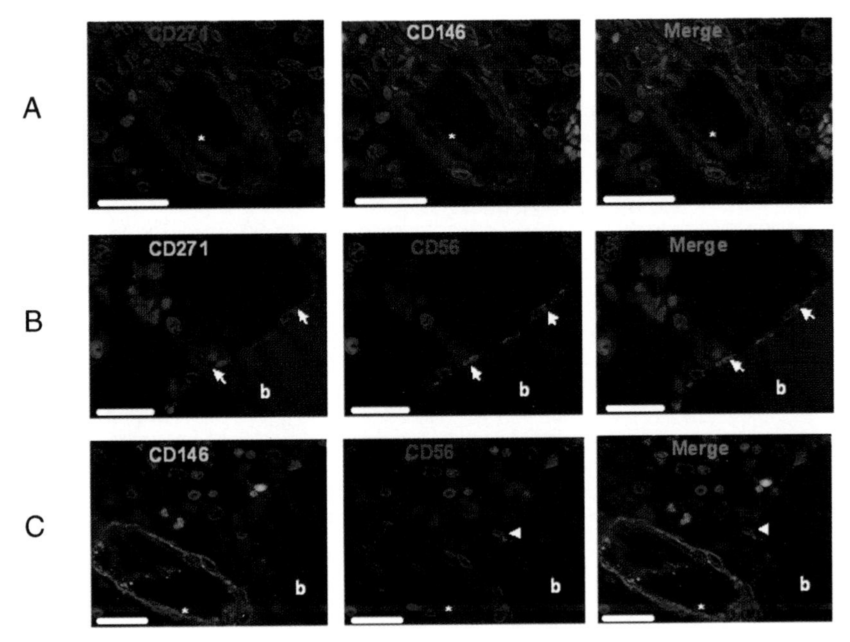

Figure 3. (A) CD271 and CD146 are coexpressed on perivascular MSCs. (B) CD271 and CD56 are coexpressed on bone-lining MSCs. (C) Perivascular MSCs (marked by *) are negative for CD56, whereas bone lining MSCs (marked by ↑) are negative for CD146. Scale bars: 20 μm.

Whereas SSEA-3 is a suitable and selective marker for the isolation of adipocyte precursors, CD56 is the more appropriate marker for isolating chondrocyte precursors.

Many of the tested markers in both subsets are either up- or downregulated during culture. Whereas CD271 is expressed at high levels in all primary MSC subsets, and CD56 and SSEA-3 are expressed in the respective subsets, these antigens are rapidly downregulated during culture (Table 2). CD166 is expressed at low levels on primary $CD56^+$ MSCs but upregulated to high levels on all cultured MSCs. Finally, CD109 and CD318 are negative on primary MSCs but highly expressed on cultured MSCs. These data suggest that the conventional definition of MSC-reactive surface markers, which is based on cultured cells, may be revised and specified.

The two identified MSC subsets not only differ in their exclusive expression of either SSEA-3 or CD56 but also in several additional surface markers. Thus, as revealed by four-color immunofluorescence analysis, $CD56^+$, but not SSEA-3^+, MSCs coexpress CD166 and the recently identified antibody-defined cell surface antigen 2B1H4 (Figs. 2C–2E). In contrast, SSEA-3^+, but not $CD56^+$, MSCs coexpress the adhesion molecule CD146 as well as the ecto-enzymes CD10 (not shown), CD26, and high levels of TNAP. A few cells in the $CD56^+$ and SSEA-3^+CD146^+ populations are positive for CD34, a marker generally regarded to be negative for bone marrow–derived MSCs. However, cloning of MSCs by single cell sorting revealed that these rare MSCs indeed exist and that CD34 expression remains stable during culture.[3] Sorting of single cells into culture plates not only provide information about the growth characteristics of individual MSC clones, it also provides information about the frequency of MSC clones with defined differentiation potential. Pittenger *et al.* described that almost 100% of colonies derived from single bone marrow cells underwent osteogenic differentiation, about 80% of the colonies revealed adipogenic differentiation potential, and that only 30% of the colonies showed chondrogenic differentiation potential.[13] Our group was successful in isolating clones with the capacity for osteoblast but not adipocyte differentiation, as well as for adipocyte but not osteoblast differentiation.[3] Further experiments are required to determine the frequency of MSCs with multipotent differentiation capacity and those with restricted differentiation potential. These analyses may

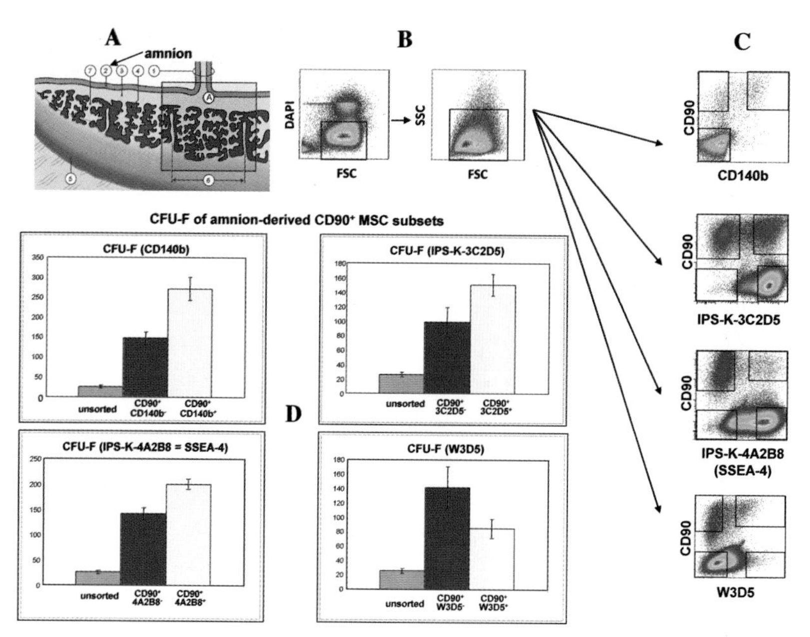

Figure 4. (A) Schematic view of placenta. Note that the amniotic sac and its filling provide a liquid that surrounds and cushions the fetus. (B) Before analysis of marker expression, enzymatically isolated cells were stained with CD90 and selected markers and gated on live (DAPI-negative) cells and by dual scatter criteria. (C) The markers shown here are suitable to distinguish between $CD90^+$ MSC subsets and between $CD90^-$ epithelial cell subsets. Note that only $CD90^+$ cells give rise to CFU-F. (D) CD140b, SSEA-4, and IPS-K-3C2D5 antigen are suitable markers to enrich CFU-F. In contrast, $CD90^+W3D5^+$ cells contain less CFU-F than $CD90^+W3D5^-$ cells. The *y*-axis shows the CFU-F number per 2,000 plated cells.

contribute to the development of customized complex models of MSC maturation and differentiation, similar to those proposed for cells of the hematopoietic system.

In a recent publication, Tormin *et al.* described the localization of two MSC subsets in distinct areas of the bone marrow: dominating perivascular MSCs, which coexpress CD271 and CD146, and bone-lining MSCs, which express CD271 but not CD146.[79] This prompted us to analyze whether the bone-lining MSCs correspond to the $CD56^+$ MSC subset (described above) and whether the perivascular $CD146^+$ MSCs lack CD56 expression. We not only confirmed that MSCs in perivascular regions coexpress CD271 and CD146, we also showed that $CD271^+$ bone-lining MSCs are negative for CD146 and positive for CD56 (Figs. 3A–3C). This suggests that, apart from the distinct surface antigen expression profile, morphology, and differentiation potential, bone-lining MSCs may also have distinct functional—yet to be identified—properties. As the niches of $CD34^+$ HSCs are supposed to be localized near the bone, it is intriguing to speculate that $CD56^+$, but not $CD56^-$, MSCs contribute to the stromal niche of $CD34^+$ HSCs.

Prospective isolation of amnion-derived MSC subsets

Four regions of the placenta can be distinguished: the amniotic epithelial, amniotic mesenchymal, chorionic mesenchymal, and chorionic trophoblastic.[80] Of these, the amniotic membrane (Fig. 4A) is the preferred source of MSC because it can be easily removed and distinguished from other placental tissue, and because it excludes the potential contamination of maternal MSCs. In the past, epithelial and mesenchymal stem cells have been separated by enzymatic procedures and by applying selective culture protocols.[81,82] As a consequence, the separation of mesenchymal and epithelial cells was incomplete and characterization solely relied on cultured cells.

To gain information on the phenotypic profile of primary amniotic cells directly after preparation of single cell suspensions, cells were stained with a large panel of antibodies against defined cell surface antigens and analyzed by flow cytometry. After multicolor staining with many antibody combinations, CD90 was found to be the best marker to separate MSCs from CD324(E-cadherin)$^+$CD326(EpCAM)$^+$ epithelial cells because of its pronounced selectivity for MSCs, and because of its high expression level on these cells. In subsequent approaches, amnion cells were screened by two-color fluorescence for their reactivity with $CD90^+$ and $CD90^-$ cells. Antibodies that divided the respective populations in at least two subsets were further examined and used for cell sorting. After exclusion of dead cells and dual scatter gating (Fig. 4B), sort windows were set as shown in Figure 4C. After sorting, cells were grown in culture to determine the CFU-F capacity and the phenotypic profile at passage 2 of culture. $CD90^-$ epithelial cells did not give rise to colonies and therefore were not analyzed for clonogenic capacity. Figure 4D shows that three of the four selected antibodies, namely the CD140b-reactive antibody 28D4, the SSEA-4-reactive antibody IPS-K-4A2B8, and the antibody with orphan specificity IPS-K-3C2D5, were suitable to enrich for clonogenic MSCs. Of these, CD140b appeared to be the best target. Indeed, CFU-F was almost two times enriched in the $CD90^+CD140b^+$ population compared with the $CD90^+CD140b^-$ subset. In contrast, the $CD90^+W3D5^+$ subset contained less CFU-F than did the $CD90^+W3D5^-$ population.

Due to their niche-forming and immunomodulatory properties, epithelial and mesenchymal cells from the amniotic membrane are not only discussed as a potent niche source for HSCs but also as attractive cells for applications in the field of regenerative medicine,[83–89] applications that have been used in many preclinical studies in animal models, including hepatic regeneration, cardiac repair, and the treatment of neurological disorders.[87,89–92] Future studies may focus on the identification of MSC or epithelial subsets that contain the most appropriate cells for clinical applications. As placenta tissue is readily available and does not elicit ethical debate, it may be in some cases the preferred source for cell replacement therapy. However, bone marrow–derived MSCs are superior in autologous settings and for the cure of spinal disk injuries because of the lack of graft rejection and the better chondrogenic differentiation potential.

Conflicts of interest

The authors declare no conflicts of interest.

References

1. Friedenstein, A.J., U. Deriglasova & N. Kulagina. 1974. Precursors for fibroblasts in different populations of hematopoietic cells as detected by the in vitro colony assay method. *Exp. Hematol.* **2:** 83–92.
2. Friedenstein, A.J., J. Gorskaja & N. Kulagina. 1976. Fibroblast precursors in normal and irradiated mouse hematopoietic organs. *Exp. Hematol.* **4:** 267–274.
3. Battula, V.L., S. Treml, P.M. Bareiss, *et al.* 2009. Isolation of functionally distinct mesenchymal stem cell subsets using antibodies against CD56, CD271, and mesenchymal stem cell antigen-1. *Haematologica* **94:** 173–184.
4. Maijenburg, M.W., M. Kleijer, K. Vermeul, *et al.* 2011. Primary bone marrow-derived MSC subsets have distinct Wnt-signatures compared to conventionally cultured MSC and differ in their capacity to support hematopoiesis in vitro. In *Proceedings of the 53rd Annual Meeting of the American Society of Hematology*, San Diego, CA.
5. Maijenburg, M.W., M. Kleijer, K. Vermeul, *et al.* 2012. The composition of the mesenchymal stromal cell compartment in human bone marrow changes during development and aging. *Haematologica* **97:** 179–183.
6. Nauta, A.J. & W.E. Fibbe. 2007. Immunomodulatory properties of mesenchymal stromal cells. *Blood* **110:** 3499–3506.
7. Amado, L.C., A.P. Saliaris, K.H. Schuleri, *et al.* 2005. Cardiac repair with intramyocardial injection of allogeneic mesenchymal stem cells after myocardial infarction. *Proc. Natl. Acad. Sci.* USA **102:** 11474–11479.
8. Rojas, M., J. Xu, C.R. Woods, *et al.* 2005. Bone marrow-derived mesenchymal stem cells in repair of the injured lung. *Am. J. Respir. Cell Mol. Biol.* **33:** 145–152.
9. Maitra, B., E. Szekely, K. Gjini, *et al.* 2004. Human mesenchymal stem cells support unrelated donor hematopoietic stem cells and suppress T-cell activation. *Bone Marrow Transplant.* **33:** 597–604.
10. Parr, A.M., C.H. Tator & A. Keating. 2007. Bone marrow-derived mesenchymal stromal cells for the repair of central nervous system injury. *Bone Marrow Transplant.* **40:** 609–619.
11. Soncini, M., E. Vertua, L. Gibelli, *et al.* 2007. Isolation and characterization of mesenchymal cells from human fetal membranes. *J. Tissue Eng. Regen. Med.* **1:** 296–305.
12. Schallmoser, K., C. Bartmann, E. Rohde, *et al.* 2007. Human platelet lysate can replace fetal bovine serum for clinical-scale expansion of functional mesenchymal stromal cells. *Transfusion* **47:** 1436–1446.
13. Pittenger, M.F., A.M. Mackay, S.C. Beck, *et al.* 1999. Multilineage potential of adult human mesenchymal stem cells. *Science* **284:** 143–147.
14. Battula, V.L., S. Treml, H. Abele & H.J. Buhring. 2008. Prospective isolation and characterization of mesenchymal stem cells from human placenta using a frizzled-9-specific monoclonal antibody. *Differentiation* **76:** 326–336.
15. Hu, Y., L. Liao, O. Wang, *et al.* 2003. Isolation and identification of mesenchymal stem cells from human fetal pancreas. *J. Lab Clin. Med.* **141:** 342–349.
16. Jin, H.J., Y. Nam, Y. Bae, *et al.* 2010. GD2 expression is closely associated with neuronal differentiation of human umbilical cord blood-derived mesenchymal stem cells. *Cell Mol. Life Sci.* **67:** 1845–1858.
17. Lin, K., Y. Matsubara, Y. Masuda, *et al.* 2008. Characterization of adipose tissue-derived cells isolated with the Celution system. *Cytotherapy* **10:** 417–426.

18. Lu, L., Y. Liu, S. Yang, *et al.* 2006. Isolation and characterization of human umbilical cord mesenchymal stem cells with hematopoiesis-supportive function and other potentials. *Haematologica* **91:** 1017–1026.
19. Martin, J., K. Helm, P. Ruegg, *et al.* 2008. Adult lung side population cells have mesenchymal stem cell potential. *Cytotherapy* **10:** 120–151.
20. Traktuev, D., S. Merfeld-Clauss, J. Li, *et al.* 2008. A population of multipotent CD34-positive adipose stromal cells share pericyte and mesenchymal surface markers, reside in a periendothelial location, and stabilize endothelial networks. *Circ. Res.* **102:** 77–85.
21. Yoshimura, K., T. Shigeura, D. Matsumoto, *et al.* 2006. Characterization of freshly isolated and cultured cells derived from the fatty and fluid portions of liposuction aspirates. *J. Cell Physiol.* **208:** 64–76.
22. Zannettino, A., S. Paton, A. Arthur, *et al.* 2008. Multipotential human adipose-derived stromal stem cells exhibit a perivascular phenotype in vitro and in vivo. *J. Cell Physiol.* **214:** 413–421.
23. Zvaifler, N., L. Marinova-Mutafchieva, G. Adams, *et al.* 2000. Mesenchymal precursor cells in the blood of normal individuals. *Arthritis Res.* **2:** 477–488.
24. Yen, B.L., H.I. Huang, C.C. Chien, *et al.* 2005. Isolation of multipotent cells from human term placenta. *Stem Cells* **23:** 3–9.
25. Sobiesiak, M., K. Sivasubramaniyan, C. Hermann, *et al.* 2010. The mesenchymal stem cell antigen MSCA-1 is identical to tissue non-specific alkaline phosphatase. *Stem Cells Dev.* **19:** 669–677.
26. Buhring, H.J., S. Treml, F. Cerabona, *et al.* 2009. Phenotypic characterization of distinct human bone marrow-derived MSC subsets. *Ann. N. Y. Acad. Sci.* **1176:** 124–134.
27. Gronthos, S., S. Fitter, P. Diamond, *et al.* 2007. A novel monoclonal antibody (STRO-3) identifies an isoform of tissue nonspecific alkaline phosphatase expressed by multipotent bone marrow stromal stem cells. *Stem Cells Dev.* **16:** 953–963.
28. Bernardo, M.E., J.A. Emons, M. Karperien, *et al.* 2007. Human mesenchymal stem cells derived from bone marrow display a better chondrogenic differentiation compared with other sources. *Connect. Tissue Res.* **48:** 132–140.
29. Battula, V.L., P.M. Bareiss, S. Treml, *et al.* 2007. Human placenta and bone marrow derived MSC cultured in serum-free, b-FGF-containing medium express cell surface frizzled-9 and SSEA-4 and give rise to multilineage differentiation. *Differentiation* **75:** 279–291.
30. Muller, I., S. Kordowich, C. Holzwarth, *et al.* 2006. Animal serum-free culture conditions for isolation and expansion of multipotent mesenchymal stromal cells from human BM. *Cytotherapy.* **8:** 437–444.
31. Bieback, K., A. Hecker, A. Kocaomer, *et al.* 2009. Human alternatives to fetal bovine serum for the expansion of mesenchymal stromal cells from bone marrow. *Stem Cells* **27:** 2331–2341.
32. Vogel, W., F. Grunebach, C.A. Messam, *et al.* 2003. Heterogeneity among human bone marrow-derived mesenchymal stem cells and neural progenitor cells. *Haematologica* **88:** 126–133.
33. Fickert, S., J. Fiedler & R.E. Brenner. 2003. Identification, quantification and isolation of mesenchymal progenitor cells from osteoarthritic synovium by fluorescence automated cell sorting. *Osteoarthritis Cartilage* **11:** 790–800.
34. Varma, M., R. Breuls, T. Schouten, *et al.* 2007. Phenotypical and functional characterization of freshly isolated adipose tissue-derived stem cells. *Stem Cells Dev.* **16:** 91–104.
35. Quirici, N., C. Scavullo, L. de Girolamo, *et al.* 2010. Anti-L-NGFR and -CD34 monoclonal antibodies identify multipotent mesenchymal stem cells in human adipose tissue. *Stem Cells Dev.* **19:** 915–925.
36. Mitchell, J., K. McIntosh, S. Zvonic, *et al.* 2006. Immunophenotype of human adipose-derived cells: temporal changes in stromal-associated and stem cell-associated markers. *Stem Cells* **24:** 376–385.
37. Martins, A., A. Paiva, J. Morgado, *et al.* 2009. Quantification and immunophenotypic characterization of bone marrow and umbilical cord blood mesenchymal stem cells by multicolor flow cytometry. *Transplant. Proc.* **41:** 943–946.
38. Deschaseaux, F., F. Gindraux, R. Saadi, *et al.* 2003. Direct selection of human bone marrow mesenchymal stem cells using an anti-CD49a antibody reveals their CD45med, low phenotype. *Br. J. Haematol.* **122:** 506–517.
39. Gindraux, F., Z. Selmani, L. Obert, *et al.* 2007. Human and rodent bone marrow mesenchymal stem cells that express primitive stem cell markers can be directly enriched by using the CD49a molecule. *Cell Tissue Res.* **327:** 471–483.
40. Letchford, J., A. Cardwell, K. Stewart, *et al.* 2006. Isolation of C15: a novel antibody generated by phage display against mesenchymal stem cell-enriched fractions of adult human marrow. *J. Immunol. Methods* **308:** 124–137.
41. Rider, D., T. Nalathamby, V. Nurcombe & S. Cool. 2007. Selection using the alpha-1 integrin (CD49a) enhances the multipotentiality of the mesenchymal stem cell population from heterogeneous bone marrow stromal cells. *J. Mol. Histol.* **38:** 449–458.
42. Stewart, K., P. Monk, S. Walsh, *et al.* 2003. STRO-1, HOP-26 (CD63), CD49a and SB-10 (CD166) as markers of primitive human marrow stromal cells and their more differentiated progeny: a comparative investigation in vitro. *Cell Tissue Res.* **313:** 281–290.
43. Baksh, D., P. Zandstra & J. Davies. 2007. A non-contact suspension culture approach to the culture of osteogenic cells derived from a CD49elow subpopulation of human bone marrow-derived cells. *Biotechnol. Bioeng.* **98:** 1195–1208.
44. Delorme, B., J. Ringe, N. Gallay, *et al.* 2008. Specific plasma membrane protein phenotype of culture-amplified and native human bone marrow mesenchymal stem cells. *Blood* **111:** 2631–2635.
45. Liu, P., D. Zhou & T. Shen. 2005. Identification of human bone marrow mesenchymal stem cells: preparation and utilization of two monoclonal antibodies against SH2, SH3. *Zhongguo Shi Yan. Xue. Ye. Xue. Za Zhi.* **13:** 656–659.
46. Odabas, S., F. Sayar, G. Guven, *et al.* 2008. Separation of mesenchymal stem cells with magnetic nanosorbents carrying CD105 and CD73 antibodies in flow-through and batch systems. *J. Chromatogr. B Analyt. Technol. Biomed. Life Sci.* **861:** 74–80.

47. Campioni, D., F. Lanza, S. Moretti, *et al.* 2008. Loss of Thy-1 (CD90) antigen expression on mesenchymal stromal cells from hematologic malignancies is induced by in vitro angiogenic stimuli and is associated with peculiar functional and phenotypic characteristics. *Cytotherapy* **10:** 69–82.
48. Schwab, K., P. Hutchinson & C. Gargett. 2008. Identification of surface markers for prospective isolation of human endometrial stromal colony-forming cells. *Hum. Reprod.* **23:** 934–943.
49. Arufe, M., A. De la Fuente, I. Fuentes-Boquete, *et al.* 2009. Differentiation of synovial CD-105(+) human mesenchymal stem cells into chondrocyte-like cells through spheroid formation. *J. Cell Biochem.* **108:** 145–155.
50. Aslan, H., Y. Zilberman, L. Kandel, *et al.* 2006. Osteogenic differentiation of noncultured immunoisolated bone marrow-derived CD105+ cells. *Stem Cells* **24:** 1728–1737.
51. Jarocha, D., E. Lukasiewicz & M. Majka. 2008. Adventage of mesenchymal stem cells (MSC) expansion directly from purified bone marrow CD105+ and CD271+ cells. *Folia Histochem. Cytobiol.* **46:** 307–314.
52. Kastrinaki, M.C., I. Andreakou, P. Charbord & H. Papadaki. 2008. Isolation of human bone marrow mesenchymal stem cells using different membrane markers: comparison of colony/cloning efficiency, differentiation potential, and molecular profile. *Tissue Eng. Part C. Methods* **14:** 333–339.
53. Alsalameh, S., R. Amin, T. Gemba & M. Lotz. 2004. Identification of mesenchymal progenitor cells in normal and osteoarthritic human articular cartilage. *Arthritis Rheum.* **50:** 1522–1532.
54. Tsuji, S., M. Yoshimoto, K. Takahashi, *et al.* 2008. Side population cells contribute to the genesis of human endometrium. *Fertil. Steril.* **90:** 1528–1537.
55. Conconi, M., P. Burra, R. Di Liddo, *et al.* 2006. CD105(+) cells from Wharton's jelly show in vitro and in vivo myogenic differentiative potential. *Int. J. Mol. Med.* **18:** 1089–1096.
56. Gronthos, S., A. Zannettino, S. Hay, *et al.* 2003. Molecular and cellular characterisation of highly purified stromal stem cells derived from human bone marrow. *J. Cell Sci.* **116:** 1827–1835.
57. Gronthos, S. & A. Zannettino. 2008. A method to isolate and purify human bone marrow stromal stem cells. *Methods Mol. Biol.* **449:** 45–57.
58. De Coppi, P., G. Bartsch, Jr., M.M. Siddiqui, *et al.* 2007. Isolation of amniotic stem cell lines with potential for therapy. *Nat. Biotechnol.* **25:** 100–106.
59. Schwab, K. & C. Gargett. 2007. Co-expression of two perivascular cell markers isolates mesenchymal stem-like cells from human endometrium. *Hum. Reprod.* **22:** 2903–2911.
60. Sorrentino, A., M. Ferracin, G. Castelli, *et al.* 2008. Isolation and characterization of CD146+ multipotent mesenchymal stromal cells. *Exp. Hematol.* **36:** 1035–1046.
61. Astori, G., F. Vignati, S. Bardelli, *et al.* 2007. "In vitro" and multicolor phenotypic characterization of cell subpopulations identified in fresh human adipose tissue stromal vascular fraction and in the derived mesenchymal stem cells. *J. Transl. Med.* **5:** 55.
62. Horn, P., S. Bork, A. Diehlmann, *et al.* 2008. Isolation of human mesenchymal stromal cells is more efficient by red blood cell lysis. *Cytotherapy* **10:** 676–685.
63. Jones, E., A. English, S. Kinsey, *et al.* 2006. Optimization of a flow cytometry-based protocol for detection and phenotypic characterization of multipotent mesenchymal stromal cells from human bone marrow. *Cytometry B Clin. Cytom.* **70:** 391–399.
64. Poloni, A., G. Maurizi, V. Rosini, *et al.* 2009. Selection of CD271(+) cells and human AB serum allows a large expansion of mesenchymal stromal cells from human bone marrow. *Cytotherapy* **11:** 153–162.
65. Quirici, N., D. Soligo, P. Bossolasco, *et al.* 2002. Isolation of bone marrow mesenchymal stem cells by anti-nerve growth factor receptor antibodies. *Exp. Hematol.* **30:** 783–791.
66. Liu, L., Z. Sun, B. Chen, *et al.* 2006. Ex vivo expansion and in vivo infusion of bone marrow-derived Flk-1+CD31-CD34- mesenchymal stem cells: feasibility and safety from monkey to human. *Stem Cells Dev.* **15:** 349–357.
67. Gentry, T., S. Foster, L. Winstead, *et al.* 2007. Simultaneous isolation of human BM hematopoietic, endothelial and mesenchymal progenitor cells by flow sorting based on aldehyde dehydrogenase activity: implications for cell therapy. *Cytotherapy* **9:** 259–274.
68. Martinez, C., T. Hofmann, R. Marino, *et al.* 2007. Human bone marrow mesenchymal stromal cells express the neural ganglioside GD2: a novel surface marker for the identification of MSCs. *Blood* **109:** 4245–4248.
69. Gronthos, S., R. McCarty, K. Mrozik, *et al.* 2009. Heat shock protein-90 beta is expressed at the surface of multipotential mesenchymal precursor cells: generation of a novel monoclonal antibody, STRO-4, with specificity for mesenchymal precursor cells from human and ovine tissues. *Stem Cells Dev.* **18:** 1253–1262.
70. Buhring, H.J., V.L. Battula, S. Treml, *et al.* 2007. Novel markers for the prospective isolation of human MSC. *Ann. N. Y. Acad.Sci.* **1106:** 262–271.
71. Masuda, H., S.S. Anwar, H.J. Buhring, *et al.* 2012. A novel marker of human endometrial mesenchymal stem-like cells. *Cell Transplant.* [Epub ahead of print] PMID: 22469435.
72. Krinner, A., M. Hoffmann, M. Loeffler, *et al.* 2010. Individual fates of mesenchymal stem cells in vitro. *BMC. Syst. Biol.* **4:** 73.
73. Lin, C.S., Z.C. Xin, C.H. Deng, *et al.* 2010. Defining adipose tissue-derived stem cells in tissue and in culture. *Histol. Histopathol.* **25:** 807–815.
74. Harichandan, A. & H.J. Buhring. 2011. Prospective isolation of human MSC. *Best. Pract. Res. Clin. Haematol.* **24:** 25–36.
75. Gronthos, S., A.C. Zannettino, S.E. Graves, *et al.* 1999. Differential cell surface expression of the STRO-1 and alkaline phosphatase antigens on discrete developmental stages in primary cultures of human bone cells. *J. Bone Miner. Res.* **14:** 47–56.
76. Wang, X., H. Hisha, S. Taketani, *et al.* 2005. Neural cell adhesion molecule contributes to hemopoiesis-supporting capacity of stromal cell lines. *Stem Cells* **23:** 1389–1399.
77. Kato, J., H. Hisha, X.L. Wang, *et al.* 2008. Contribution of neural cell adhesion molecule (NCAM) to hemopoietic system in monkeys. *Ann. Hematol.* **87:** 797–807.

78. Wang, X., H. Hisha, T. Mizokami, *et al.* 2010. Mouse mesenchymal stem cells can support human hematopoiesis both in vitro and in vivo: the crucial role of neural cell adhesion molecule. *Haematologica* **95:** 884–891.
79. Tormin, A., O. Li, J.C. Brune, *et al.* 2011. CD146 expression on primary non-hematopoietic bone marrow stem cells correlates to in situ localization. *Blood.* **19:** 5067–5077.
80. Parolini, O., F. Alviano, G.P. Bagnara, *et al.* 2008. Concise review: isolation and characterization of cells from human term placenta: outcome of the first international Workshop on Placenta Derived Stem Cells. *Stem Cells* **26:** 300–311.
81. Marongiu, F., R. Gramignoli, Q. Sun, *et al.* 2010. Isolation of amniotic mesenchymal stem cells. *Curr. Protoc. Stem Cell Biol.* **12:** 1E.5.1–1E.5.11.
82. Miki, T., F. Marongiu, K. Dorko, *et al.* 2010. Isolation of amniotic epithelial stem cells. *Curr. Protoc. Stem Cell Biol.* **3:** 1E.3.1–1E.3.9.
83. Subrahmanyam, M. 1995. Amniotic membrane as a cover for microskin grafts. *Br. J. Plast. Surg.* **48:** 477–478.
84. Diaz-Prado, S., E. Muinos-Lopez, T. Hermida-Gomez, *et al.* 2011. Human amniotic membrane as an alternative source of stem cells for regenerative medicine. *Differentiation* **81:** 162–171.
85. Parolini, O., M. Soncini, M. Evangelista & D. Schmidt. 2009. Amniotic membrane and amniotic fluid-derived cells: potential tools for regenerative medicine? *Regen. Med.* **4:** 275–291.
86. Toda, A., M. Okabe, T. Yoshida & T. Nikaido. 2007. The potential of amniotic membrane/amnion-derived cells for regeneration of various tissues. *J. Pharmacol. Sci.* **105:** 215–228.
87. Tsuji, H., S. Miyoshi, Y. Ikegami, *et al.* 2010. Xenografted human amniotic membrane-derived mesenchymal stem cells are immunologically tolerated and transdifferentiated into cardiomyocytes. *Circ. Res.* **106:** 1613–1623.
88. Gomes, J.A., A. Romano, M.S. Santos & H.S. Dua. 2005. Amniotic membrane use in ophthalmology. *Curr. Opin. Ophthalmol.* **16:** 233–240.
89. Sakuragawa, N., H. Misawa, K. Ohsugi, *et al.* 1997. Evidence for active acetylcholine metabolism in human amniotic epithelial cells: applicable to intracerebral allografting for neurologic disease. *Neurosci. Lett.* **232:** 53–56.
90. Elwan, M.A. & N. Sakuragawa. 1997. Evidence for synthesis and release of catecholamines by human amniotic epithelial cells. *Neuroreport* **8:** 3435–3438.
91. Zhao, P., H. Ise, M. Hongo, *et al.* 2005. Human amniotic mesenchymal cells have some characteristics of cardiomyocytes. *Transplantation* **79:** 528–535.
92. Strom, S.C., P. Bruzzone, H. Cai, *et al.* 2006. Hepatocyte transplantation: clinical experience and potential for future use. *Cell Transplant.* **15**(Suppl. 1): S105–S110.

Ann. N.Y. Acad. Sci. ISSN 0077-8923

ANNALS OF THE NEW YORK ACADEMY OF SCIENCES
Issue: *Hematopoietic Stem Cells VIII*

Safety and efficacy of mesenchymal stromal cell therapy in autoimmune disorders

Maria Ester Bernardo[1] and Willem E. Fibbe[2]

[1]Department of Pediatric Hematology–Oncology, IRCCS Bambino Gesù Children's Hospital, Rome, Italy. [2]Department of Immunohematology and Blood Transfusion, Center for Stem Cell Therapy, Leiden University Medical Center, Leiden, the Netherlands

Address for correspondence: Willem E. Fibbe, M.D., Ph.D., Department of Immunohematology and Blood Transfusion, Center for Stem Cell Therapy, E3-Q, Leiden University Medical Center, Albinusdreef 2, 2333 ZA Leiden, the Netherlands. W.E.Fibbe@lumc.nl

Mesenchymal stromal cells (MSCs) are being employed in clinical trials to facilitate engraftment and to treat steroid-resistant acute graft-versus-host disease after hematopoietic stem cell transplantation, as well as to repair tissue damage in inflammatory/degenerative disorders, in particular, in inflammatory bowel diseases (IBDs). When entering the clinical arena, a few potential risks of MSC therapy have to be taken into account: (i) immunogenicity of the cells, (ii) biosafety of medium components, (iii) risk of ectopic tissue formation, and (iv) potential *in vitro* transformation of the cells during expansion. This paper analyzes the main risks connected with the use of MSCs in cellular therapy approaches, and reports on some of the most intriguing findings on the use of MSCs in the context of regenerative medicine. Experimental studies in animal models and phase I/II clinical trials on the use of MSCs for the treatment of IBDs and other inflammatory/degenerative conditions are reviewed.

Keywords: mesenchymal stromal cells; biosafety; tissue repair; paracrine mechanism; autoimmune disorders; inflammatory bowel diseases

Introduction

Mesenchymal stromal cells (MSCs) are multipotent cells that can be isolated from several human tissues and expanded *ex vivo* for clinical use.[1,2] They are recognized as pivotal contributors to the creation of the hematopoietic stem cell (HSC) niche and are important players in the development and differentiation of the lymphohematopoietic system.[1,3] Moreover, MSCs are endowed with immunomodulatory properties that make them capable of affecting, both *in vitro* and *in vivo,* the function of all cells involved in the immune response.[4,5] MSCs have also been reported to promote graft survival and engraftment of HSCs in experimental animal models, as well as to display the ability to home to sites of injury, where they may promote tissue repair through the production of trophic factors and anti-inflammatory molecules.[6–8]

Because of these properties, the role of MSCs in the clinical arena has been explored in phase I/II trials both in the context of hematopoietic stem cell transplantation (HSCT), with the aim to facilitate engraftment and to treat steroid-resistant acute graft-versus-host disease (GvHD),[9,10] and in the field of regenerative medicine, in particular, for the treatment of gastrointestinal disorders such as inflammatory bowel diseases (IBDs).[11,12]

Safety issues

When considering the use of *ex vivo* expanded MSCs for clinical application, several potential risks should be considered: (i) the immunogenicity of the cells, (ii) the biosafety of medium components, (iii) the risk of ectopic tissue formation, and (iv) the potential *in vitro* transformation of the cells during expansion.

Immunogenicity

Concerning the immunogenicity of MSCs, it has been demonstrated that MSCs are not intrinsically immunoprivileged. Infusion of allogeneic

doi: 10.1111/j.1749-6632.2012.06667.x

MSCs into immunocompetent and major histocompatibility complex (MHC)-mismatched mice may induce an immune response, resulting in their rejection.[13] On the contrary, the infusion of syngeneic host–derived MSCs resulted, in the same mouse model, in enhanced engraftment of allogeneic stem cells.[13] Moreover, it has been demonstrated that IL-2–activated autologous and allogeneic natural killer (NK) cells are capable of effectively lysing MSCs *in vitro*.[14] Although MSCs express normal levels of MHC class I, which should protect against NK-mediated killing, they display ligands that are recognized by activating NK receptors that, in turn, trigger NK alloreactivity.[14]

The majority of the clinical reports on MSC therapeutic application have suggested low immunogenicity of MSCs in humans.[9–12] However, when gene-marked MSCs were employed to treat children with osteogenesis imperfecta, the gene-marked cells were not detected in the treated patients, indicating their potential recognition and rejection by the host immune system.[15]

On the basis of these experimental and clinical findings, some fundamental issues should be taken into consideration when determining the clinical application of MSCs, including whether autologous or allogeneic MSCs should be employed, the state of immune competence of the patient at time of infusion, and the number of infusions needed to treat the patient.

The safety of the culture medium

Although current standard conditions for *ex vivo* expansion of MSCs are based on the presence of fetal calf serum (FCS), the use of bovine proteins might be associated with the risk of transmission of zoonoses and potential immune reactions in the host, resulting in rejection of the cells especially after repeated treatments.[16,17] For these reasons, various animal-free additives have been considered for clinical-grade expansion of MSCs: autologous and allogeneic human serum,[18] cytokines and growth factors,[19] and platelet lysate (PL)/platelet rich plasma (PRP).[20] In particular, PL/PRP has been proposed as a suitable and efficacious candidate substitute for FCS in the near future.[20–22] Several research groups have demonstrated that the growth factors contained in PL/PRP are able to promote MSC expansion, and that a concentration of PL/PRP of 5% is sufficient to guarantee the optimal growth of MSCs while substantially preserving their biological and functional properties, including those relating to modulation of immune responses.[20–22] However, clinical trials conducted so far have mainly employed FCS-expanded cells, and available data *in vivo* on PL/PRP-cultured MSCs are still scarce. Therefore, these latter cells need to be extensively tested *in vivo* before being considered as a safe and effective substitute for MSCs generated in the presence of FCS-based media.

Ectopic tissue formation

One of the potential risks of MSC treatment involves the formation of mesenchymal tissues at ectopic sites. In a rat myocardial infarction model, it has been reported that MSCs may form bone following local injection into the myocardium.[23] Similarly, formation of adipose tissue in kidneys has been observed in a rat model of experimental glomerulonephritis.[24] Moreover, it has been recently reported in a murine model of GvHD that local implantation of MSCs resulted in ectopic bone formation in syngeneic recipients, whereas it lead to transplant rejection in allogeneic mice.[25] These studies underline the potential danger of ectopic tissue formation in patients treated with MSCs for myocardial infarction and other diseases; however, in clinical trials so far, no ectopic tissue or tumor formation *in vivo* has been observed. Few studies have attempted to specifically address this concern, and factors governing the postinfusion fate of MSCs and the influence of the local environment on MSC behavior are still largely unknown and need further investigation. Therefore, a strict and long-term follow-up of patients treated with MSCs is recommended to monitor the potential formation of mesenchymal tissues at ectopic sites.

Malignant transformation

It has been shown by a few groups that long-term manipulation *in vitro* of both adipose tissue (AT) and bone marrow (BM)–derived MSCs may alter their functional and biological properties, leading to the accumulation of genetic alterations and malignant transformation.[26–28] By contrast, other researchers have suggested that human MSCs of various tissue origin can be cultured *in vitro* long term without losing their usual phenotypical/functional characteristics and without developing chromosomal aberrations.[29–31] In particular, genetic studies performed through

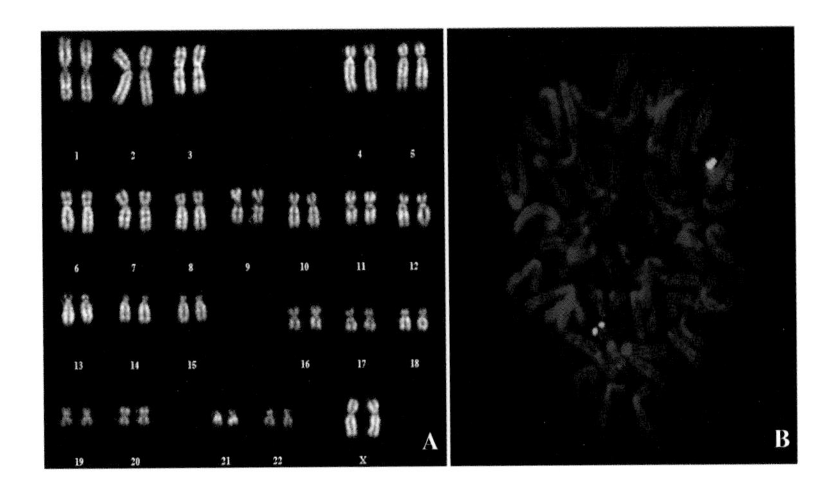

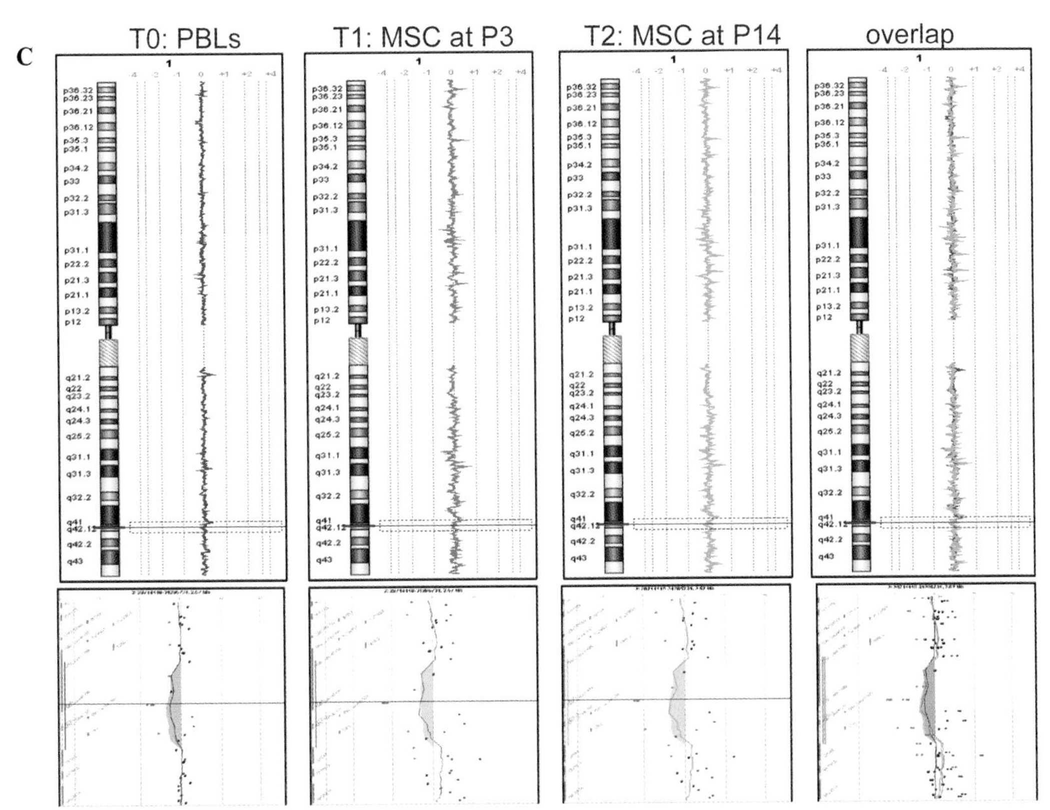

Figure 1. (A) Normal Q-banding karyotype (360–400 band) of MSCs from a healthy donor at passage 2 (P2). (B) FISH analysis with chromosome 16 subtelomeric-specific probes on a metaphase from a healthy donor at P18. Green signals represent subtelomeric regions of the short arm of chromosome 16; red signals indicate subtelomeric regions of the long arm. (C) Array-CGH profiles of chromosome 1 from a healthy donor at T0: peripheral blood lymphocytes (PBLs); T1: MSCs at P3; T2: MSCs at P14; overlap: three overlapping experiments (blue line, PBLs; red line, P3; green line, P14). Modified from Bernardo *et al.*[29]

both conventional karyotyping and molecular techniques, such as array-comparative genomic hybridization (array-CGH), have been employed to document the absence of chromosomal abnormalities in cultured MSCs (see Fig. 1).[29–31] Telomere length and telomerase activity analyses, together with the study of several proteins and genes involved in the regulation of cell cycle, senescence, and tumorigenesis have been also tested, confirming the absence of transformation.[29,31] Moreover, French researchers have demonstrated that the occasional presence of anueploidy in some MSC preparations may be related to the occurrence of senescence, but not to the development of transformation.[31]

More importantly, the reports on MSC malignant transformation have been recently retracted because it was found that the tumor cells in MSC cultures were unrelated to the original MSCs; rather, they derived from contaminating tumor cell lines.[32,33] Together, these data indicate that malignant transformation in *ex vivo* expanded human MSCs is likely to be an extremely uncommon event, estimated to be in the frequency of $<10^{-9}$ (Ref. 34). As a general recommendation, phenotypic, functional and genetic assays, although known to have limited sensitivity, should be routinely performed on MSCs before *in vivo* use; in particular, a genetic characterization of MSCs through conventional/molecular karyotyping may be considered before release of MSCs for clinical application, in particular for patient-derived MSCs.

MSCs in animal models for tissue repair

MSCs have been tested in a variety of animal models to facilitate engraftment of HSCs,[6] to treat/cure acute GvHD,[35] and to promote repair of damaged tissues.[8,35] With respect to the latter application, MSCs have been reported to secrete soluble factors capable of stimulating survival and functional recovery of injured cells.[8,36]

Inflammatory bowel disorders

MSCs have been recently shown to migrate to sites of inflammation in the 2,4,6-trinitrobonzene sulfonic acid (TNBS)–induced model of colitis[36] and to restore epithelial barrier integrity in the experimental model of dextran sulphate sodium-induced colitis.[37] In a rat model of colitis induced by intraluminal instillation of TNBS, topical implantation of BM-derived MSCs into the colonic submucosa was associated with accelerated healing of the damaged tissue.[36] In similar models of experimental colitis, the intravenous and intraperitoneal administration of MSCs also alleviated the signs of the disease by ameliorating inflammation-related tissue destruction.[38–40] When directly injected in the tissue surrounding a gastric ulcer in a rat model, BM-derived MSCs were able to stimulate rapid healing and this was associated with expression in the engrafted cells of angiogenic factors such as vascular endothelial growth factor (VEGF) and hepatocyte growth factor (HGF)[41] (see also Table 1).

Radiation and abdominal sepsis models

Tissue regeneration after radiation injury has been obtained with BM-derived MSCs that were intravenously infused into irradiated immunodeficient mice; MSCs were able to specifically home to radiation-injured tissues, to increase self-renewal of the gut epithelium, and to accelerate structural recovery of small intestine.[42,43] In a recent study, BM-derived MSCs were infused intravenously in an animal model of abdominal sepsis before and after its induction; this was associated with a decrease in animal mortality and an improvement in animal organic functions.[44]

Liver disorders

With respect to liver diseases, MSC-derived hepatocytes and undifferentiated MSCs were injected intrasplenically or intravenously into immunodeficient mice with carbon tetrachloride–induced lethal fulminant hepatic failure. The administration of both types of cells rescued the animals from hepatic failure; this was associated with a reduction of oxidative stress and an accelerated repopulation of hepatocytes.[45] In a rat model, MSC-derived conditioned medium proved effective in reversing fulminant hepatic failure, suggesting that soluble factors released by MSCs might be capable of promoting regeneration of hepatocytes.[8,46] In a small-for-size liver graft model in rats, implantation of MSCs overexpressing HGF through the portal vein prevented liver failure and reduced animal mortality[47] (see Table 1). This reparative effect was not confirmed in a rat model of severe chronic liver injury, that is, MSC administration was not associated with an improvement of liver function and fibrosis.[48]

Autoimmune disorders

MSCs have been also infused in animal models of neurodegenerative disorders, with very promising experimental results. In a model of experimental autoimmune encephalomyelitis (EAE), murine MSCs were able to ameliorate the signs of the disease through the induction of peripheral T cell tolerance against central nervous system (CNS)-restricted antigens.[49,50] By inhibiting autoreactive T and B cells, the infusion of MSCs mitigated the clinical manifestations of systemic lupus erythematosus

Table 1. MSCs in animal models of tissue repair[a]

Context	Outcome	Refs.
Rat, dextran sulphate sodium–induced colitis	Favored and restored epithelial barrier integrity	36
Rat, TNBS-induced experimental colitis	Stimulated intestinal mucosa healing	35, 37–39
Rat, gastric ulcer	Accelerated organ damage healing via angiogenesis	40
Immunodeficient mice, radiation injury model	Homing to radiation-injured tissues, accelerated structural recovery of small intestine	41, 42
Mouse, abdominal sepsis model	Decreased mortality, improved organic animal function	43
Immunodeficient mice, CT-induced hepatic failure	Rescued animals, accelerated hepatocyte repopulation	44
Rat, acute hepatic failure	MSC-derived CM: reversed fulminant hepatic failure	8, 45
Rat, small-for-size liver graft model	Prevented liver failure, reduced mortality	46
Mouse, EAE	Prevention of EAE development	48, 49
Mouse, SLE	Ameliorated signs and symptoms of SLE	50
Mouse, CIA	No beneficial effect; accentuation of Th1 response	51
Mouse, STZ diabetes	Ameliorated diabetes and kidney disease	52–54
Rat, glomerulonephritis	Stimulated glomerular healing	55

TNBS, 2,4,6-trinitrobonzene sulfonic acid; MSCs, mesenchymal stromal cells; CT, carbon tetrachloride; CM, conditioned medium; EAE, experimental autoimmune encephalomyelitis; SLE, systemic lupus erythematosus; CIA, collagen-induced arthritis; STZ, streptozotocin.
[a]This table has been adapted from Ref. 80.

(SLE) in a murine model.[51] However, in a model of collagen-induced arthritis (CIA), the administration of MSCs did not confer any benefit and was associated with an enhanced Th1 response. Because MSCs were not detected in the articular environment, the worsening symptoms was unlikely due to the homing of MSCs to the joints.[52]

In the context of other autoimmune disorders, MSCs have been employed in mouse models of diabetes where their infusion was associated with the suppression of T cell proliferation and the generation of myeloid/inflammatory dendritic cells (DCs), resulting in long-term reversal of hyperglycemia.[53,54] In another mouse model of diabetes, MSC treatment resulted in an increase of the number of insulin-producing β cells and in the repair of renal glomeruli[55] (see Table 1).

In addition, the infusion of rat MSCs in an experimental model of glomerulonephritis was able to stimulate glomerular healing.[56]

Mechanisms of repair

The potential mechanisms through which MSCs display their reparative/regenerative effects after tissue damage include the capacity to home to sites of injury, the ability to release anti-inflammatory soluble factors, and the capacity to modulate immune responses, all of which have to be taken into account.[4,5,8,36,38,42] Although the reparative effect of MSCs has been rarely ascribed to their differentiation into resident cells, their anti-inflammatory/antiproliferative properties are displayed through the production of paracrine molecules capable of stimulating the recovery of damaged cells.[8,11,12,46,47] It has been demonstrated that MSCs release several soluble molecules and chemokines, either constitutively or following cross-talk with other cells;[4,5] these include transforming growth factor-β (TGF-β), prostaglandin-E2 (PGE2), heme oxygenase-1 (HO-1), and indoleamine 2,3-dioxygenase (IDO) for human MSCs, and nitric oxide (NO) for murine MSCs.[4,5] Release of interferon-γ (IFN-γ) by damaged target cells is able to induce the release of IDO by human MSCs, which, through the depletion of tryptophan, results in an antiproliferative effect.[57,58] It has been recently shown that MSCs can be activated by macrophage-derived TNF-α and other proinflammatory cytokines to secrete TNF-α stimulated gene/protein 6 (TSG-6), a multifunctional anti-inflammatory protein that is able to modulate the cascade of proinflammatory cytokines.[59,60]

The anti-inflammatory/antiproliferative effect of MSCs might also be due to their ability to stimulate the generation/differentiation of regulatory T cells (T_{reg} cells). In an experimental murine model of Crohn's disease (CD), the infusion of MSCs was associated with the induction of FoxP3$^+$ T_{reg} cells, which was efficacious in both preventing and curing colitis.[38] Moreover, it has been recently demonstrated that HO-1 produced by human MSCs is able to promote the formation of Tr1 and Th3 T_{reg} cells *in vitro*, and that this process is influenced by the environment in which MSCs and target cells interact.[61] It has been suggested that MSCs acquire their immunosuppressive functions after being exposed to an inflammatory environment;[59] the presence and the level of inflammatory cytokines, such as IFN-γ, TNF-α, and IL-1β, can activate MSCs to deliver their immunosuppressive effect.[62]

MSC survival and sustained engraftment do not seem to be required for their reparative effect;[8,46] chemokines and growth factors released in an inflamed environment may be able to guide MSCs to the damaged tissue because of the interaction of these latter molecules with integrins and selectins expressed on MSCs.[63–65]

Biological and functional properties of "diseased" MSCs

Although MSCs isolated form healthy donors have been shown to display consistent and rather uniform properties,[4,5] biological and functional characteristics of MSCs derived from patients affected by degenerative/inflammatory disorders may display different features. These "diseased" MSCs might be impaired in some of their intrinsic properties and may therefore be unsuitable for treatment.

Although one group reported that MSCs isolated from patients with systemic sclerosis (SS) are functionally impaired *in vitro*,[66] other researchers suggested that MSCs derived from patients affected by various autoimmune diseases, including SS, exhibit comparable phenotypical and functional properties as their healthy counterparts.[67]

BM-derived MSCs isolated from patients affected by refractory CD have been demonstrated to show similar biological and functional properties compared with MSCs expanded from healthy donors.[11,68] Moreover, MSCs isolated from CD patients (CD-MSCs) did not show a propensity to undergo spontaneous transformation after long-term *in vitro* culture, as assessed by both conventional and molecular karyotyping.[68] CD-MSCs were also able to inhibit *in vitro* lymphocyte proliferation after stimulation with phytoemagglutinin,[11,68] and MSC proliferation was not affected by coculture with immunosuppressive drugs commonly employed in CD patients.[11]

The demonstration that MSCs derived from diseased patients display the typical MSC properties and functions provides the experimental background for considering their use as a therapeutic strategy. When MSCs are infused in nonprofoundly immunodepressed patients, such as those affected by inflammatory and degenerative disorders, autologous rather than allogeneic or third-party cells might be preferred, owing to the possible recognition of allogeneic or third-party cells by the host immune system and consequent rejection. Therefore, the functionality of MSCs derived from diseased patients should be investigated before considering their infusion in the patients for therapeutic purposes.

Clinical applications of MSCs in regenerative medicine

MSCs as therapeutic strategy in gastrointestinal disorders

MSCs are being tested as a therapeutic strategy in gastrointestinal disorders, in particular, in IBDs (see Table 2). The encouraging results obtained with experimental models[36–40] led to the systemic administration of autologous BM–derived MSCs that have been employed in a phase I study for the treatment of 10 patients affected by refractory luminal CD.[11] Although feasibility and safety of the approach was demonstrated, a clinical response, defined as a significant drop in CD activity index (CDAI), was seen in three patients. Moreover, an improvement in mucosal inflammation was noted in two patients, together with an increase in CD4$^+$ CD127$^+$ T_{reg} cells and a decrease of proinflammatory cytokines in mucosal biopsies.[11]

In another phase I study, five patients affected by fistulizing CD and treated with local administration of autologous, AT-derived MSCs showed promising results.[69] In a subsequent phase II, randomized, controlled trial, patients with complex perianal fistulas of either criptoglandular origin or associated with CD were randomized to receive local

Table 2. MSCs as therapeutic strategy in gastrointestinal and autoimmune disorders[a]

Context	Outcome	Refs.
Luminal CD		
Phase I, 10 pts, autologous BM-MSC i.v.	Feasibility and safety, clinical response in 3 pts	11
Fistulizing CD		
Phase I, 5 pts, autologous AT-MSC l.i.	Promoted fistula repair	66
Phase II randomized controlled trial, 49 pts, autologous AT-MSC l.i.	70% response in fibrin glue + MSC versus 16% in fibrin glue alone (also fistulas of cryptoglandular origin)	67
1 pt, rectovaginal fistula, allogeneic AT-MSC l.i.	Partial healing of the fistula, no MSC rejection	68
Phase I-II trial, 10 pts, autologous BM-MSC l.i.	Feasibility and safety, CR in 7 pts, reduction of CDAI/PDAI	12
Ulcerative colitis		
Ulcerative colitis, 39 pts, allogeneic BM-MSC i.v.	Improved inflammation indices, increased remission duration	70
Neurodegenerative disorders		
MS, 10 pts, autologous MSC i.v.	Feasibility and safety	71
MS (15 pts) and ALS (19 pts), intrathecal and/or i.v. MSCs	Demonstration of safety + increase in T_{reg} cells	72
SAA		
1 pt, coinfusion HSC-MSC i.v.	Hematopoietic engraftment and sustained remission	75
2 pediatric pts, coinfusion HSC-MSC i.v.	Hematopoietic engraftment and sustained remission	76

CD, Crohn's disease; CDAI, CD Activity Index; pts, patients; BM, bone marrow; i.v., intravenous infusion; AT, adipose tissue; l.i., local injection; CR, complete response; PDAI, perianal disease activity index; MS, multiple sclerosis; ALS, amyotrophic lateral sclerosis; T_{reg} cells, regulatory T cells; SAA, severe aplastic anemia; HSC, hematopoietic stem cell; MSCs, mesenchymal stromal cells.
[a]This table has been adapted from Ref. 80.

treatment with fibrin glue or fibrin glue + autologous AT–derived MSCs. Fistula healing was observed in the majority of the patients receiving MSCs in addition to fibrin glue, independently of the origin of the fistulas.[70] In a recent case report, a CD patient with a rectovaginal fistula was treated with a local injection of allogeneic AT–derived MSCs and showed partial healing of the fistula in the absence of adverse events.[71]

In a phase I/II study, 10 CD patients with refractory and actively draining complex perianal fistulas were treated with repeated intrafistular injections of autologous BM-derived MSCs.[12] All patients showed a clinical response; seven patients benefited from complete closure of the fistulas, whereas the remaining three had a partial response. Moreover, a significant reduction of both CDAI and perianal disease activity index (PDAI) scores was achieved in all patients. As already shown in the study by Duijvenstein *et al.*,[11] in this case, the percentage of mucosal, as well as circulating, T_{reg} cells increased during MSC treatment, confirming the observation made in an animal model resembling human colitis and suggesting a possible role of T_{reg} cells in MSC-mediated repair of inflamed tissues in IBDs.[38]

Together, these studies demonstrate the feasibility and safety of both intravenous and local treatment with MSCs in refractory CD, and promising preliminary results for efficacy can be observed. These findings need to be further substantiated in larger cohorts of CD patients and in a randomized trial to define the role of MSC therapy in the management of this disabling condition. The possibility of coating MSCs with anti-addressin antibodies, with the aim of increasing their delivery to inflamed colon and therefore their therapeutic efficacy, has been tested in a mouse model with interesting preliminary results.[72]

Systemic infusion of allogeneic BM–derived MSCs has also been applied in patients with ulcerative colitis; an improvement of inflammation indexes and an increased duration of remission was noted in MSC-treated patients compared with patients receiving conventional therapy.[73]

MSCs as a therapeutic strategy in autoimmune disorders and aplastic anemia

In a phase I study, 10 patients affected by multiple sclerosis (MS) have been treated with intravenous infusion of autologous MSCs. Although 6 of the 10 patients showed some degree of functional improvement, a demonstration of feasibility and safety was obtained.[74] In another phase I/II study, 15 patients affected by MS and 19 with amyotrophic lateral sclerosis (ALS) were treated with intrathecal and/or intravenous MSC infusion. As seen in the previous study, the MSC treatment was safe, and an increase in the proportion of $CD4^+$ $CD25^+$ T_{reg} cells in the peripheral blood of the patients was reported.[75] Despite the promising experimental results in animal models, clear evidence of the beneficial effect of MSCs for the treatment of neurodegenerative disorders is lacking.[76]

On the basis of interesting experimental findings[53–55] MSC therapy is being considered as a promising possibility to be explored in the treatment of type 1 diabetes.[77] Both the regenerative capacities (to restore pancreatic islets) and the immunomodulatory properties of MSCs are stimulating researchers to launch initial studies to test MSC therapeutic efficacy in treating diabetes.

Recently, Jaganathan *et al.*[78] reported on a patient with severe aplastic anemia (SAA) who failed to respond to three allogeneic HSCTs but displayed hematopoietic engraftment and sustained remission after coadministration of HSCs and third-party MSCs. Although donor MSC DNA was not found in the recipient, 25 days after the coinfusion the BM histology showed normal hematopoietic structures. In another paper, two pediatric patients with SAA were cotransplanted with haploidentical MSCs and HLA-identical peripheral blood stem cells (PBSCs) after graft failure. They both showed hematopoietic engraftment and remained transfusion-independent more than two years after treatment.[79] It is reasonable to speculate that coinfused MSCs are capable of modifying the recipient BM microenvironment via either a paracrine effect or a direct cell–cell-mediated reparative function, allowing recipient MSC to support hematopoietic engraftment; however, these single observations need to be corroborated in a larger cohort of patients affected by SAA (see also Table 2).

Conclusions and future directions

The currently available experimental and clinical data indicate that, similar to previously obtained data in the setting of HSCT, MSC treatment for autoimmune disorders (i.e., IBDs) is feasible and safe. Neither early toxicity nor later side effects have been registered in treated patients, although a longer follow-up is necessary to draw definitive conclusions on potential long-term adverse events.[11,12,66–68] Although patient improvements have been reported in most studies, a clear demonstration of efficacy of MSC therapy in this context is lacking. This needs to be further demonstrated in large randomized clinical trials, which should enroll homogeneous patients who receive homogeneous cellular products, in terms of tissue source, medium components, and route of administration.

More insight into the mechanisms by which MSCs mediate their anti-inflammatory/reparative effect is also warranted. One hypothesis proposes that MSCs act through paracrine mechanisms that prevent apoptosis and promote endogenous repair of injured tissues.[8,36,39,46,49] The immune modulatory effect of MSC may involve the generation/differentiation of T_{reg} cells and the interplay between MSCs and T_{reg} cells, although evidence for the *in vivo* role of T_{reg} cells in MSC-mediated repair is still lacking. It has been suggested that MSCs undergo *in vivo* activation in an inflammatory environment before becoming effective. However, no clinical data have yet been generated with activated MSCs, and further (pre) clinical data are required to further substantiate this hypothesis.

On the clinical side, the simultaneous administration of other immunosuppressive treatments that could potentiate or perhaps reduce the therapeutic benefit of MSCs needs to be addressed in future clinical studies. Additional studies are also required to define the role of host-related factors, including optimal timing of MSC administration, as well as product-related factors, including cell dose and schedule of administration.

Conflicts of interest

The authors declare no conflicts of interest.

References

1. Friedenstein, A.J., K.V. Petrakova, A.I. Kurolesova & G.P. Frolova. 1968. Heterotopic of bone marrow. Analysis of precursor cells for osteogenic and hematopoietic tissues. *Transplantation* **6:** 230–247.
2. Pittenger, M.F., A.M. Mackay, S.C. Beck, *et al.* 1999. Multilineage potential of adult human mesenchymal stem cells. *Science* **284:** 143–147.
3. Zhang, J., C. Niu, L. Ye, *et al.* 2003. Identification of the haematopoietic stem cell niche and control of the niche size. *Nature* **425:** 836–841.
4. Nauta, A.J. & W.E. Fibbe. 2007. Immunomodulatory properties of mesenchymal stromal cells. *Blood* **110:** 3499–3506.
5. Locatelli, F., R. Maccario & F. Frassoni. 2007. Mesenchymal stromal cells, from indifferent spectators to principal actors. Are we going to witness a revolution in the scenario of allograft and immune-mediated disorders? *Haematologica* **92:** 872–877.
6. Almeida-Porada, G., A.W. Flake, H.A. Glimp & E.D. Zanjani. 1999. Cotransplantation of stroma results in enhancement of engraftment and early expression of donor hematopoietic stem cells in utero. *Exp. Hematol.* **27:** 1569–1575.
7. Bartholomew, A., C. Sturgeon, M. Siatskas, *et al.* 2002. Mesenchymal stem cells suppress lymphocyte proliferation in vitro and prolong skin graft survival in vivo. *Exp. Hematol.* **30:** 42–48.
8. Parekkadan, B., D. van Poll, K. Suganuma, *et al.* 2007. Mesenchymal stem cell-derived molecules reverse fulminant hepatic failure. *PLoS One* **2:** e941.
9. Ball, L.M., M.E. Bernardo, H. Roelofs, *et al.* 2007. Cotransplantation of ex-vivo expanded mesenchymal stem cells accelerates lymphocyte recovery and may reduce the risk of graft failure in haploidentical hematopoietic stem cell transplantation. *Blood* **110:** 2764–2767.
10. Le Blanc, K., F. Frassoni, L.M. Ball, *et al.* 2008. Mesenchymal stem cells for treatment of steroid-resistant, severe, acute graft-versus-host disease: a phase II study. *Lancet* **371:** 1579–1586.
11. Duijvenstein, M., A.C. Vos, H. Roelofs, *et al.* 2010. Autologous bone marrow-derived mesenchymal stromal cell treatment for refractory luminal Crohn's disease: results of a phase I study. *Gut* **59:** 1662–1669.
12. Ciccocioppo, R., M.E. Bernardo, A. Sgarella, *et al.* 2011. Autologous bone marrow-derived mesenchymal stromal cells in the treatment of fistulising Crohn's disease. *Gut* **60:** 788–798.
13. Nauta, A.J., G. Westerhuis, A.B. Kruisselbrink, *et al.* 2006. Donor-derived mesenchymal stem cells are immunogenic in an allogeneic host and stimulate donor graft rejection in a non-myeloablative setting. *Blood* **108:** 2114–2120.
14. Spaggiari, G.M., A. Capobianco, S. Becchetti, *et al.* 2006. Mesenchymal stem cell-natural killer cell interactions: evidence that activated NK cells are capable of killing MSCs, whereas MSCs can inhibit IL-2-induced NK-cell proliferation. *Blood* **107:** 1484–1490.
15. Horwitz, E.M., P.L. Gordon, W.K. Koo, *et al.* 2002. Isolated allogeneic bone marrow-derived mesenchymal cells engraft and stimulate growth in children with osteogenesis imperfecta: implications for cell therapy of bone. *Proc. Natl. Acad. Sci. U. S. A.* **99:** 8932–8937.
16. Spees, J.L., C.A. Gregory, H. Singh, *et al.* 2004. Internalized antigens must be removed to prepare hypoimmunogenic mesenchymal stem cells for cell and gene therapy. *Mol. Ther.* **9:** 747–756.
17. Horwitz, E.M., D.J. Prockop, L.A. Fitzpatrick, *et al.* 1999. Transplantability and therapeutic effects of bone marrow-derived mesenchymal cells in children with osteogenesis imperfecta. *Nat. Med.* **5:** 309–313.
18. Shahdadfar, A., K. Fronsdal, T. Haug, *et al.* 2005. In vitro expansion of human mesenchymal stem cells: choice of serum is a determinant of cell proliferation, differentiation, gene expression, and transcriptosome stability. *Stem Cells* **23:** 1357–1366.
19. Jung, S., A. Sen, L. Rosenberg & L.A. Behie. 2010. Identification of growth and attachment factors for the serum-free isolation and expansion of human mesenchymal stromal cells. *Cytotherapy* **12:** 637–657.
20. Doucet, C., I. Ernou, Y. Zhang, *et al.* 2005. Platelet lysates promote mesenchymal stem cell expansion: a safety substitute for animal serum in cell-based therapy applications. *J. Cell Physiol.* **205:** 228–236.
21. Bernardo, M.E., M.A. Avanzini, C. Perotti, *et al.* 2007. Optimization of in vitro expansion of human multipotent mesenchymal stromal cells for cell-therapy approaches: further insights in the search for a fetal calf serum substitute. *J. Cell Physiol.* **211:** 121–130.
22. Schallmoser, K., C. Bartmann, E. Rohde, *et al.* 2007. Human platelet lysate can replace fetal bovine serum for clinical-scale expansion of functional mesenchymal stromal cells. *Transfusion* **47:** 1436–1446.
23. Breitbach, M., T. Bostani, W. Roell, *et al.* 2007. Potential risks of bone marrow cell transplantation into infarcted hearts. *Blood* **110:** 1362–1369.
24. Kunter, U., S. Rong, P. Boor, *et al.* 2007. Mesenchymal stem cells prevent progressive experimental renal failure but maldifferentiate into glomerular adipocytes. *J. Am. Soc. Nephrol.* **18:** 1754–1764.
25. Prigozhina, T.B., S. Khitrin, G. Elkin, *et al.* 2008. Mesenchymal stromal cells lose their immunosuppressive potential after allotransplantation. *Exp. Hematol.* **36:** 1370–1376.
26. Rubio, D., J. Garcia-Castro, M.C. Martin, *et al.* 2005. Spontaneous human adult stem cell transformation. *Cancer Res.* **65:** 3035–3039.
27. Wang, Y., D.L. Huso, J. Harrington, *et al.* 2005. Outgrowth of a transformed cell population derived from normal human BM mesenchymal stem cell culture. *Cytotherapy* **7:** 509–519.
28. Røsland, G.V., A. Svendsęn, A. Torsvik, *et al.* 2009. Long-term cultures of bone marrow-derived human mesenchymal stem cells frequently undergo spontaneous malignant transformation. *Cancer Res.* **69:** 5331–5339.
29. Bernardo, M.E., N. Zaffaroni, F. Novara, *et al.* 2007. Human bone marrow-derived mesenchymal stem cells do not undergo transformation after long-term *in vitro* culture and do not exhibit telomere maintenance mechanisms. *Cancer Res.* **67:** 9142–9149.
30. Avanzini, M.A., M.E. Bernardo, A.M. Cometa, *et al.* 2009. Generation of mesenchymal stromal cells in the presence of platelet lysate: a phenotypical and functional comparison

between umbilical cord blood- and bone marrow-derived progenitors. *Haematologica.* **94:** 1649–1660.
31. Tarte, K., J. Gaillard, J. Lataillade, *et al.* 2010. Clinical-grade production of human mesenchymal stromal cells: occurrence of aneuploidy without transformation. *Blood* **115:** 1549–1553.
32. Vogel, G. 2010. To scientists's dismay, mixed-up cell lines strike again. *Science* **329:** 1004.
33. Torsvik, A., G.V. Røsland, A. Svendsen, *et al.* 2010. Spontaneous malignant transformation of human mesenchymal stem cells reflects cross-contamination: putting the research field on track – letter. *Cancer Res.* **70:** 6393–6396.
34. Prockop, D.J., M. Brenner, W.E. Fibbe, *et al.* 2010. Defining the risks of mesenchymal stromal cell therapy. *Cytotherapy* **2:** 576–578.
35. Tisato, V., K. Naresh, J. Girdlestone, *et al.* 2007. Mesenchymal stem cells of cord blood origin are effective at preventing but not treating graft-versus-host disease. *Leukemia* **21:** 1992–1999.
36. Hayashi, Y., S. Tsuji, M. Tsujii, *et al.* 2008. Topical implantation of mesenchymal stem cells has beneficial effects on healing of experimental colitis in rats. *J. Pharmacol. Exp. Ther.* **326:** 523–531.
37. Yabana, T., Y. Arimura, H. Tanaka, *et al.* 2009. Enhancing epithelial engrafment of rat mesenchymal stem cells restores epithelial barrier integrity. *J. Pathol.* **218:** 350–359.
38. González, M.A., E. Gonzalez-Rey, L. Rico, *et al.* 2009. Adipose-derived mesenchymal stem cells alleviate experimental colitis by inhibiting inflammatory and autoimmune responses. *Gastroenterology* **136:** 978–989.
39. Tanaka, F., K. Tominaga, M. Ochi, *et al.* 2008. Exogenous administration of mesenchymal stem cells ameliorates dextran sulphate-sodium-induced colitis via anti-inflammatory action in damaged tissue in rats. *Life Science* **83:** 771–779.
40. Zhang, Q., S. Shi, Y. Liu, *et al.* 2009. Mesenchymal stem cells derived from human gingiva are capable of immunomodulatory functions and ameliorate inflammation-related tissue destruction in experimental colitis. *J. Immunol.* **183:** 7787–7798.
41. Hayashy, Y., S. Tsuji, M. Tsujii, *et al.* 2008. Topical transplantation of mesenchymal stem cells accelerates gastric ulcer healing in rats. *Am. J. Physiol. Gastrointest. Liver Physiol.* **294:** G778–G786.
42. Hayashy, Y., J. Zhang, J.F. Gong, *et al.* 2008. Effects of transplanted bone marrow mesenchymal stem cells on the irradiated intestine of mice. *J. Biomed. Sci.* **15:** 585–594.
43. Mouiseddine, M., S. François, A. Semont, *et al.* 2007. Human mesenchymal stem cells home specifically to radiation-injured tissues in a non-obese diabetes/severe combined immunodeficiency mouse model. *Br. J. Radiol.* **80:** S49–S55.
44. Németh, K., A. Leelahavanichkul, P.S. Yuen, *et al.* 2009. Bone marrow stromal cells attenuate sepsis via prostaglandin E(2)-dependent reprogramming of host macrophages to increase their interleukin-10 production. *Nat. Med.* **15:** 42–49.
45. Kuo, T.K., S.P. Hung, C.H. Chuang, *et al.* 2008. Stem cell therapy for liver disease: parameters governing the success of using bone marrow mesenchymal stem cells. *Gastroenterology* **134:** 2111–2121.
46. van Poll, D., B. Parekkadan, C.H. Cho, *et al.* 2008. Mesenchymal stem cell-derived molecules directly modulate hepatocellular death and regeneration in vitro and in vivo. *Hepatology* **47:** 1634–1643.
47. Yu, Y., A.H. Yao, N. Chen, *et al.* 2007. Mesenchymal stem cells over-expressing hepatocyte growth factor improve small-for-size liver grafts regeneration. *Mol. Ther.* **15:** 1382–1389.
48. Carvalho, A.B., L.F. Quintanilha, J.V. Dias, *et al.* 2008. Bone marrow multipotent mesenchymal stromal cells do not reduce fibrosis or improve function in a rat model of severe chronic liver injury. *Stem. Cells* **26:** 1307–1314.
49. Zappia, E., S. Casazza, E. Pedemonte, *et al.* 2005. Mesenchymal stem cells ameliorate experimental autoimmune encephalomyelitis inducing T-cell anergy. *Blood* **106:** 1755–1761.
50. Zhang, J., Y. Li, J. Chen, *et al.* 2005. Human bone marrow stromal cell treatment improves neurological functional recovery in EAE mice. *Exp. Neurol.* **195:** 16–26.
51. Deng, W., Q. Han, L. Liao, *et al.* 2005. Effect of allogeneic bone marrow-derived mesenchymal stem cells on T and B lymphocytes from BXSB mice. *DNA Cell. Biol.* **24:** 458–463.
52. Djouad, F., V. Fritz, F. Apparailly, *et al.* 2005. Reversal of the immunesuppressive properties of mesenchymal stem cells by tumor necrosis factor alpha in collagen-induced arthritis. *Arthritis Rheum.* **52:** 1595–1603.
53. Jurewicz, M., S. Yang, A. Augello, *et al.* 2010. Congenic mesenchymal stem cell therapy reverses hyperglycemia in experimental type 1 diabetes. *Diabetes* **59:** 3139–3147.
54. Fiorina, P., M. Jurewicz, A. Augello, *et al.* 2009. Immunomodulatory function of bone marrow-derived mesenchymal stem cells in experimental autoimmune type 1 diabetes. *J. Immunol.* **183:** 993–1004.
55. Lee, R.H., M.J. Seo, A.A. Pulin, *et al.* 2006. Multipotent stromal cells from human marrow home to and promote repair of pancreatic islets and renal glomeruli in diabetic NOD/SCID mice. *Proc. Natl. Acad. Sci. U. S. A.* **103:** 17438–17443.
56. Kunter, U., S. Rong, Z. Djuric, *et al.* 2006. Transplanted mesenchymal stemm cells accelerate glomerular healing in experimental glomerulonephritis. *J. Am. Soc. Nephrol.* **17:** 2202–2212.
57. Krampera, M., L. Cosmi, R. Angeli, *et al.* 2006. Role of interferon-gamma in the immunomodulatory activity of human bone marrow mesenchymal stem cells. *Stem Cells* **24:** 386–398.
58. Ryan, J.M., F. Barry, J.M. Murphy & B.P. Mahon. 2007. Interferon-γ does not break, but promotes the immunosuppressive capacity of adult human mesenchymal stem cells. *Clin. Exp. Immunol.* **149:** 353–363.
59. Prockop, D.J. & J.Y. Oh. 2012. Mesenchymal stem/stromal cells (MSCs): role as guardians of inflammation. *Mol. Ther.* **20:** 14–20.
60. Roddy, G.W., J.Y. Oh, R.H. Lee, *et al.* 2011. Action at a distance: systemically administered adult stem/progenitor cells (MSCs) reduce inflammatory damage to the cornea without engraftment and primarily by secretion of TNF- α stimulated gene/protein 6. *Stem Cells* **29:** 1572–1579.
61. Mougiakakos, D., R. Jitschin, C.C. Johansson, *et al.* 2011. The impact of inflammatory licensing on heme oxygenase-1-mediated induction of regulatory T cells by human mesenchymal stem cells. *Blood* **117:** 4826–4835.

62. Dazzi, F. & M. Krampera. 2011. Mesenchymal stem cells and autoimmune diseases. *Best Pract. Res. Clin. Haematol.* **24:** 49–57.
63. Ren, G., L. Zhang, X. Zhaon, *et al.* 2008. Mesenchymal stem cell-mediated immunosuppression occurs via concerted action of chemokines and nitric oxide. *Cell Stem Cell* **2:** 141–150.
64. Wynn, R.F., C.A. Hart, C. Corradi-Perini, *et al.* 2004. A small proportion of mesenchymal stem cells strongly expresses functionally active CXCR4 receptor capable of promoting migration to bone marrow. *Blood* **104:** 2643–2645.
65. Sordi, V., M.L. Malosio, F. Marchesi, *et al.* 2005. Bone marrow mesenchymal stem cells express a restricted set of functionally active chemokine receptors capable of promoting migration to pancreatic islets. *Blood* **106:** 419–427.
66. Larghero, J., D. Farge, A. Braccini, *et al.* 2008. Phenotypical and functional characteristics of in vitro expanded bone marrow mesenchymal stem cells from patients with systemic sclerosis. *Ann. Rheum. Dis.* **67:** 443–449.
67. Bocelli-Tyndall, C., L. Bracci, G. Spagnoli, *et al.* 2007. Bone marrow mesenchymal stromal cells (BM-MSCs) from healthy donors and auto-immune disease patients reduce the proliferation of autologous- and allogeneic-stimulated lymphocytes in vitro. *Rheumatology* **46:** 403–408.
68. Bernardo, M.E., M.A. Avanzini, R. Ciccocioppo, *et al.* 2009. Phenotypical/functional characterization of in vitro expanded mesenchymal stromal cells from Crohn's disease patients. *Cytotherapy* **11:** 825–836.
69. García-Olmo, D., M. García-Arranz, D. Herreros, *et al.* 2005. A phase I clinical trial of the treatment of Crohn's fistula by adipose mesenchymal stem cell transplantation. *Dis. Colon. Rectum.* **48:** 1416–1423.
70. Garcia-Olmo, D., D. Herreros, I. Pascual, *et al.* Expanded adipose-tissue derived stem cells for the treatment of complex perianal fistula: a phase II clinical trial. *Dis. Colon. Rectum.* **52:** 79–86.
71. Garcia-Olmo, D., D. Herreros, P. De-La-Quintana, *et al.* 2010. Adipose-derived stem cells in Crohn's rectovaginal fistula. *Case Report Med.* **2010:** 961758.
72. Ko, I.K., B.G. Kim, A. Awadallah, *et al.* 2010. Targeting improves MSC treatment of inflammatory bowel disease. *Mol. Ther.* **18:** 1365–1372.
73. Lazebnik, L.B., A.G. Konopliannikov, O.V. Kniazev, *et al.* 2010. Use of allogeneic mesenchymal stem cells in the treatment of intestinal inflammatory diseases. *Ter. Arkh.* **82:** 38–43.
74. Mohyeddin Bonab, M., S. Yazdanbakhsh, J. Lotfi, *et al.* 2007. Does mesenchymal stem cell therapy help multiple sclerosis patients? Report of a pilot study. *Iran. J. Immunol.* **4:** 50–57.
75. Karussis, D., C. Karageorgiou, A. Vaknin-Dembinsky, *et al.* 2010. Safety and immunological effect of mesenchymal stem cell transplantation in patients with multiple sclerosis and amyotrophic lateral sclerosis. *Arch. Neurol.* **67:** 1187–1194.
76. Freedman, M.S., A. Bar-Or, H.L. Atkins, *et al.* 2010. The therapeutic potential of mesenchymal stem cell transplantation as a treatment for multiple sclerosis: consensus report of the International MSCT Study Group. *Mult. Scler.* **16:** 503–510.
77. Mabed, M. & M. Shahin. 2012. Mesenchymal stem cell-based therapy for the treatment of type 1 Diabetes Mellitus. *Curr. Stem. Cell. Res. Ther.* **7:** 179–190.
78. Jaganathan, B.G., V. Tisato, T. Vulliamy, *et al.* 2010. Effects of MSC co-injection on the reconstitution of aplastic anemia patient following hematopoietic stem cell transplantation. *Leukemia* **24:** 1791–1795.
79. Fang, B., N. Li, Y. Song, *et al.* 2009. Cotransplantation of haploidentical mesenchymal stem cells to enhance engraftment of hematopoietic stem cells and to reduce the risk of graft failure in two children with severe aplastic anemia. *Pediatr. Transplant* **13:** 499–502.
80. Hernatti, P. & A. Keating. "Mesenchymal stromal cells: Basic Biology and Clinical Applications." Section II, chapter 6: "MSCs for gastrointestinal disorders."

Ann. N.Y. Acad. Sci. ISSN 0077-8923

ANNALS OF THE NEW YORK ACADEMY OF SCIENCES
Issue: *Hematopoietic Stem Cells VIII*

TIM-3 as a therapeutic target for malignant stem cells in acute myelogenous leukemia

Yoshikane Kikushige and Koichi Akashi
Department of Medicine and Biosystemic Sciences, Kyushu University Graduate School of Medicine, Fukuoka, Japan

Address for correspondence: Koichi Akashi, Department of Medicine and Biosystemic Sciences, Kyushu University, 3-1-1 Maidashi, Higashi-Ku, Fukuoka 812-8582, Japan. akashi@med.kyushu-u.ac.jp

Acute myeloid leukemia (AML) originates from self-renewing leukemic stem cells (LSCs), an ultimate therapeutic target for AML. Recent studies have shown that many AML LSC–specific surface antigens could be such candidates. T cell immunoglobulin mucin-3 (TIM-3) is expressed on LSCs in most types of AML, except for acute promyelocytic leukemia, but not on normal hematopoietic stem cells (HSCs). In mouse models reconstituted with human AML LSCs or human hematopoietic stem cells, a human TIM-3 mouse IgG2a antibody with complement-dependent and antibody-dependent cellular cytotoxic activities eradicates AML LSCs *in vivo* but does not affect normal human hematopoiesis. Thus, TIM-3 is one of the promising targets to eradicate AML LSCs.

Keywords: leukemic stem cell; AML; TIM-3

Introduction

Acute myeloid leukemia (AML) is a clonal malignant disorder derived from a small number of leukemic stem cells (LSCs). The concept of LSCs has been proposed based on the finding in the 1980s that only a small fraction of AML blasts can form colonies *in vitro*. In 1996, Dick *et al.* showed that the $CD34^+CD38^-$, but not other fractions of bone marrow cells from AML patients, can reconstitute human AML in immunodeficient mice, demonstrating direct evidence for the presence of LSCs.[1] LSCs can self-renew and generate a large number of clonogenic leukemic blasts.[1–3] Although recent studies have suggested that LSCs are present in either in the $CD34^+CD38^+$ fraction[4] or the $CD34^-$ blastic fraction, at least in some types of AML[5,6] the $CD34^+CD38^-$ population represents highly-enriched LSCs in the vast majority of cases.[7] The $CD34^+CD38^-$ phenotype is, however, shared with normal human hematopoietic stem cells (HSCs) that have long-term reconstitution activity.[8,9]

In the clinic, conventional chemotherapies can currently achieve complete remission in ~90% of AML cases. However, a considerable fraction (~60%) of AML patients still relapse after intensive chemotherapy. The recurrence of AML in these patients could be caused by regrowth of the remaining LSCs. Therefore, LSCs could be the ultimate therapeutic target to achieve cure in AML patients. Thus, to selectively kill AML LSCs, sparing normal HSCs, one of the most practical approaches is to target the AML LSC–specific surface or functionally indispensable molecules. To achieve specificity for LSCs, the target molecule should be expressed on LSCs at a high level and not on normal HSCs.[10] It should not matter whether the molecule is expressed in normal blood cells or normal progenitor cells, because normal HSCs that are spared would replenish all mature blood cells after treatment.

Recently, two papers have reported the T cell immunoglobulin mucin-3 (TIM-3) as a surface molecule expressed in LSCs of most AML types.[11,12] Here, we discuss the potential usefulness of TIM-3 in eradicating AML LSCs, leaving normal HSCs intact.

Expression and functions of TIM-3 in normal hematopoiesis

TIM-3 was originally identified as a surface molecule expressed in interferon (IFN)-γ–producing $CD4^+$ Th1 cells and in $CD8^+$ T cytotoxic type 1 (Tc1) cells[13] in mouse hematopoiesis. TIM-3 plays an important role in limiting and controlling

doi: 10.1111/j.1749-6632.2012.06550.x

Th1-dependent immune responses and in inducing immune tolerance.[13–15] The ligand of TIM-3 in lymphocytes is galectin-9, an S-type lectin with two distinct carbohydrate recognition domains that can bind to carbohydrate chains on the TIM-3 IgV domain. Engagement of TIM-3 by galectin-9 induces Th1 cells to undergo apoptosis and inhibits their production of IFN-γ.[16] Thus, TIM-3 is a negative regulator of Th1- and Tc1-driven immune responses.

TIM-3 is also expressed in myeloid cells, including $CD11b^+$ macrophages, $CD11c^+$ dendritic cells (DCs), and mast cells, and it recognizes apoptotic cell–expressed phosphatidylserine (PS) through the TIM-3 IgV domain.[17–20] The binding of PS to TIM-3 does not interfere with binding of galectin-9 to TIM-3, as the binding sites of these molecules are located at opposite sides of the IgV domain. In TIM-3–expressing DCs, recognition of PS by TIM-3 enhanced phagocytosis of apoptotic cells and cross-presentation of apoptotic cell–associated antigens to $CD8^+$ T cells.[18]

In human steady-state hematopoiesis, TIM-3 is expressed mainly in monocytes and a fraction of NK

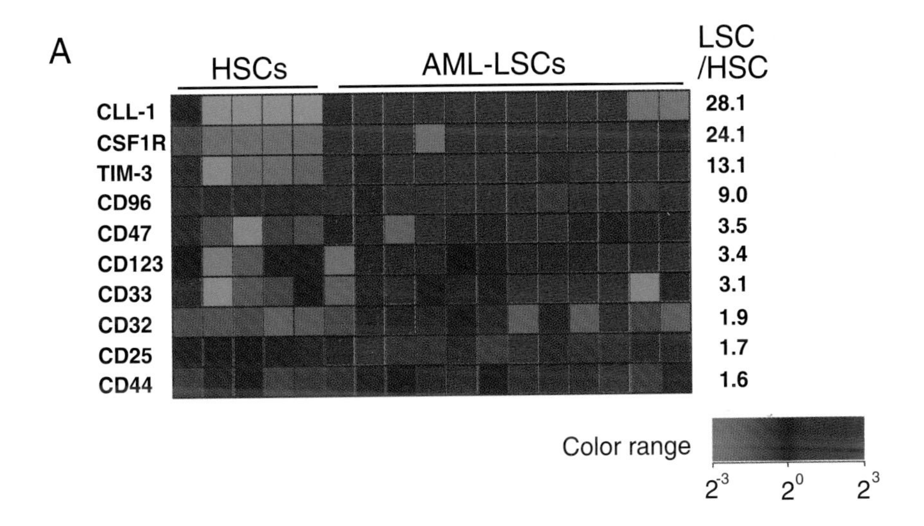

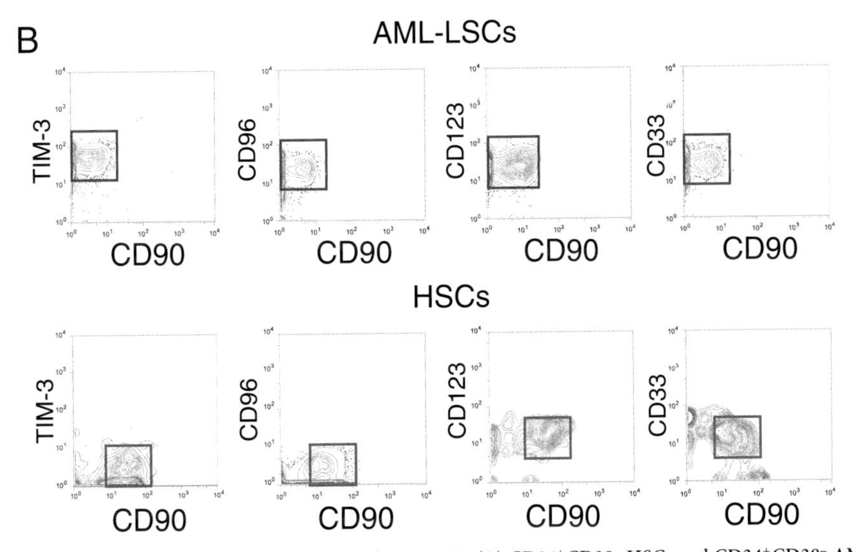

Figure 1. The expression of LSC-specific surface molecules in AML. (A) $CD34^+CD38^-$ HSCs and $CD34^+CD38^-$ AML LSCs were purified and analyzed on cDNA microarray. The surface molecules that are expressed strongly in AML LSCs compared to normal HSCs are listed. (B) The expression of representative surface markers in AML LSCs and normal HSCs.

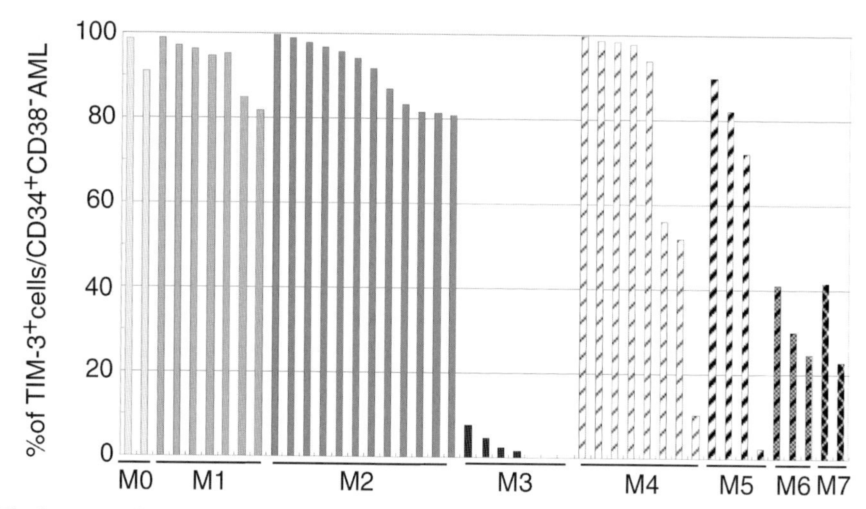

Figure 2. The frequency of TIM-3^{+} cells in the CD34^{+}CD38^{-} fraction in each AML subtype. TIM-3 protein is expressed in the vast majority of the CD34^{+}CD38^{-} LSC population in AML types M0, M1, M2, and M4.

cells, but not in granulocytes, T cells, or B cells.[11] In the bone marrow TIM-3 is not expressed in normal HSCs or the vast majority of the CD34^{+}CD38^{+} progenitor population. But within the CD34^{+}CD38^{+} fraction, TIM-3 was expressed in a fraction of granulocyte/monocyte progenitors (GMPs) at a low level, while it is not in common myeloid progenitors (CMPs), megakaryocyte/erythrocyte progenitors (MEPs), or common lymphoid progenitors (CLPs). The vast majority of purified TIM-3^{+} GMPs gave rise to CFU-M, suggesting that upregulation of TIM-3 mainly occurs in concert with the monocyte lineage commitment at the GMP stage.[11]

TIM-3 expression in human AML

TIM-3 has been identified as an AML LSC–specific marker based on the comparison between expression profiles of CD34^{+}CD38^{-} AML cells and normal HSCs (Fig. 1A). While TIM-3 protein is not expressed in normal HSCs (Fig. 1B), it is expressed at a high level in the vast majority of CD34^{+}CD38^{-} LSCs and the CD34^{+}CD38^{+} progenitor fraction in AML M0, M1, M2, and M4 types in virtually all cases (Fig. 2) and in all cytogenic subgroups.[11,12] In AML M5, M6, and M7, a considerable fraction of CD34^{+}CD38^{-} cells express TIM-3. However, TIM-3 expression was not seen in the CD34^{+}CD38^{-} population in M3 cases (Fig. 2). TIM-3 expression tends to decline at the CD34^{-} leukemic blast stage. Of note, the expression level of TIM-3 was found to be high in AML with core-binding factor translocations or with mutations in CEBPA.[12]

Strikingly, the TIM-3^{+} population in the bone marrow contains all AML LSCs, and normal HSCs are always included in patients' TIM-3^{-} populations. TIM-3^{+} and TIM-3^{-} AML populations have been transplanted into sublethally-irradiated immunodeficient mice and only mice transplanted with TIM-3^{+} AML cells developed human AML.[11] In another experiment, the TIM-3^{-} fraction did not include LSCs but normal HSCs, as evidenced by the fact that normal human hematopoiesis was frequently reconstituted in mice transplanted with the TIM-3^{-} fraction.[12]

Targeting AML LSCs by TIM-3-specific antibodies in xenograft models

To utilize TIM-3 to target AML LSCs, it is critical to establish human TIM-3 antibodies that can kill TIM-3–expressing cells *in vivo*. In terms of an antibody-based treatment, knowing the antibody-dependent cell-mediated cytotoxicity (ADCC) and the complement-dependent cytotoxicity (CDC) activities is critical in order to eliminate target cells.[21] A TIM-3 monoclonal antibody (IgG2b) was obtained by immunizing Balb/c mice with L929 cells stably-expressing human TIM-3 and soluble TIM-3 protein.[11] For this antibody, the variable portion of the VH regions of the cloned hybridoma that

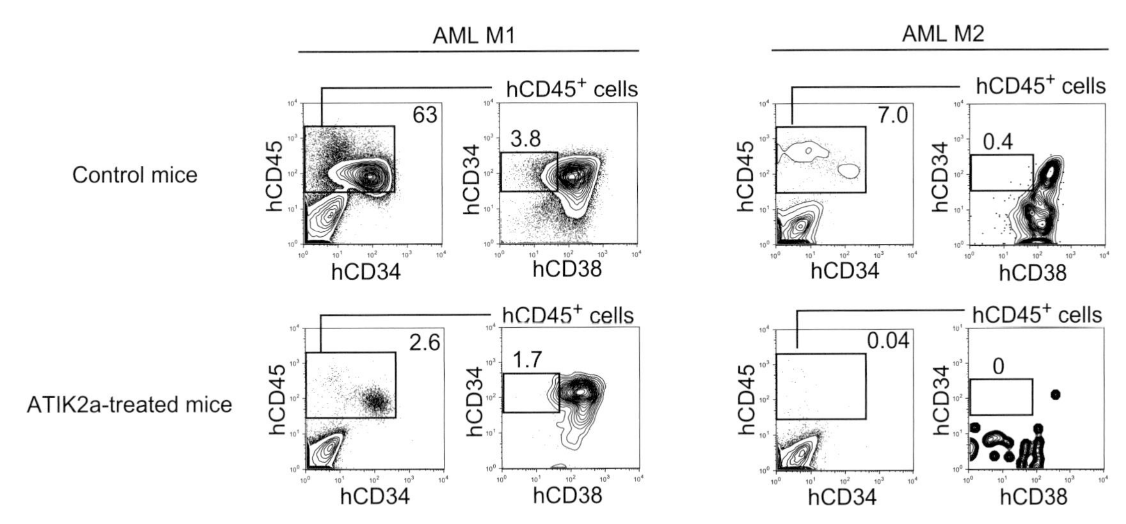

Figure 3. Potent antileukemic effects of the TIM-3 monoclonal antibody in a xenograft model. Mice were transplanted with LSCs purified from patients with AML M1 (left) or M2 (right) and were then treated with ATIK2a or control antibodies. In both experiments, ATIK2a treatment significantly reduced the human $CD45^+$ AML cells and $CD34^+CD38^-$ AML LSCs *in vivo*. Representative results of bone marrow analysis four weeks after treatment are shown.

recognize TIM-3 were grafted onto IgG2a Fc regions because the IgG2a subclass is the most efficient at inducing ADCC activity in mice.[22,23] The established clone, ATIK2a, was effective at killing TIM-3–expressing cell lines via both CDC and ADCC.[11]

The effect of ATIK2a on normal and malignant AML hematopoiesis was tested in xenograft models. NOD-SCID mice transplanted with 10^5 $CD34^+$ cord blood cells with or without ATIK2a injection developed almost equal percentages of human cells. In mice injected with ATIK2a, however, human $TIM\text{-}3^+$ mature monocytes were not found, suggesting that while targeting TIM-3 does not affect hematopoiesis, it does eliminate normal monocytes.

In contrast, ATIK2a exerted profound effects on leukemia development. In mice transplanted with human AML of M0, M1, and M4 types, ATIK2a treatment significantly reduced the human $CD45^+$ AML cell burden, as well as the $CD34^+CD38^-$ LSC cell numbers *in vivo* (Fig. 3). Retransplantation into secondary recipients of the remaining AML cells from primary recipients treated with ATIK2a never gave rise to human AML, indicating that the ATIK2a treatment successfully eradicated the LSCs in the primary recipients. These data suggest that eliminating AML LSCs by using TIM-3–"killing" antibodies may be a practical approach to curing human AML.

Perspective

To use surface markers for targeting AML LSCs, specificity as well as sensitivity are critical. TIM-3 has several advantages over other candidate markers. First, TIM-3 protein is not detectable in normal HSCs or in other myelo-erythroid or lymphoid progenitors, although monocyte-lineage committed progenitors begin to upregulate TIM-3. Second, TIM-3 could mark all LSCs that can reconstitute human AML in immunodeficient mice in the majority of M0, M1, M2, and M4 AML cases, and its expression level is sufficient to eradicate LSCs by antibody-based treatment. The expression level of other candidate molecules, including CD25,[24] CD32,[24] CD44,[25] and CD47,[26] in LSCs was only two- to threefold higher at the mRNA level compared with normal HSCs (Fig. 1A), and in some AML cases, LSCs did not express these molecules. CD33 and CD123 proteins are expressed at relatively high levels in normal HSCs (Fig. 1B) and myeloid progenitors, including CMPs and GMPs,[27] suggesting that targeting these molecules would harm normal hematopoiesis. In fact, prolonged cytopenia has been observed in AML patients treated with gemtuzumab, a recombinant humanized anti-CD33 monoclonal antibody conjugated with the cytotoxic antibiotic calicheamicin.

In contrast, CLL-1,[28] CSF1R,[29] and CD96[30] are molecules specifically expressed by LSCs. CLL-1 is a transmembrane glycoprotein;[31] the proportion of CLL-1–expressing CD34$^+$CD38$^-$ AML cells is highly diverse in cases.[28] CD96 is a member of the Ig gene superfamily; it is expressed on activated T cells.[32] Of note, similar to the case of TIM-3, CD96$^+$, but not CD96$^-$, AML cells efficiently reconstitute AML in the immunodeficient mice,[30] suggesting that CD96 can mark all functional AML LSCs. The expression level of CD96 protein is also high enough to clearly separate AML LSCs and normal HSCs (Fig. 1B). However, the sensitivity of TIM-3 is likely to be the highest among these molecules, at least for AML M0, M1, M2, and M4.

To apply our findings to clinic, we have developed a chimeric TIM-3 monoclonal antibody by fusing the variable regions of ATIK2a to the human IgG constant region. Thus far we have injected this antibody to cynomolgus monkeys, which showed no significant adverse effects (unpublished data). A clinical study with further intensive treatment is being planned.

It is important to understand the function of molecules in the maintenance and/or reconstitution capabilities of LSCs. For example, it was shown that anti-CD44 monoclonal antibodies reduced leukemic burden and blocked secondary engraftment in a NOD-SCID model.[25] This effect on LSCs was mediated in part by the disruption of LSC–niche interactions.[25] Anti-CD47 antibodies can block LSC reconstitution in a NOD-SCID model,[26] and this might be due to the activation of phagocytosis by macrophages through inhibition of interaction of CD47 with SIRPA. Since the pathway for eradication of LSCs by anti-CD44 or anti-CD47 treatment is different from that of anti-TIM-3, the combination of these antibodies might be critical for future treatments targeting AML LSCs.

Conflicts of interest

The authors declare no conflicts of interest.

References

1. Lapidot, T. *et al.* 1994. A cell initiating human acute myeloid leukaemia after transplantation into SCID mice. *Nature* **367:** 645–648.
2. Bonnet, D. & J.E. Dick. 1997. Human acute myeloid leukemia is organized as a hierarchy that originates from a primitive hematopoietic cell. *Nat. Med.* **3:** 730–737.
3. Hope, K.J., L. Jin & J.E. Dick. 2004. Acute myeloid leukemia originates from a hierarchy of leukemic stem cell classes that differ in self-renewal capacity. *Nat. Immunol.* **5:** 738–743.
4. Taussig, D.C. *et al.* 2008. Anti-CD38 antibody-mediated clearance of human repopulating cells masks the heterogeneity of leukemia-initiating cells. *Blood* **112:** 568–575.
5. Martelli, M.P. *et al.* 2010. CD34+ cells from AML with mutated NPM1 harbor cytoplasmic mutated nucleophosmin and generate leukemia in immunocompromised mice. *Blood* **116:** 3907–3922.
6. Taussig, D.C. *et al.* 2010. Leukemia-initiating cells from some acute myeloid leukemia patients with mutated nucleophosmin reside in the CD34(-) fraction. *Blood* **115:** 1976–1984.
7. Ishikawa, F. *et al.* 2007. Chemotherapy-resistant human AML stem cells home to and engraft within the bone-marrow endosteal region. *Nat. Biotechnol.* **25:** 1315–1321.
8. Bhatia, M. *et al.* 1997. Purification of primitive human hematopoietic cells capable of repopulating immune-deficient mice. *Proc. Natl. Acad. Sci. USA* **94:** 5320–5325.
9. Ishikawa, F. *et al.* 2005. Development of functional human blood and immune systems in NOD/SCID/IL2 receptor {gamma} chain(null) mice. *Blood* **106:** 1565–1573.
10. Krause, D.S. & R.A. Van Etten. 2007. Right on target: eradicating leukemic stem cells. *Trends Mol. Med.* **13:** 470–481.
11. Kikushige, Y. *et al.* 2010. TIM-3 is a promising target to selectively kill acute myeloid leukemia stem cells. *Cell Stem Cell.* **7:** 708–717.
12. Jan, M. *et al.* 2011. Prospective separation of normal and leukemic stem cells based on differential expression of TIM3, a human acute myeloid leukemia stem cell marker. *Proc. Natl. Acad. Sci. USA* **108:** 5009–5014.
13. Monney, L. *et al.* 2002. Th1-specific cell surface protein Tim-3 regulates macrophage activation and severity of an autoimmune disease. *Nature* **415:** 536–541.
14. Sanchez-Fueyo, A. *et al.* 2003. Tim-3 inhibits T helper type 1-mediated auto- and alloimmune responses and promotes immunological tolerance. *Nat. Immunol.* **4:** 1093–1101.
15. Sabatos, C.A. *et al.* 2003. Interaction of Tim-3 and Tim-3 ligand regulates T helper type 1 responses and induction of peripheral tolerance. *Nat. Immunol.* **4:** 1102–1110.
16. Zhu, C. *et al.* 2005. The Tim-3 ligand galectin-9 negatively regulates T helper type 1 immunity. *Nat. Immunol.* **6:** 1245–1252.
17. Anderson, A.C. *et al.* 2007. Promotion of tissue inflammation by the immune receptor Tim-3 expressed on innate immune cells. *Science* **318:** 1141–1143.
18. Nakayama, M. *et al.* 2009. Tim-3 mediates phagocytosis of apoptotic cells and cross-presentation. *Blood* **113:** 3821–3830.
19. Nakae, S. *et al.* 2007. TIM-1 and TIM-3 enhancement of Th2 cytokine production by mast cells. *Blood* **110:** 2565–2568.
20. Dekruyff, R.H. *et al.* 2010. T cell/transmembrane, Ig, and mucin-3 allelic variants differentially recognize phosphatidylserine and mediate phagocytosis of apoptotic cells. *J. Immunol.* **184:** 1918–1930.
21. Nimmerjahn, F. & J.V. Ravetch. 2007. Antibodies, Fc receptors and cancer. *Curr. Opin. Immunol.* **19:** 239–245.
22. Nimmerjahn, F. & J.V. Ravetch. 2005. Divergent immunoglobulin g subclass activity through selective Fc receptor binding. *Science* **310:** 1510–1512.
23. Uchida, J. *et al.* 2004. The innate mononuclear phagocyte network depletes B lymphocytes through Fc

receptor-dependent mechanisms during anti-CD20 antibody immunotherapy. *J. Exp. Med.* **199:** 1659–1669.
24. Saito, Y. *et al.* 2010. Identification of therapeutic targets for quiescent, chemotherapy-resistant human leukemia stem cells. *Sci. Transl. Med.* **2:** 17ra19.
25. Jin, L. *et al.* 2006. Targeting of CD44 eradicates human acute myeloid leukemic stem cells. *Nat. Med.* **12:** 1167–1174.
26. Majeti, R. *et al.* 2009. CD47 is an adverse prognostic factor and therapeutic antibody target on human acute myeloid leukemia stem cells. *Cell* **138:** 286–299.
27. Taussig, D.C. *et al.* 2005. Hematopoietic stem cells express multiple myeloid markers: implications for the origin and targeted therapy of acute myeloid leukemia. *Blood* **106:** 4086–4092.
28. van Rhenen, A. *et al.* 2007. The novel AML stem cell associated antigen CLL-1 aids in discrimination between normal and leukemic stem cells. *Blood* **110:** 2659–2666.
29. Aikawa, Y. *et al.* PU.1-mediated upregulation of CSF1R is crucial for leukemia stem cell potential induced by MOZ-TIF2. *Nat. Med.* **16:** 580–585, 581p following 585.
30. Hosen, N. *et al.* 2007. CD96 is a leukemic stem cell-specific marker in human acute myeloid leukemia. *Proc. Natl. Acad. Sci. USA* **104:** 11008–11013.
31. Bakker, A.B. *et al.* 2004. C-type lectin-like molecule-1: a novel myeloid cell surface marker associated with acute myeloid leukemia. *Cancer Res.* **64:** 8443–8450.
32. Wang, P.L. *et al.* 1992. Identification and molecular cloning of tactile. A novel human T cell activation antigen that is a member of the Ig gene superfamily. *J Immunol.* **148:** 2600–2608.

Ann. N.Y. Acad. Sci. ISSN 0077-8923

ANNALS OF THE NEW YORK ACADEMY OF SCIENCES
Issue: *Hematopoietic Stem Cells VIII*

Zebrafish xenografts as a tool for *in vivo* studies on human cancer

Martina Konantz,[1] Tugce B. Balci,[2] Udo F. Hartwig,[3] Graham Dellaire,[4,5] Maya C. André,[6,7] Jason N. Berman,[2,5,8,9] and Claudia Lengerke[1]

[1]Department of Hematology and Oncology, University of Tübingen Medical Center II, Tübingen, Germany. [2]IWK Health Centre, Halifax, Nova Scotia, Canada. [3]3rd Department of Medicine—Hematology, Oncology and Pneumology, University Medical Center, Johannes Gutenberg-University, Mainz, Germany. [4]Departments of Biochemistry and Molecular Biology, Dalhousie University, Halifax, Nova Scotia, Canada. [5]Department of Pathology, Dalhousie University, Halifax, Nova Scotia, Canada. [6]Department of Pediatric Hematology/Oncology, University Children's Hospital, Eberhard Karls University, Tübingen, Germany. [7]Department of Pediatric Intensive Care Medicine, University Children's Hospital (UKBB), Basel, Switzerland. [8]Department of Microbiology and Immunology, Dalhousie University, Halifax, Nova Scotia, Canada. [9]Department of Pediatrics, Dalhousie University, Halifax, Nova Scotia, Canada

Address for correspondence: Claudia Lengerke, M.D., University of Tübingen Medical Center II, Otfried-Mueller-Strasse 10, 72076 Tübingen, Germany. claudia.lengerke@med.uni-tuebingen.de

The zebrafish has become a powerful vertebrate model for genetic studies of embryonic development and organogenesis and increasingly for studies in cancer biology. Zebrafish facilitate the performance of reverse and forward genetic approaches, including mutagenesis and small molecule screens. Moreover, several studies report the feasibility of xenotransplanting human cells into zebrafish embryos and adult fish. This model provides a unique opportunity to monitor tumor-induced angiogenesis, invasiveness, and response to a range of treatments *in vivo* and in real time. Despite the high conservation of gene function between fish and humans, concern remains that potential differences in zebrafish tissue niches and/or missing microenvironmental cues could limit the relevance and translational utility of data obtained from zebrafish human cancer cell xenograft models. Here, we summarize current data on xenotransplantation of human cells into zebrafish, highlighting the advantages and limitations of this model in comparison to classical murine models of xenotransplantation.

Keywords: human xenografts; zebrafish; organogenesis; cancer biology; xenotransplantation

Introduction

Over the past decades, the zebrafish has developed into a powerful vertebrate model for genetic studies of embryonic development and organogenesis. The high fecundity and short generation times of the zebrafish embryo facilitate the generation of transgenic lines and allow performance of large-scale mutagenesis screens that can identify novel genetic pathways.[1–6] Transient genetic modification can be readily achieved in the early embryo via direct microinjection of messenger ribonucleic acid (mRNA) to obtain gene overexpression or morpholino oligonucleotides, inducing gene knockdown through antisense technology. Importantly, high interspecies conservation of molecular pathways has been shown between zebrafish and mammals.[7,8]

In addition to developmental studies, zebrafish have increasingly been employed to analyze cancer and genetic diseases through diverse forward and reverse genetic approaches. Chemical carcinogenesis screens providing cancer models resembling human disease and transgenic zebrafish expressing human and mouse oncogenes and tumor suppressors have enabled studies on tumor formation and maintenance in zebrafish.[9–15] Despite general conservation of oncogenic pathways between zebrafish and humans, exceptions exist, such as the apparent absence of the *INK4a/ARF* tumor suppressor gene in zebrafish.[16] Thus, full recapitulation of the genetic complexity of human tumors may not be possible

doi: 10.1111/j.1749-6632.2012.06575.x

using zebrafish animal models and certain genetic diseases may be restricted to humans making verification of results in primary human cells mandatory. Due to obvious ethical and practical limitations, *in vivo* studies on human cells are limited to xenografts. While these assays are so far best established in immunopermissive mice, several reports document the feasibility of xenografting human cells to zebrafish. This technique has the potential of marrying the clinically relevant context of human-derived cells with the unique opportunities for imaging and genetic manipulation provided by the zebrafish. Here, we summarize available data on zebrafish as hosts for human xenotransplant assays and discuss the potential of this model to complement and, in some cases, surpass murine systems.

The classical approach: xenotransplantation of human cells into immunopermissive mouse strains

To overcome the system-immanent limitations of biomedical research in humans, various immunopermissive mouse models have been developed in the last decades. Initially, the gold standard, for *in vivo* xenotransplantation was defined as reconstitution of human immunity or malignancy in immunodeficient mice displaying the severe combined immunodeficiency (SCID) or recombination activating gene (Rag)(null) mutation. These models have been then improved by the generation of $Rag2^{-/-}$ $\gamma c^{-/-}$ and, alternatively, NOD/SCID/$IL2rg^{tmWjl}$/Sz (NSG) mice lacking T, B, and natural killer (NK) cells that, due to a deficient common cytokine receptor γ-chain (γc), have defective signaling of the γc-cytokine family and sustained immune suppression.[17–19] Accordingly, superior engraftment of human healthy and malignant cells has been demonstrated in NSG mice.[19–26]

The engraftment efficacy of individual tissue specimens is highly variable and influenced by inherent parameters. An obvious explanation could be the differential response to murine niches and cytokines; human cells less reliant on human-specific microenvironmental cues may show better engraftment upon xenotransplantation in mice. Consistent with this notion, constitutive expression of human cytokines was shown to augment engraftment and differentiation capacity of xenotransplanted healthy and malignant human myeloid cells in NSG mice.[25,27–31] Furthermore, murine xenografts derived from clinically aggressive leukemia show superior engraftment and higher frequencies of putative leukemia stem cells in, for example, pediatric B cell precursors ALL, AML, and T-ALL (Hartwig and Andre, personal communication).[32,33] Although inherently connected to latency periods of several weeks, monitoring engraftment efficacy and expansion of acute leukemia samples from patients in NSG mice might serve as an additional prognostic marker. Xenotransplantation assays in immunopermissive mice have also been used to explore treatment strategies such as the administration of cytokines with antitumor activity (e.g., IL-27, see Ref. 34), monoclonal antibodies targeting cancer stem cells, or *in vitro* engineered T cell grafts used for adoptive cellular immunotherapy.[35,36] Recent advances in the field of murine xenograft models have been extensively reviewed elsewhere.[37,38]

Despite being the gold standard, there are several major drawbacks of using the mouse as a xenograft model. These include the long duration of time to visible engraftment of human cells, the difficulties in imaging of single or low numbers of engrafting cells and the engraftment heterogeneity between samples requiring laborious experiments with high numbers of animals. The costs of xenograft experiments in murine hosts are furthermore enhanced by the requirement of immune-permissive strains that need specific housing and care and the often performed cotreatment with subcutaneously or intravenously injected cytokines or drugs.

The zebrafish model overcomes some of the requirements mentioned above but obviously cannot supplant the use of mammalian model systems such as the mouse, dog, or monkey that will provide a more evolutionarily similar biological environment for xenotransplanted human cells when used as hosts. However, dog or monkey transplant models are even more laborious and less amenable for genetic studies than mouse models.

Advantages and limitations of zebrafish as hosts for human xenografts

First transplants of human cells into zebrafish were reported in 2005, when Lee *et al.* showed engraftment of human metastatic melanoma cells in zebrafish embryos at the blastula stage.[39] This pioneering work was followed by several studies demonstrating engraftment of a diverse range of human tumors into zebrafish.[40,41]

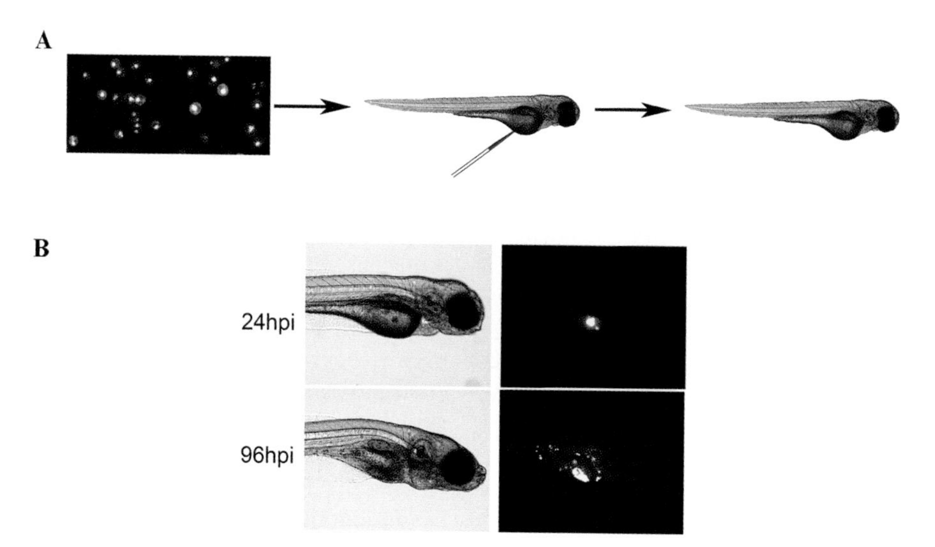

Figure 1. Schematic representation of xenotransplantation in zebrafish. (A) Fluorescently labeled human leukemia cells (THP1) are transplanted into the yolk sac of zebrafish embryos at 48 h postfertilization using a microinjection needle. (B) Proliferation can be visualized and quantified *in vivo* in a matter of days. hpi = hours post injection. (5× objective, Axiocam Rev 3.0 CCD camera and Axiovision Rel 4.0 software by Carl Zeiss Microimaging Inc., Oberkochen, Germany.) Modified from Corkery, D. *et al.*[60]

The zebrafish model provides unique tools for visualization of tumor cell behavior and interaction with host cells. Zebrafish embryos develop *ex utero* and up to one month of age, their larvae are transparent, allowing direct imaging of development, organogenesis, and cancer progression.[42,43] In adults, transparent zebrafish lines enable tracking of transplanted cells down to the single cell level.[44] Several other transgenic lines are available and can facilitate studies on interactions between human cells and specific host factors, as well as exposure to genetically modified host tissue niches.[45] For example, *Tg(fli1-eGFP)* embryos in which the *fli1* promoter is able to drive the expression of EGFP in all blood vessels throughout embryogenesis can be used for analyzing tumor-induced angiogenesis.[46] Other available transgenic lines labeling blood vessels are *Tg(flk1:mCherry)* or *Tg(vegfr2:g-rGFP)*.[47–49] More recently, the *Tg(mpx:GFP)*[i114] line was used to examine neutrophil-mediated experimental metastasis.[50,51] Other transgenic lines that specifically label macrophages *Tg(mpeg1:eGFP)* or platelets *Tg(cd41:eGFP)* could similarly be employed to study the impact of these components of the host's inflammatory response on tumor cell behavior.[52,53]

In addition to providing unique visualization tools, the external development of the zebrafish embryo also allows transplantations at specific developmental stages. If performed at early embryonic stages (blastula stage to 48 h-postfertilization (hpf)), when the zebrafish adaptive immune response has not yet been established, the xenotransplantation procedure does not require immune suppression.[40,41,54] At these early developmental stages, transplantation can be performed via mechanical injection, analogous to the delivery procedures for mRNA or morpholino oligonucleotides, making the procedure amenable for high-throughput screens (Fig. 1A). In contrast, transplantation into adult fish has to be performed manually and requires immune suppression (e.g., by irradiation or dexamethasone treatment).[43,55–57]

Engraftment rates of human cells in mice are assessable following a latency of several weeks, making this assay less suitable for prognostication in the clinical practice. If tumor engraftment rates in zebrafish prove to be similar surrogate markers for clinical disease aggressiveness they can provide readout within a few days. In leukemia, differential engraftment of putative $CD34^+$ stem cells versus $CD34^-$ primary myeloid blasts was observed only four days after xenotransplantation into zebrafish embryo.[58,59] Similarly, human chronic myeloid leukemia (CML) K562 and acute promyelocytic leukemia (APL) NB-4 cell lines that were injected into 48-hpf embryos could be tracked for up to seven days. During this time, engraftment and circulation of these leukemia cells were monitored by live-cell microscopy and proliferation was quantified by enumerating fluorescently labeled human cancer cells at 24- and 72-h postinjection (hpi). Using this proliferation assay, we observed a reproducible increase in leukemia cell numbers within the embryo of approximately threefold after 48 hours (see also Fig. 1B).[60] These developments hearken at the potential for the zebrafish system as experimental readout for leukemia proliferation and progression in real time.

Successful engraftment in zebrafish can be achieved from fewer cells and host numbers can be scaled up easily, improving the validity of statistical analyses. As such, the zebrafish model offers advantages for studies using primary material available in limited quantities.[61,62] Fish can absorb small molecular weight compounds directly from the water, compared to injection or feeding in mice, enabling high-throughput drug screening.[3,4,6,13] Treatments can further be enhanced by incorporating robotic injection and quantifying responses using high-content bioimaging platforms. These additions provide fast and automated whole-animal imaging of xenotransplanted zebrafish embryos in a 96-well format, further supporting expedited screening procedures.[63,64]

However, there are limitations to the use of zebrafish as hosts for human xenotransplants. An obvious hurdle is the difference in temperature requirement for the maintenance of human cells compared with zebrafish embryos, since the latter are usually maintained at 28 °C.[65] Human cells are usually incubated at 37 °C, reflecting human body temperature. Although consistent comparative studies on specific cell behaviors in response to modulating temperatures between 31 °C and 37 °C are not available to our knowledge, several groups have shown that human cells kept at temperatures as low as 31 °C are still able to proliferate and form colonies in soft agar clonogenic assays.[39,58,66] Moreover, results from different groups including our laboratory suggest that zebrafish embryos kept at 35 °C develop normally, while xenotransplanted human (melanoma and leukemia) cells still readily proliferate.[40,58,60] Thus, higher zebrafish maintenance temperatures around 34–35 °C should be considered when evaluating biological characteristics of xenotransplated human cells.

Another concern relates to cellular size. Zebrafish cells, as well as zebrafish vessels and other anatomic structures are smaller than the corresponding human structures.[67] However, circulation of leukemic cells and migration of even larger human solid tumor cells, such as sarcoma cells, has been repeatedly demonstrated in zebrafish indicating that passage is possible (Fig. 2A and unpublished data).[68] Moreover, we have used the nuclear stain, DRAQ5, to confirm that the human cells under observation are intact and not damaged due to injection or passage through the embryo.[60] Thus, zebrafish vessels, even

A

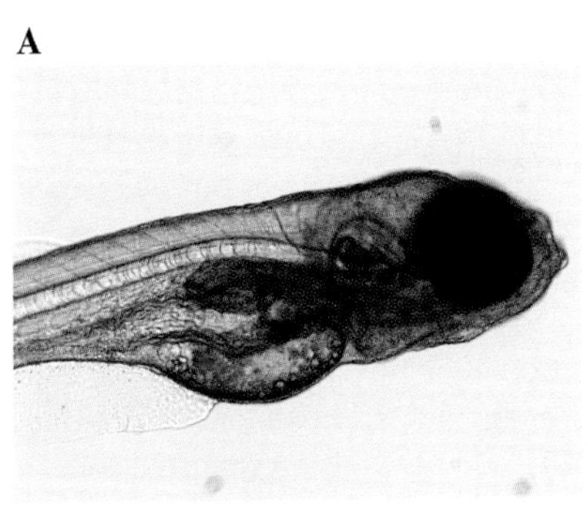

24hpi, 5x

B

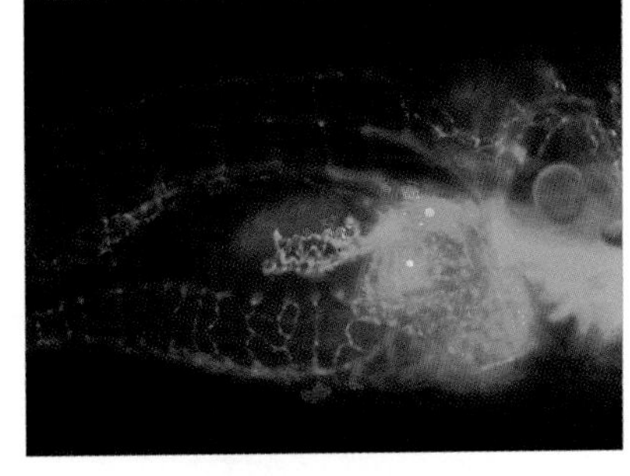

96hpi, 10x

Figure 2. Transgenic zebrafish lines can be used to monitor angiogenic response to solid tumor proliferation. (A) Brightfield image of sarcoma cells (TC32) inside the *Tg(fli1a:eGFP)* transgenic zebrafish embryo at 24 hpi. (B) Fluorescence imaging of the same embryos at 96 hpi showing interaction between the vasculature (labeled green) and the tumor cells (labeled red). hpi = hours post injection. (5× and 10× objective, Axiocam Rev 3.0 CCD camera and Axiovision Rel 4.0 software by Carl Zeiss Microimaging Inc.) Modified from Corkery, D. *et al.*[60]

Table 1. Advantages and limitations of performing human xenograft experiments in zebrafish

Advantages	Limitations
• Large numbers of offspring • Transplantation at embryonic stages possible • Permeability of zebrafish to small molecules delivered by the water enables drug screens • External development and existence of transparent lines allow visualization and bioluminescence readouts • No immune rejection in early transplantation settings • Small numbers of cells per animal required for xenotransplantation • Visualization (transparency, transgenic lines, measurable bioluminescence) • Fast readout (hours to days)	• Little knowledge about niche structures and microenvironmental cues • Different biological environment in developing zebrafish for transplanted adult cells • Differences in size (small zebrafish organs/vessels—large human cells) • Different maintenance temperatures (overcome at 35 °C) • Absent organs (e.g., breast, lung) (may be overcome using analogous structures, e.g., gills) • Fewer possibilities for orthotopic transplantation • Limited numbers of zebrafish antibodies available • No adult immuno-permissive zebrafish lines available yet

though narrower in caliber, appear large enough to allow passage of most human cells. Whether in some regions vessel elasticity may aid this process is not known. Interpretation of results obtained from zebrafish xenografts are challenged by virtue of differences in host niche and environmental factors. Cells transplanted into the growing embryo are exposed to an entirely different biological environment. To enhance specificity, results from screens performed on zebrafish embryos could be reevaluated in adult fish recipients and shared findings would suggest the efficacy of a given compound is more robust and extends beyond a specific structurally developmental context. However, missing supportive cytokines and/or adequate protein interactions due to lack of protein homology between zebrafish and humans may impair *in vivo* behavior of human cells even when experiments are performed in adult fish. Such biases have been demonstrated in mice and have led to the generation of immunopermissive animals with knock-in human cytokines.[28,38] Equivalent approaches may be considered in zebrafish, where to date, there are no reports on successful engraftment of healthy hematopoietic $CD34^+$ stem/progenitor cells.

Advantages and limitations of human xenografts in zebrafish are summarized in Table 1.

Engraftment of primary cells

Most of the xenotransplantation studies in zebrafish have been performed using cancer cell lines and transplantation at various sites and different developmental stages (e.g., directly at the blastula stage using 1–200 cancer cells,[39,66,69–73] in two days postfertilization (dpf) embryos using 50–2000 human cells, transplanted in the yolk close to the duct of Cuvier or in 30-day-old fish using 50–300 cells transplanted into the peritoneal cavity).[43] Successful engraftment has also been reported by injection into the hindbrain ventricle, the perivitelline cavity, or the posterior cardinal vein.[40,41,58,60,74–80] Orthotopic transplantation, as reported in mice, is less well established in fish and in certain instances may not be possible due to the absence of the corresponding organ (e.g., breast, lung). In these cases, the functionally equivalent organ, such as the gills for the lungs, may approximate orthotopic transplantation.[78,81,82] Details on reported transplant procedures are summarized in Table 2.

To date, only a few reports demonstrate zebrafish xenografts from primary human cells. Bagowski *et al.* for example report successful engraftment of small pieces or dissociated cells from human pancreas, colon, and stomach carcinoma as well as from

Table 2. Transplant modalities of human cells into zebrafish

Tissue/Cell line	Injection timepoint	Location	Number of injected cells	Label	Reference
Metastatic melanoma line C8161	Blastula stage		$<$ 100 cells	1	39
Epidermal melanocytes		Near blastodisk			
Dermal fibroblasts					
Aggressive and non-aggressive melanoma cell lines	Blastula stage		50–100 cells	1	69
Neuroblastoma cell line U87-L	6 hpf	Yolk	~100 cells	1	71
Adenocarcinoma cell line MDA-MB435	30 dpf	Peritoneal cavity	0–300 cells	1	43
Fibrosarcoma cell line HT1080					
Melanoma cell line B16					
Breast cancer cell line MD-MB-435	48 hpf	Yolk close to the duct of cuvier	1,000–2,000 cells in matrigel	1	41
Ovarial carcinoma cell line A2780					
Melanoma cell line WM-266–4	48 hpf	Yolk, (hindbrain ventricle, circulation)	50–200 cells	2	40
Colorectal cancer cell line SW620					
Pancreas cancer cell line FG CAS/Crk					
Fibroblast cell line CCD-1092Sk					
Breast cancer cell line MDA MB-231	48 hpf	Perivitelline cavity	100–00 cells	2	74
Ovarial carcinoma cell line OVCAR 8					
PaTu-S, PaTu-T, and other pancreas tumor cell lines	48 hpf	Yolk	200 cells	2	78, 79
Cells of pancreatic tumors	48 hpf	Yolk	Tissue pieces and dissociated cells		
Human pancreatic cancer lines PaTu-S and Panc-1	48 hpf	Yolk	1,000–2,000 cells	2	75
Lung adenocarcinoma cell line A549	Blastula stage	Blastocyst		1	72
Glioma/glioblastoma cell line U251	Blastula	Close to blastodisk	1–100 cells	1	70
Breast cancer cell line MDA-MB-231	48 hpf	Yolk	500 cells in matrigel	1, 2	76

Continued

Table 2. Continued

Tissue/Cell line	Injection timepoint	Location	Number of injected cells	Label	Reference
Prostate metastatic cell line C4–2B	Blastula stage			2	73
Ovarialcarcinoma cell line OVCA-433	48 hpf	Yolk	~100 cells	2	80
Glioma/glioblastoma cell line U251	Blastula stage		50–200 cells	1	66
Fibrocsarcoma cell line 1080	48 hpf	Duct of cuvier	1,000–2,000 cells in matrigel	-	77
Leukemia cell lines K562, Jurkat, NB-4	48 hpf	Posterior cardinal vein	50–200 cells	2	58
$CD34^{+/-}$ cells from AML patient blasts					
Leukemia cell lines K562 und, NB-4	48 hpf	Yolk	200 cells	2	60
Human prostate cancer cell line PC3	48 hpf	Yolk	~100 cells	1	63
Breast carcinoma cell line MDA-MB-23	48 hpf	Duct of cuvier	40–400 cells	2	51
Human prostate cancer cell line PC3					

1 = Cells are stably transduced with a reporter construct. 2 = Cells are labelled with a cell tracker. 3 = Cells are visible through bioluminescence.

chronic pancreatitis.[78,79] Upon transplantation into the yolk of 48-hpf embryos, tumor cells invade the embryo and form micrometastases as early as 24-hpi. Importantly, this approach allowed discrimination between infiltrating pancreatic adenocarcinoma and noninvasive chronic pancreatitis.[78,79]

Zebrafish engraftment of human leukemia cell lines and primary cells has also been reported. Upon transplantation of 50–200 leukemic cells into the yolk of 48-hpf zebrafish embryos, leukemic cells were shown to proliferate and migrate throughout the fish and could be followed *in vivo* for several days.[58,60] Studies using primary leukemia patient-derived bone marrow samples furthermore demonstrate the feasibility of determining drug response of primary samples in real time with therapy (Balci, Dellaire, and Berman, unpublished results). Importantly, leukemic cell circulation was shown to be an active process, requiring functional living cells, since fixed controls remained in the yolk. Furthermore, in acute myeloid leukemia, engraftment was demonstrated from $CD34^+$ putative leukemia stem cells but not from $CD34^-$ cells, indicating that zebrafish models may reflect the biology of the disease in a similar way as mouse models do and enable studies on tumorigenicity and tumor stem cells.[59,83,84] Zebrafish may in fact be ideally suited for limiting dilution cell transplantation assays to determine self-renewal cell frequency, as these experiments can be easily performed with high numbers of fish, and single tumor-initiating cells can be tracked *in vivo* in this system. Although it is attractive to postulate that the zebrafish could be used to determine leukemia-initiating cell populations from human bone marrow or peripheral blood, to date limiting dilution assays have only been performed in syngeneic zebrafish.[61,85]

While successful human tumor engraftment into zebrafish has been reported by several groups, engraftment of healthy human cells appears, as in mice, to be more cumbersome. Unlike leukemic cells, healthy human $CD34^+$ hematopoietic stem and progenitor cells from cord blood origin rapidly disappear after injection into zebrafish embryos, indicating that essential microenvironmental cues are missing in this transplant setting.[58] As mentioned above, it remains to be established if supplementation with human cytokines can complement this deficiency. In the embryo, in addition to creation of knock-in transgenic lines, cytokine addition to the water could be performed, since molecules including chemicals or peptides dissolved in water freely diffuse into the zebrafish in the presence of a carrier at various embryonic stages.[6,13] Another strategy would be to transplant hematopoietic cells orthotopically into the zebrafish kidney marrow, the place where adult hematopoietic stem cells reside in the fish. Engraftment efficiency shows variation depending on the transplanted tissue type. For example, human healthy melanocytes can be retrieved from the skin of xenotransplanted zebrafish hosts,[39] indicating that zebrafish embryo contain niches supporting outgrowth of human melanocytic cells.

Thus, as previously reported in mice, xenotransplantation of healthy cells to zebrafish appears more difficult in comparison to malignant ones. As such, this is a fertile area for future research requiring further optimization with respect to transplantation techniques, host age effects, supplementation of cytokines and growth factors, and other species specific and developmental factors yet to be determined.

From metastasis to angiogenesis: zebrafish xenograft studies and cancer cell biology

Several reports show successful engraftment of human malignant cells of different origins into zebrafish, and *in vivo* assessment of, for example, invasiveness, metastatic behavior,[39,69,71,78] angiogenesis,[40,41,43,75,86] and response to anticancer therapies.[58,60,66,70,72,73,75–80] Xenotransplated tumor cells can be retrieved from the fish and used for molecular analysis, as shown among others by Weiss *et al.* in their analysis of the microRNA miR-10a during metastasis of pancreas carcinoma cells.[79] Thus, the zebrafish xenograft model provides a unique assay for certain areas of tumor biology and may complement existing murine models.

Studies of tumor invasiveness and metastasis

As about 90% of all cancer deaths arise from metastatic expansion of primary tumors, it is of great importance to understand the metastatic behavior of cancer cells.[87] This is often limited by the inability to visualize tumor cell behavior in real time in animal models. Due to its transparency, the zebrafish is ideally suited for this purpose. Prior to transplantation, cells can be stably transduced with a luciferase reporter construct enabling *in vivo* bioluminescent tracking of migratory cells.[71] The validity of this model for the assessment of invasion and metastasis has been validated by Bagowski *et al.*, who transplanted two sister human pancreatic tumor cell lines showing high (PaTu8988-T cells lacking E-cadherin) and low migratory capacity (PaTu8988-S cells expressing E-cadherin) respectively into 48-hpf zebrafish embryos. While PaTu8988-T cells revealed invasiveness and strong formation of micrometastasis, PaTu8988-S cells remained in the yolk.[78] Continuing this line of work, Weiss *et al.* identified the microRNA miR-10a to be a key mediator of metastatic behavior in pancreatic cancer.[79] While miR-10a was expressed in highly migratory PaTu8988-T cells, it was not detected in the low migratory sister cell line PaTu8988-S. Exposure to morpholino oligonucleotides specifically inhibiting miR-10a blocked tissue invasion and metastasis formation of xenotransplanted PaTu8988-T cells. In contrast, PaTuT-8988-S cells became highly invasive upon miR-10a overexpression. Moreover, inhibition of retinoic acid (RA), an upstream regulator of miR-10a, suppresses metastasis from pancreatic cancer cells in zebrafish xenografts. Additional genes implicated in metastasis, such as Y-box binding protein 1 (YB-1)[88] are also currently being examined using zebrafish xenograft approaches.[68]

By performing a siRNA kinome library screen Pardo *et al.* identified ribosomal S6 kinase 1 (RSK1) as a key regulator of metastasis.[72] Downregulation of RSK1 in the lung adenocarcinoma cell line A549-enhanced cell migration in an *in vitro* cell motility assay. Patients with RSK1 negative lung tumors showed increased number of metastases in comparison to patients with RSK1 positive tumors. The authors confirmed their findings in zebrafish xenograft experiments, where siRNA-mediated RSK1 downregulation strongly enhanced invasion in A549 xenotransplanted cells.

Evaluation of tumor-induced angiogenesis

A solid body of evidence has emerged showing that the zebrafish is a suitable model system to study angiogenesis and neovascularization occurring with cancer development in live animals.[40,41,43,75,86] First attempts to study tumor-induced angiogenesis have been performed by Haldi *et al.*, who transplanted WM-266-4 melanoma cells into the yolk of 2-dpf zebrafish. The xenotransplanted cells rapidly formed masses and recruited zebrafish endothelial cells that were able to infiltrate these masses and lead to vessel formation.[40] These results were confirmed by follow-up studies from Nicoli *et al.*, which showed angiogenic responses upon transplantation of human breast adenocarcinoma MDA-MB-435 and melanoma B16 cells into 2-dpf zebrafish embryos.[41,86] The newly formed vessels expressed known markers like FLI-1, VEGFR2/KDR, and VE-cadherin. Interestingly, responses could be modulated by genetic modification of tumor cells with human FGF or VEGF prior to transplantation. Strong suppression of vessel formation was observed from cells defective in the production of these growth factors.

The *Tg(fli1:eGFP)* zebrafish line, which expresses green fluorescence in the vasculature, provides an additional valuable tool to monitor vascular recruitment in the metastatic process. Recruitment of new blood vessels to the vicinity of the xenografted tumor mass can be visualized in real time (Fig. 2B and unpublished data),[68] similar to the more established chicken chorioallantoic membrane (CAM) angiogenesis model, but rivaling this system for more mechanistic studies due to the ease by which the zebrafish host can be genetically modified.[89] In another study, silencing of the two LIM domain kinases LIMK1 and LIMK2 was shown to reduce tumor cell-induced neovascularization in human pancreatic tumor cell fish xenografts.[75] Furthermore, using the so-called "late zebrafish xenotransplantation model," Stoletov *et al.* studied microtumor formation and angiogenesis after transplantation of cells from the adenocarcinoma cell line MDA-MB435, fibrosarcoma cell line HT1080, and melanoma cell line B16 in the peritoneal cavity of 30-day-old zebrafish.[43] They showed that indeed tumor cells survive, invade, and remodel the fish vasculature and that these effects are mediated by *RhoC*. Modulation of VEGF levels induces however more complex responses,[90] as more recently demonstrated in zebrafish xenograft studies by the group of Snaar-Jagalska, which showed that administration of VEGFR inhibitors blocked tumor vascularization but enhanced migration of neutrophils, the consequence of which was the promotion of tumor invasion and the formation of micrometastasis.[51]

Clearly, the zebrafish embryo represents an important new tool in studying tumor-induced angiogenesis and is the subject of several excellent reviews.[91–93]

Zebrafish xenografts as tools for drug discovery and *in vivo* assessment of anticancer therapies

Evaluation of new therapeutic agents requires a long and complex process that involves validation in animal models to study drug efficacy and toxicity. In recent years, the zebrafish has become a cost-effective alternative to mammals.[3] The first experiments exposing human cancer xenografts to anticancer therapies in zebrafish have been performed in 2007.[76] Using zebrafish xenografts from breast cancer and melanoma cell lines, Harfouche *et al.* showed that nanoparticle-enabled targeting of the PI3K pathway by LY294002 (NP-LY) inhibited downstream Akt phosphorylation and thereby suppressed tumor-induced angiogenesis. In melanoma, further research by Topczewska *et al.* demonstrated that transplantation of metastatic cells disrupts zebrafish development and identified Nodal as a key regulator of disease aggressiveness.[69] These observations triggered a series of follow-up studies on Nodal as a novel therapeutic or prognostic marker in melanoma.[94–97] Human glioma cells (U87MG) also showed robust engrafting upon transplantation into the yolk of 2-dpf zebrafish embryos, which could be inhibited by treatment with specific drugs such as the γ-secretase inhibitors DAPT and compound E.[71]

To evaluate a nanotechnological approach using plasmonic nanobubbles (PNBs) as an *in vivo* adjustable theranostic agent, human prostate cancer cells (C4–2B) were labeled with gold nanoparticles and transplanted into zebrafish. PNBs were then selectively generated by short laser pulses leading to mechanic ablation of cancer cells surrounded by the PNBs without damage of nearby tissue.[73] Further studies showed efficacy and biocompatibility of single-walled carbon nanotubes (SWCNT) for specific delivery of antiangiogenetic drugs such as thalidomide against transplanted human HT1080 sarcoma cells.[77]

In ovarian carcinoma, cisplatin treatment has been reported to activate ERK and thereby induce epithelial-to-mesenchymal transition (EMT), generating chemoresistant mesenchymal stem cell-like cells.[80] In 48-hpf zebrafish embryos, cisplatin treatment of xenotransplanted human OVCA 433 ovarian cancer cells enhanced cellular migration, an effect that could be inhibited by cotreatment with U0126, a selective MEK1/2 inhibitor suppressing ERK2. Together, these findings suggest that cotreatment of cisplatin and ERK2-inhibitors may reduce the development of chemoresistance to platin-based therapies, which is a crucial mechanism of disease progression in ovarian carcinoma.

Finally, in leukemia xenografts, drugs like imatinib (to target BCR-ABL1 in engrafted K562 cells),[58,60] all-trans retinoic acid (ATRA) (to target PML-RARA in engrafted NB-4 cells),[58,60] cyclophosphamide and mafosfamide (to target Jurkat cell growth)[58] were shown to reduce the leukemic burden with surprisingly little effect on animal development during the reported follow-up window. Importantly, reciprocal experiments, i.e., using ATRA on K562 cells and imatinib on NB4 cells, failed to show a significant inhibition in proliferation, confirming that the reduction in observed cell numbers was due to conserved targeted tumor–drug interactions in the context of the zebrafish xenotransplantation model, rather than nonspecific cellular toxicity.[60]

However, drug screening in the zebrafish is not without its caveats. For example, care must be taken in employing drugs with teratogenic effects on vertebrate development such as ATRA, which at high and prolonged doses results in embryonic lethality.[58] In addition, drug doses active against engrafted human cells within the zebrafish, when compounds are added to the water, may be many times higher than tolerated by human cells in tissue culture assays and could have little relationship to the final pharmacological dose in humans. Nonetheless, the relative efficacy of compounds can be determined rapidly once the maximum tolerated dose range for the zebrafish embryo has been determined for individual drugs (or drug classes) and embryo stage.[60] It is also possible that pro-drugs or other compounds that function as metabolites may not be metabolized the same way in zebrafish, which express fish-specific subclasses of p450 cytochrome genes of the CYP3 family.[98] Finally, compounds may also be actively transported or detoxified by zebrafish embryos before they can reach active concentrations *in vivo*. Nonetheless, despite these caveats the conservation of drug response between human cells and the zebrafish is high, with an estimated 50–70% of drugs known to be active against human cells in tissue culture retaining their activity when screened against zebrafish (e.g., drugs directed against the cell cycle).[3]

Zebrafish xenografts as a model for radiotherapy approaches

Zebrafish models have been also used for *in vivo* radiotherapy studies. For example, efficacy and toxicity of treatment with the radiosensitizer 4′-bromo-3′nitropropiophenone (NS-123) was tested on U251 neuroblastoma cells transplanted in zebrafish embryos.[70] Cotreatment with NS-123 and irradiation therapy resulted in markedly reduced numbers of surviving tumor cells in zebrafish xenografts. Importantly, no toxic effects were observed through NS-123 treatment. These effects were successfully reproduced in murine xenograft models, underscoring the value of the zebrafish model. Following a similar strategy, Geiger *et al.* showed that sensitization with the DNA-methylating agent temozolomide enhances the inhibitory effects of radiotherapy on proliferation and angiogenesis in U251-RFP human glioma cells transplanted in zebrafish embryo, without grossly affecting animal development.[66]

Conclusions and perspectives

The zebrafish has evolved as a powerful vertebrate model for genetic studies of embryonic development and oncogenesis. Some of the advantages of using the zebrafish compared to other animal models includes the feasibility of reverse and forward genetic approaches,[99] including mutagenesis screens, and high-throughput screens of small molecules that can be applied through the water. Another key advantage of the zebrafish is the ease by which developmental and oncogenic processes can be visualized due to the transparency of their larvae and the availability of adult transparent zebrafish.[44]

Human malignant cells from a variety of malignancies have been successfully xenotransplanted into zebrafish embryo or adult fish, providing models for genetic studies of tumor-induced angiogenesis, invasiveness, and of therapeutic drug response, which together provide a host of options for antitumor drug screening. Furthermore, the incorporation of specific transgenic zebrafish lines expressing fluorescently labeled immune cells or tissues into these xenograft assays will provide an unprecedented window into tumor cell and microenvironmental interactions during cancer development and in response to treatment. In the future, these strategies can be enhanced through the genetic manipulation of both the graft and the host, to explore critical factors in cancer cell behavior, such as metastasis, drug resistance, and apoptosis. Serial transplants have not yet been undertaken for human cells engrafted in zebrafish, but represent an important next step for evaluating cancer cell self-renewal and could provide a means of determining the frequency and potency of cancer initiating cells from patient biopsies.

However, xenotransplantation of human cells into zebrafish is associated with certain limitations, including differences in tissue-niche and microenvironmental cues between the fish and engrafted human cells, and difficulty or impossibility of orthotopic transplantation due to the absence of certain organs (e.g., lungs, mammary tissue, prostate, etc). Through the recent growth and technical advancements in this field, the efficiency and robustness of the zebrafish human cancer cell xenotransplantation holds much promise for providing new insights into cancer cell and microenvironmental interactions and as a tool for novel drug discovery. Finally, by optimizing techniques for primary cell engraftment, the zebrafish may provide for the first time, a live animal model system for evaluating individual patient drug responses in a clinically relevant setting.

Acknowledgments

This study was supported by grants from the Deutsche Krebshilfe (Max-Eder Program), the Deutsche Forschungsgemeinschaft (SFB773), the University of Tuebingen (Fortüne Program) and the Boehringer Ingelheim Foundation (Exploration Grant) for C.L., postdoctoral fellow awards from the Beatrice Hunter Cancer Research Institute Cancer Research Training Program and the IWK Health Centre for T.B.B., U.F.H., and M.C.A. were funded by the Deutsche Forschungsgemeinschaft KFO 183/TP 4 and 6. G.D. is Canadian Institutes of Health Research (CIHR) New Investigator and Senior Scientist of the Beatrice Hunter Cancer Research Institute and his work is supported in part by an operating grant from the Canadian Breast Cancer Foundation Atlantic. J.N.B is the Cancer Care Nova Scotia Peggy Davison Clinician Scientist and Senior Scientist of the Beatrice Hunter Cancer Research Institute. We thank Dale Corkery, Department of Biochemistry and Molecular Biology, and Chansey Veinotte, Department of Microbiology and Immunology, Dalhousie University, Halifax, NS, Canada, for providing the figures.

Conflicts of interest

The authors declare no conflicts of interest.

References

1. Mullins, M.C. *et al.* 1994. Large-scale mutagenesis in the zebrafish: in search of genes controlling development in a vertebrate. *Curr. Biol.* **4:** 189–202.
2. Stern, H.M. & L.I. Zon. 2003. Cancer genetics and drug discovery in the zebrafish. *Nat. Rev. Cancer* **3:** 533–539.
3. Zon, L.I. & R.T. Peterson. 2005. In vivo drug discovery in the zebrafish. *Nat. Rev. Drug Discov.* **4:** 35–44.
4. den Hertog, J. 2005. Chemical genetics: drug screens in zebrafish. *Biosci. Rep.* **25:** 289–297.
5. Amsterdam, A. & N. Hopkins. 2006. Mutagenesis strategies in zebrafish for identifying genes involved in development and disease. *Trends Genet.* **22:** 473–478.
6. Kari, G., U. Rodeck & A.P. Dicker. 2007. Zebrafish: an emerging model system for human disease and drug discovery. *Clin. Pharmacol. Ther.* **82:** 70–80.
7. Granato, M. & C. Nusslein-Volhard. 1996. Fishing for genes controlling development. *Curr. Opin. Genet. Dev.* **6:** 461–468.
8. Chen, J.N. & M.C. Fishman. 1996. Zebrafish tinman homolog demarcates the heart field and initiates myocardial differentiation. *Development* **122:** 3809–3816.
9. Amatruda, J.F. *et al.* 2002. Zebrafish as a cancer model system. *Cancer Cell* **1:** 229–231.
10. Berghmans, S. *et al.* 2005. Making waves in cancer research: new models in the zebrafish. *Biotechniques* **39:** 227–237.
11. Goessling, W., T.E. North & L.I. Zon. 2007. New waves of discovery: modeling cancer in zebrafish. *J. Clin. Oncol.* **25:** 2473–2479.
12. Liu, S. & S.D. Leach. 2011. Zebrafish models for cancer. *Annu. Rev. Pathol.* **6:** 71–93.
13. Parng, C. *et al.* 2002. Zebrafish: a preclinical model for drug screening. *Assay Drug Dev. Technol.* **1:** 41–48.
14. Amatruda, J.F. & E.E. Patton. 2008. Genetic models of cancer in zebrafish. *Int. Rev. Cell Mol. Biol.* **271:** 1–34.

15. Payne, E. & T. Look. 2009. Zebrafish modelling of leukaemias. *Br. J. Haematol.* **146:** 247–256.
16. Sharpless, N.E. 2005. INK4a/ARF: a multifunctional tumor suppressor locus. *Mutat. Res.* **576:** 22–38.
17. Goldman, J.P. *et al.* 1998. Enhanced human cell engraftment in mice deficient in RAG2 and the common cytokine receptor gamma chain. *Br. J. Haematol.* **103:** 335–342.
18. Ito, M. *et al.* 2002. NOD/SCID/gamma(c)(null) mouse: an excellent recipient mouse model for engraftment of human cells. *Blood* **100:** 3175–3182.
19. Shultz, L.D. *et al.* 2005. Human lymphoid and myeloid cell development in NOD/LtSz-scid IL2R gamma null mice engrafted with mobilized human hemopoietic stem cells. *J. Immunol.* **174:** 6477–6489.
20. Ishikawa, F. *et al.* 2007. Chemotherapy-resistant human AML stem cells home to and engraft within the bone-marrow endosteal region. *Nat. Biotechnol.* **25:** 1315–1321.
21. Agliano, A. *et al.* 2008. Human acute leukemia cells injected in NOD/LtSz-scid/IL-2Rgamma null mice generate a faster and more efficient disease compared to other NOD/scid-related strains. *Int. J. Cancer* **123:** 2222–2227.
22. Simpson-Abelson, M.R. *et al.* 2008. Long-term engraftment and expansion of tumor-derived memory T cells following the implantation of non-disrupted pieces of human lung tumor into NOD-scid IL2Rgamma(null) mice. *J. Immunol.* **180:** 7009–7018.
23. Sanchez, P.V. *et al.* 2009. A robust xenotransplantation model for acute myeloid leukemia. *Leukemia* **23:** 2109–2117.
24. Diamanti, P., C.V. Cox & A. Blair. 2012. Comparison of childhood leukemia initiating cell populations in NOD/SCID and NSG mice. *Leukemia.* **26:** 376–380.
25. Martin-Padura, I. *et al.* 2010. Sex-related efficiency in NSG mouse engraftment. *Blood* **116:** 2616–2617.
26. Bankert, R.B. *et al.* 2011. Humanized mouse model of ovarian cancer recapitulates patient solid tumor progression, ascites formation, and metastasis. *PLoS One* **6:** e24420.
27. Feuring-Buske, M. *et al.* 2003. Improved engraftment of human acute myeloid leukemia progenitor cells in beta 2-microglobulin-deficient NOD/SCID mice and in NOD/SCID mice transgenic for human growth factors. *Leukemia* **17:** 760–763.
28. Wunderlich, M. *et al.* 2010. AML xenograft efficiency is significantly improved in NOD/SCID-IL2RG mice constitutively expressing human SCF, GM-CSF and IL-3. *Leukemia* **24:** 1785–1788.
29. Rongvaux, A. *et al.* 2011. Human thrombopoietin knockin mice efficiently support human hematopoiesis in vivo. *Proc. Natl. Acad. Sci. USA* **108:** 2378–2383.
30. Malaise, M. *et al.* 2011. Stable and reproducible engraftment of primary adult and pediatric acute myeloid leukemia in NSG mice. *Leukemia* **25:** 1635–1639.
31. Rathinam, C. *et al.* 2011. Efficient differentiation and function of human macrophages in humanized CSF-1 mice. *Blood* **118:** 3119–3128.
32. Morisot, S. *et al.* 2010. High frequencies of leukemia stem cells in poor-outcome childhood precursor-B acute lymphoblastic leukemias. *Leukemia* **24:** 1859–1866.
33. Meyer, L.H. *et al.* 2011. Early relapse in all is identified by time to leukemia in NOD/SCID mice and is characterized by a gene signature involving survival pathways. *Cancer Cell* **19:** 206–217.
34. Canale, S. *et al.* 2011. Interleukin-27 inhibits pediatric B-acute lymphoblastic leukemia cell spreading in a preclinical model. *Leukemia* **25:** 1815–1824.
35. Distler, E. *et al.* 2008. Acute myeloid leukemia (AML)-reactive cytotoxic T lymphocyte clones rapidly expanded from CD8(+) CD62L((high)+) T cells of healthy donors prevent AML engraftment in NOD/SCID IL2Rgamma(null) mice. *Exp. Hematol.* **36:** 451–463.
36. Stevanovic, S. *et al.* 2012. Human allo-reactive CD4+ T cells as strong mediators of anti-tumor immunity in NOD/scid mice engrafted with human acute lymphoblastic leukemia. *Leukemia.* **26:** 312–322.
37. Manz, M.G. & J.P. Di Santo. 2009. Renaissance for mouse models of human hematopoiesis and immunobiology. *Nat. Immunol.* **10:** 1039–1042.
38. Willinger, T. *et al.* 2011. Improving human hemato-lymphoid-system mice by cytokine knock-in gene replacement. *Trends Immunol.* **32:** 321–327.
39. Lee, L.M. *et al.* 2005. The fate of human malignant melanoma cells transplanted into zebrafish embryos: assessment of migration and cell division in the absence of tumor formation. *Dev. Dyn.* **233:** 1560–1570.
40. Haldi, M. *et al.* 2006. Human melanoma cells transplanted into zebrafish proliferate, migrate, produce melanin, form masses and stimulate angiogenesis in zebrafish. *Angiogenesis* **9:** 139–151.
41. Nicoli, S. *et al.* 2007. Mammalian tumor xenografts induce neovascularization in zebrafish embryos. *Cancer Res.* **67:** 2927–2931.
42. Spitsbergen, J. 2007. Imaging neoplasia in zebrafish. *Nat. Methods* **4:** 548–549.
43. Stoletov, K. *et al.* 2007. High-resolution imaging of the dynamic tumor cell vascular interface in transparent zebrafish. *Proc. Natl. Acad. Sci. USA* **104:** 17406–17411.
44. White, R.M. *et al.* 2008. Transparent adult zebrafish as a tool for in vivo transplantation analysis. *Cell Stem Cell* **2:** 183–189.
45. Kawakami, K. 2005. Transposon tools and methods in zebrafish. *Dev. Dyn.* **234:** 244–254.
46. Lawson, N.D. & B.M. Weinstein. 2002. In vivo imaging of embryonic vascular development using transgenic zebrafish. *Dev. Biol.* 248: 307–318.
47. Chi, N.C. *et al.* 2008. Foxn4 directly regulates tbx2b expression and atrioventricular canal formation. *Genes Dev.* **22:** 734–739.
48. Jin, S.W. *et al.* 2005. Cellular and molecular analyses of vascular tube and lumen formation in zebrafish. *Development* **132:** 5199–5209.
49. Cross, L.M. *et al.* 2003. Rapid analysis of angiogenesis drugs in a live fluorescent zebrafish assay. *Arterioscler Thromb. Vasc. Biol.* **23:** 911–912.
50. Renshaw, S.A. *et al.* 2006. A transgenic zebrafish model of neutrophilic inflammation. *Blood* **108:** 3976–3978.
51. He, S. *et al.* 2012. Neutrophil-mediated experimental metastasis is enhanced by VEGFR inhibition in a zebrafish

xenograft model. *J. Pathol.* doi: 10.1002/path.4013 [Epub ahead of print].
52. Ellett, F. *et al.* 2011. mpeg1 promoter transgenes direct macrophage-lineage expression in zebrafish. *Blood* **117:** e49–e56.
53. Lin, H.F. *et al.* 2005. Analysis of thrombocyte development in CD41-GFP transgenic zebrafish. *Blood* **106:** 3803–3810.
54. Lam, S.H. *et al.* 2004. Development and maturation of the immune system in zebrafish, Danio rerio: a gene expression profiling, in situ hybridization and immunological study. *Dev. Comp. Immunol.* **28:** 9–28.
55. Patton, E.E. *et al.* 2005. BRAF mutations are sufficient to promote nevi formation and cooperate with p53 in the genesis of melanoma. *Curr. Biol.* **15:** 249–254.
56. Traver, D. *et al.* 2003. Transplantation and in vivo imaging of multilineage engraftment in zebrafish bloodless mutants. *Nat. Immunol.* **4:** 1238–1246.
57. Langenau, D.M. *et al.* 2004. In vivo tracking of T cell development, ablation, and engraftment in transgenic zebrafish. *Proc. Natl. Acad. Sci. USA* **101:** 7369–7374.
58. Pruvot, B. *et al.* 2011. Leukemic cell xenograft in zebrafish embryo for investigating drug efficacy. *Haematologica* **96:** 612–616.
59. Wang, J.C. *et al.* 1998. High level engraftment of NOD/SCID mice by primitive normal and leukemic hematopoietic cells from patients with chronic myeloid leukemia in chronic phase. *Blood* **91:** 2406–2414.
60. Corkery, D.P., G. Dellaire & J.N. Berman. 2011. Leukaemia xenotransplantation in zebrafish–chemotherapy response assay in vivo. *Br. J. Haematol.* **153:** 786–789.
61. Smith, A.C. *et al.* 2010. High-throughput cell transplantation establishes that tumor-initiating cells are abundant in zebrafish T-cell acute lymphoblastic leukemia. *Blood* **115:** 3296–3303.
62. Eguiara, A. *et al.* 2011. Xenografts in zebrafish embryos as a rapid functional assay for breast cancer stem-like cell identification. *Cell Cycle* **10:** 3751–3757.
63. Ghotra, V.P. *et al.* 2012. Automated whole animal bio-imaging assay for human cancer dissemination. *PLoS One* **7:** e31281.
64. Carvalho, R. *et al.* 2011. A high-throughput screen for tuberculosis progression. *PLoS One* **6:** e16779.
65. Detrich, H.W., 3rd, M. Westerfield & L.I. Zon. 2010. The zebrafish: cellular and developmental biology, part A. Preface. *Methods Cell Biol.* **100:** xiii.
66. Geiger, G.A., W. Fu & G.D. Kao. 2008. Temozolomide-mediated radiosensitization of human glioma cells in a zebrafish embryonic system. *Cancer Res.* **68:** 3396–3404.
67. Davidson, A.J. & L.I. Zon. 2004. The 'definitive' (and 'primitive') guide to zebrafish hematopoiesis. *Oncogene* **23:** 7233–7246.
68. Veinotte, C.J. *et al.* Zebrafish and murine renal sub-capsule xenografts identify a role of Y-Box binding protein (YB-1) in the metastasis of Ewing family tumors. Unpublished work.
69. Topczewska, J.M. *et al.* 2006. Embryonic and tumorigenic pathways converge via Nodal signaling: role in melanoma aggressiveness. *Nat. Med.* **12:** 925–932.
70. Lally, B.E. *et al.* 2007. Identification and biological evaluation of a novel and potent small molecule radiation sensitizer via an unbiased screen of a chemical library. *Cancer Res.* **67:** 8791–8799.
71. Zhao, H. *et al.* 2009. A screening platform for glioma growth and invasion using bioluminescence imaging. Laboratory investigation. *J. Neurosurg.* **111:** 238–246.
72. Lara, R. *et al.* 2011. An siRNA screen identifies RSK1 as a key modulator of lung cancer metastasis. *Oncogene* **30:** 3513–3521.
73. Wagner, D.S. *et al.* 2010. The in vivo performance of plasmonic nanobubbles as cell theranostic agents in zebrafish hosting prostate cancer xenografts. *Biomaterials* **31:** 7567–7574.
74. Lee, S.L. *et al.* 2009. Hypoxia-induced pathological angiogenesis mediates tumor cell dissemination, invasion, and metastasis in a zebrafish tumor model. *Proc. Natl. Acad. Sci. USA* **106:** 19485–19490.
75. Vlecken, D.H. & C.P. Bagowski. 2009. LIMK1 and LIMK2 are important for metastatic behavior and tumor cell-induced angiogenesis of pancreatic cancer cells. *Zebrafish* **6:** 433–439.
76. Harfouche, R. *et al.* 2009. Nanoparticle-mediated targeting of phosphatidylinositol-3-kinase signaling inhibits angiogenesis. *Angiogenesis* **12:** 325–338.
77. Cheng, J. *et al.* 2011. Nanotherapeutics in angiogenesis: synthesis and in vivo assessment of drug efficacy and biocompatibility in zebrafish embryos. *Int. J. Nanomed.* **6:** 2007–2021.
78. Marques, I.J. *et al.* 2009. Metastatic behaviour of primary human tumours in a zebrafish xenotransplantation model. *BMC Cancer* **9:** 128.
79. Weiss, F.U. *et al.* 2009. Retinoic acid receptor antagonists inhibit miR-10a expression and block metastatic behavior of pancreatic cancer. *Gastroenterology* **137:** 2136–2145 e2131–2137.
80. Latifi, A. *et al.* 2011. Cisplatin treatment of primary and metastatic epithelial ovarian carcinomas generates residual cells with mesenchymal stem cell-like profile. *J. Cell Biochem.* **112:** 2850–2864.
81. Dobson, J.T. *et al.* 2008. Carboxypeptidase A5 identifies a novel mast cell lineage in the zebrafish providing new insight into mast cell fate determination. *Blood* **112:** 2969–2972.
82. Da'as, S. *et al.* 2011. Zebrafish mast cells possess an FcvarepsilonRI-like receptor and participate in innate and adaptive immune responses. *Dev. Comp. Immunol.* **35:** 125–134.
83. Lapidot, T. *et al.* 1994. A cell initiating human acute myeloid leukaemia after transplantation into SCID mice. *Nature* **367:** 645–648.
84. Bonnet, D. & J.E. Dick. 1997. Human acute myeloid leukemia is organized as a hierarchy that originates from a primitive hematopoietic cell. *Nat. Med.* **3:** 730–737.
85. Blackburn, J.S., S. Liu & D.M. Langenau. 2011. Quantifying the frequency of tumor-propagating cells using limiting dilution cell transplantation in syngeneic zebrafish. *J. Vis. Exp.* **53:** e2790.
86. Nicoli, S. & M. Presta. 2007. The zebrafish/tumor xenograft angiogenesis assay. *Nat. Protoc.* **2:** 2918–2923.
87. Jemal, A. *et al.* 2007. Cancer statistics, 2007. *CA Cancer J. Clin.* **57:** 43–66.
88. Evdokimova, V. *et al.* 2009. Translational activation of snail1 and other developmentally regulated transcription factors by

YB-1 promotes an epithelial-mesenchymal transition. *Cancer Cell* **15:** 402–415.
89. Ribatti, D. 2012. Chicken chorioallantoic membrane angiogenesis model. *Methods Mol. Biol.* **843:** 47–57.
90. Ebos, J.M. & R.S. Kerbel. 2011. Antiangiogenic therapy: impact on invasion, disease progression, and metastasis. *Nat. Rev. Clin. Oncol.* **8:** 210–221.
91. Stoletov, K. & R. Klemke. 2008. Catch of the day: zebrafish as a human cancer model. *Oncogene* **27:** 4509–4520.
92. Moshal, K.S., K.F. Ferri-Lagneau & T. Leung. 2010. Zebrafish model: worth considering in defining tumor angiogenesis. *Trends Cardiovasc. Med.* **20:** 114–119.
93. Tobia, C., G. De Sena & M. Presta. 2011. Zebrafish embryo, a tool to study tumor angiogenesis. *Int. J. Dev. Biol.* **55:** 505–509.
94. Postovit, L.M. *et al.* 2007. Targeting Nodal in malignant melanoma cells. *Expert Opin. Ther. Targets* **11:** 497–505.
95. Strizzi, L. *et al.* 2009. Development and cancer: at the crossroads of Nodal and Notch signaling. *Cancer Res.* **69:** 7131–7134.
96. Strizzi, L. *et al.* 2009. Nodal as a biomarker for melanoma progression and a new therapeutic target for clinical intervention. *Expert Rev. Dermatol.* **4:** 67–78.
97. Hooijkaas, A.I. *et al.* 2011. Expression of the embryological morphogen Nodal in stage III/IV melanoma. *Melanoma Res.* **21:** 491–501.
98. Yan, J. & Z. Cai. 2010. Molecular evolution and functional divergence of the cytochrome P450 3 (CYP3) family in Actinopterygii (ray-finned fish). *PLoS One* **5:** e14276.
99. Thisse, C. & L.I. Zon. 2002. Organogenesis—heart and blood formation from the zebrafish point of view. *Science* **295:** 457–462.

Ann. N.Y. Acad. Sci. ISSN 0077-8923

ANNALS OF THE NEW YORK ACADEMY OF SCIENCES
Issue: *Hematopoietic Stem Cells VIII*

Hematopoietic stem cell expansion: challenges and opportunities

Marta A. Walasek, Ronald van Os, and Gerald de Haan

Department of Biology of Aging, Section Stem Cell Biology, European Research Institute for the Biology of Aging (ERIBA), University Medical Center Groningen, University of Groningen, Groningen, the Netherlands

Address for correspondence: Gerald de Haan, Department of Biology of Aging, Section Stem Cell Biology, University Medical Center Groningen, University of Groningen, Groningen 9700 AD, the Netherlands. g.de.haan@umcg.nl

Attempts to improve hematopoietic reconstitution and engraftment potential of *ex vivo*–expanded hematopoietic stem and progenitor cells (HSPCs) have been largely unsuccessful due to the inability to generate sufficient stem cell numbers and to excessive differentiation of the starting cell population. Although hematopoietic stem cells (HSCs) will rapidly expand after *in vivo* transplantation, experience from *in vitro* studies indicates that control of HSPC self-renewal and differentiation in culture remains difficult. Protocols that are based on hematopoietic cytokines have failed to support reliable amplification of immature stem cells in culture, suggesting that additional factors are required. In recent years, several novel factors, including developmental factors and chemical compounds, have been reported to affect HSC self-renewal and improve *ex vivo* stem cell expansion protocols. Here, we highlight early expansion attempts and review recent development in the extrinsic control of HSPC fate *in vitro*.

Keywords: HSC expansion; cell-extrinsic factors; small molecules

Introduction

The two defining features of hematopoietic stem cells (HSCs), self-renewal and multilineage differentiation, render these cells an attractive source for stem cell-based therapies. HSC transplantation can be a life-saving procedure in the treatment of a broad spectrum of disorders, including hematologic, immune, and genetic diseases. Bone marrow (BM) and mobilized peripheral blood stem cells are the most common HSC sources for transplantation; however, their use is restricted by the low availability of suitable human leukocyte antigen (HLA)–matched donors. An alternative approach, allowing the use of more rapidly available partly mismatched donors, involves the use of umbilical cord blood (UCB), which has been shown to be a rich source of HSCs. However, the stem cell number in a single UCB unit is not sufficient for transplantation into an adult.[1] These two issues, the lack of HLA-matched donors and the low number of stem cells available from common HSC sources for transplantation, especially in the case of UCB, present serious limitations to HSC transplantations. Clinical experience has shown that the stem cell dose (measured as $CD34^+$ cell number) is related to patient survival and time required for engraftment,[1–3] indicating that amplification of hematopoietic stem and progenitor cells (HSPCs) is expected to be highly beneficial for their clinical application.

HSCs are a rare population of cells, representing less than 0.01% of all cells in the BM. During homeostasis, the stem cell pool is maintained at a relatively constant level. In contrast, several studies have shown that during hematopoietic stress, such as serial transplantations, HSCs can and will self-renew extensively. Studies by Pawliuk *et al.*, Iscove, and Nawa have shown that following injection into lethally irradiated recipients, HSC numbers increased by 10- to 20-fold.[4,5] When these cells were serially transplanted, the initial number of transplanted HSCs increased cumulatively by 8,400-fold after four successive passages.[5] Furthermore, it has been shown that a single purified HSC can reconstitute long-term hematopoiesis of lethally irradiated recipients, indicating robust *in vivo*

doi: 10.1111/j.1749-6632.2012.06549.x

self-replication capacity.[6] These studies have unequivocally demonstrated that HSCs can massively expand, and suggest that *in vivo* stem cells are exposed to specific factors/signals that promote their self-renewal and amplification.

Although HSC self-renewal divisions *in vivo* clearly occur, induction of self-renewal *in vitro* has been difficult. Even after several decades of research, the quest for factors that stimulate self-renewal *in vitro* is still continuing. The aim of these studies is to define the culture conditions that will support unlimited *ex vivo* expansion of HSCs from any source. To achieve this goal, stimulation of symmetrical self-renewal divisions over unlimited differentiation divisions is required (Fig. 1). Although many efforts have been made to expand stem cells, extensive *in vitro* amplification of HSCs without loss of their repopulating potential has not yet been achieved. Nonetheless, several cell-intrinsic and cell-extrinsic factors have been identified and shown to have potential to expand HSCs in culture. Here, we summarize how *ex vivo* HSC expansion protocols have evolved, briefly reviewing the early expansion attempts and then focusing on recent expansion approaches involving small molecules.

Regulation of HSC self-renewal by intrinsic factors

One of the remaining fundamental questions of stem cell biology is how self-renewal is regulated. Answers to this question would also substantially contribute to the success of *ex vivo* HSC expansion protocols. Valuable knowledge about factors influencing self-renewal has been gained from gene manipulation studies, which have identified multiple proteins that play an important role in the regulation of HSC self-renewal, including transcription factors, epigenetic modifiers, and cell cycle regulators (Fig. 2) (reviewed in Refs. 7 and 8). Here, we will only highlight a few examples, representing different classes of proteins, such as DNA-binding, chromatin-binding, and RNA-binding factors.

One of the first genes described to play a role in HSC fate determination was HoxB4. Ectopic expression of this transcription factor resulted in robust (40-fold) expansion of transplantable murine HSCs, but did not coincide with the development of leukemia.[9,10] HSC amplification could also be stimulated by an extrinsically delivered TAT-HoxB4 fusion protein, although the HSC expansion levels achieved with this fusion protein were much lower compared with ectopic protein expression.[11] Additionally, other members of HoxA and HoxB clusters (e.g., HoxA4, HoxA9, HoxB6) have also been shown to regulate fate determination of adult HSCs.[12–15] Overexpression of chromatin remodelers, such as members of the Polycomb group family of proteins (PcG), especially Ezh2 or Bmi1, have been shown to modulate HSC activity by preventing stem cell exhaustion or augmenting self-renewal, respectively.[16–18] The recent discovery of noncoding RNAs, including microRNAs, added an additional level of regulation to the network controlling HSC fate determination.[19,20] Several microRNAs have been reported to modulate HSC fates, including the miR-125 family and miR-29a.[21–25] Genetic studies aimed to decipher the mechanism of self-renewal have indicated that self-renewal is regulated at various levels and by a complex network of interacting stimuli.

Although the highest expansion levels of HSCs have been reported after ectopic expression of cell-intrinsic genes, introduction of genetic material is undesired in clinical protocols since continuous activation of self-renewal may lead to potential malignant transformation or to stem cell exhaustion. Therefore, mild and/or transient activation of self-renewal by extrinsic factors or molecules that perturb the activity of intrinsic HSC self-renewal regulators (described above) might be a preferred tool for *ex vivo* stem cell expansion.

Extrinsic regulators of HSC fates

The first attempts to *ex vivo* amplify HSCs with extrinsic factors focused on the use of hematopoietic cytokines, many of which are produced by the *in vivo* HSC microenvironment. Although the role of cytokines in blood lineage development has been indisputably established, their regulatory role on HSC self-renewal has been questioned.[26–28] Multiple cytokines, including SCF, Tpo, Flt-3L, IL-11, IL-3, IL-6, and GM-CSF, and combinations of these, have been studied in *in vitro* HSC expansion protocols of mouse and human cells. For details of the effects of distinct cytokines on HSC characteristics in culture, we refer to the excellent review by Sauvageau *et al.*[28] Knowledge gained from cytokine studies indicates that whereas *in vitro* cell survival and proliferation can be efficiently stimulated by several cytokines (especially SCF and Tpo), these cytokines by themselves are usually not

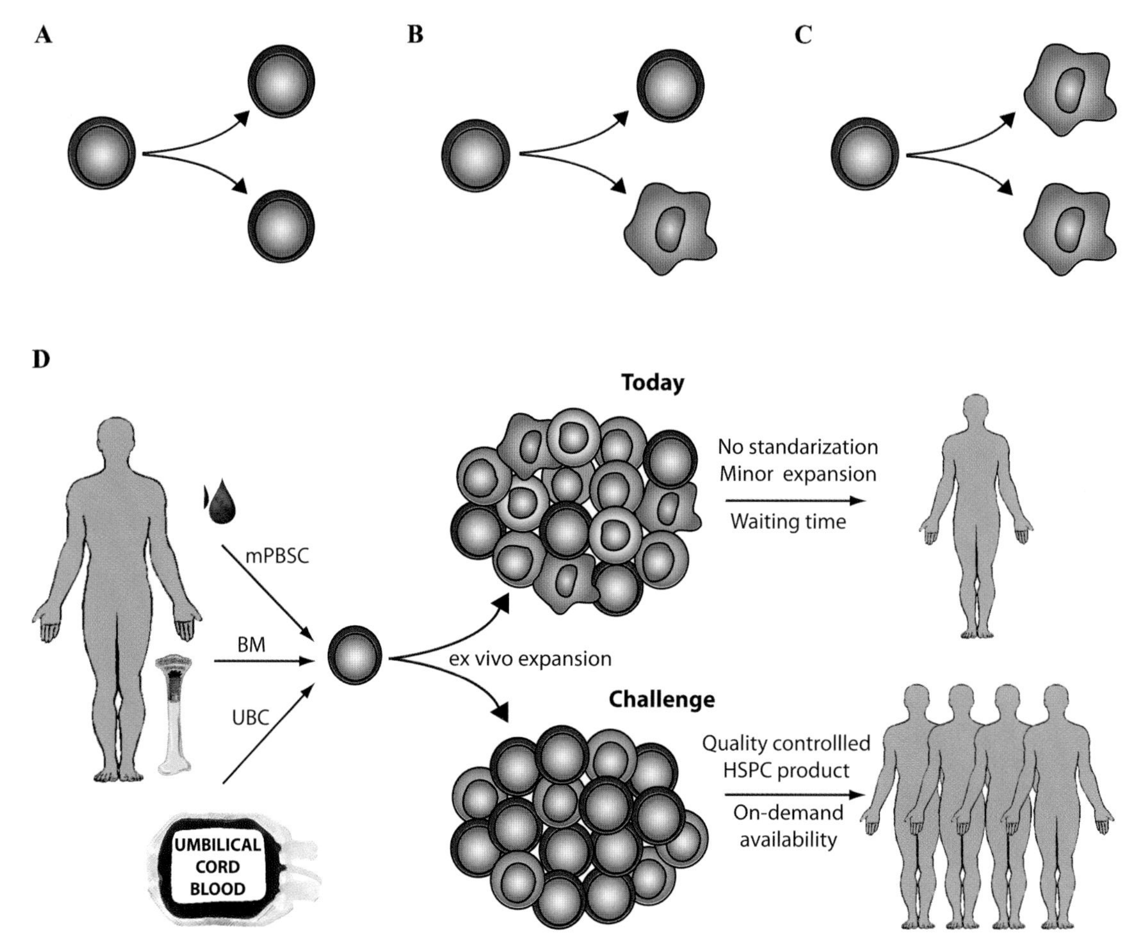

Figure 1. HSC fate outcomes define the composition of *ex vivo* expanded product. (A) Symmetrical HSC self-renewal division. As a result of this division, the parent stem cell gives rise to two daughter cell with the identical properties as the parent cell, increasing the stem cell pool. This occurs during development of the hematopoietic system in fetal liver and upon BM transplantation but remains difficult to sustain during *ex vivo* stem cell culture. (B) Asymmetrical HSC self-renewal division. Following asymmetrical self-renewal, the parent stem cell generates one identical daughter cell and a progeny committed to differentiation, thus preserving stem cell numbers. This occurs during hematopoietic homeostasis in adulthood and during *ex vivo* HSC expansion culture supporting stem cell maintenance. (C) Symmetrical HSC division. The outcome of symmetrical cell division is the generation of two daughter cells committed to differentiation, leading to stem cell depletion. This occurs in pathological conditions characterized by stem cell exhaustion and usually during *ex vivo* HSC culture. (D) HSC expansion today and the "Holy Grail" of expansion research. *Ex vivo* HSC expansion protocols available today cannot efficiently support symmetrical self-renewal divisions of HSCs, resulting (optimistically) in maintenance or minor amplification of stem cells and expansion of differentiated progeny. It remains unclear whether stem cells from any source benefit from up-to-date expansion protocols, resulting in a lack of standardized on-demand expanded product. The challenge remains to expand undifferentiated HSCs (via stimulation of symmetrical self-renewal divisions) in numbers sufficient for therapy of adult patients. This would allow development of clinically relevant and quality-controlled HSPC-expanded products that can be supplied upon demand.

sufficient to support self-renewal and they typically induce HSC differentiation, resulting at best in maintenance or modest stem cell amplification, but usually leading to progressive depletion of long-term repopulating cells.[29,30] Single-cell cytokine cultures demonstrated that *in vivo* repopulating ability of mouse and human HSCs is gradually but inevitably lost, starting from the first *in vitro* cell division.[31–33] Nonetheless, the cocktail of SCF and Flt3L with IL-11 (mouse) or Tpo (mouse and human) was defined as the cytokine combination supporting the best HSPC survival, proliferation, and maintenance during *in vitro* culture, and is often used as the core cytokine mix to which

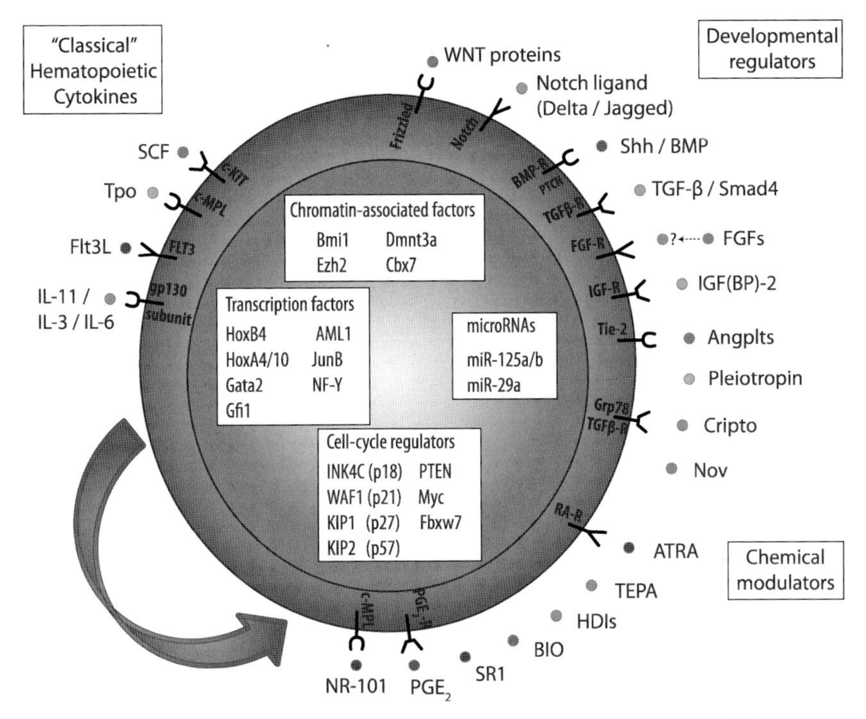

Figure 2. Cell-intrinsic and cell-extrinsic factors involved in HSC self-renewal. Self-renewal can be driven intrinsically by gene expression and can be regulated by extrinsic factors from environment. Cell-intrinsic regulation of HSC fate includes interplay between specific transcription factors, RNA/DNA-binding proteins, and chromatin-associated factors. That network can be modulated by cell-extrinsic cues such as cytokines, developmental/growth factors, and chemical compounds. These distinct stimuli create a complex matrix of interactions that defines the result of HSC fate, suggesting that combination of distinct stimuli could be required for effective stimulation of self-renewal divisions and stem cell expansion.

other cytokines or expansion factors can be added (Table 1).[55–57] To conclude, efforts to improve the rate of HSPC engraftment by *ex vivo* cytokine-based expansion protocols has been largely unsuccessful, indicating the need of additional factors/molecules in order to support HSC self-renewal and amplification *in vitro.*

Developmental regulators and HSC expansion

Recently, factors that do not necessarily qualify as typical hematopoietic growth factors but rather play an instructive role during ontogeny (i.e., at a time when the hematopoietic system is first developed and is rapidly expanding) have been shown to also regulate the adult stem cell compartment. These factors stimulate several developmentally conserved pathways, such as wingless-type (Wnt), Notch, Sonic hedgehog (Shh), and fibroblast growth factor (FGF) signaling. Adult HSCs have been shown to express receptors that can activate all of the aforementioned pathways, indicating that these pathways may be employed to extrinsically control HSC fate in culture.

Wnt proteins

The Wnt signaling pathway, known by its role in embryogenesis and cancer, regulates cell-to-cell interactions in a wide variety of tissues. A role for Wnt has been convincingly demonstrated in homeostasis of the intestinal stem cell compartment,[58] and a role for Wnt has been explored in HSC fate regulation.[59] Wnts are secreted glycoproteins, expressed by the fetal and adult HSC microenvironment. Addition of soluble Wnt proteins, such as Wnt3a or Wnt5a, to liquid cultures or cocultures with Wnt-transduced stroma, enhanced mouse and human HSC survival, proliferation, and self-renewal, as measured by engraftment rate after transplantation into lethally irradiated recipients.[60–63] Interestingly, it has recently been demonstrated that the level of Wnt signaling is critically important in regulating HSC self-renewal.[64] Binding of Wnt ligands to its receptor Frizzled triggers activation of the canonical signal transducer β-catenin and its

Table 1. Summary of recent expansion protocols discussed

Molecule	Species	Input cells	Supplements	Treatment time	Effect: Assay	Effect: Fold over control	Effect: Fold over input	Mechanism	Reference
FGFs	M	BM		3w	CAFC d-35		↑		de Haan *et al.*[34]
					HSC frequency		↑		Yeoh *et al.*[35]
IGF-2	M	BM SP CD45$^+$Sca-1$^+$	SCF Tpo FGF-1	10d	CRU (HSC frequency)		↑ 24 (WBM) ↑ 7.8 (SP)		Zhang and Lodish[36]
Angplts (2,3,5,7)	M	SP CD45$^+$Sca-1$^+$	SCF Tpo FGF-1 IGF-2	10d	CRU (HSC frequency)		↑ 24 (Angptl2) ↑ 30 (Angptl3)	Angplts expansion effects are dependent on mammalin cell-specific posttranslational modifications	Zhang *et al.*[37]
Pleiotrophin	M	LSK CD34$^-$	SCF Flt3L Tpo	7d	LSK numbers	↑ 17			Himburg *et al.*[38]
					CRU frequency (12 weeks p-tpx)	↑ 6	↑ 4		
					% engraftment secondary tpx	↑ 10	↑ 10		
	H	Lin-34$^+$38$^-$ UCB			CFU	↑ 4			
					% engraftment (4 weeks)	↑ 3	↑ 3		
ATRA	M	LSK	Serum SCF Flt3L IL-6 IL-11	7d	% Lin$^-$/Sca1$^+$ cells	↑ 1.7/4	↓ 0.5		Purton *et al.*[39]
					CFU-S (d8)	↑ 12	↑ 60		Purton *et al.*[40]
					Pre-CFU-S	↓ 0.2	↓ 0.4		
					% engraftment	↔	↓		
					Donor reconstitution (per 10^5 cells)	↑ 5			
TEPA	H	CD34$^+$ CD133$^+$	Serum SCF	3w	Total CD34$^+$CD38$^-$ cells	↑ 15			Peled *et al.*[41]
				8w	Total CD34$^+$ cells	↑ 10			Peled *et al.*[42]
		UCB	Tpo Flt3L IL-6		% engraftment	↑	↑		Peled *et al.*[43]
VPA	H	CD34$^+$ UCB MPB BM	Serum SCF Flt3L Tpo	1–3w	% CD34$^+$CD90$^+$ cells	↓ 1.5-6		Slower cell cycle	De Felice *et al.*[44]
					% CD34$^+$CD38$^-$ cells	↑ 20-140	↑ 16		
					CFU activity (plating efficiency)	↑ 85	↑ 7		
					Replating CFU activity	↓ 0.3-0.08			
			IL-3			↑ 5			
	H	CD34$^+$ UCB BM	SCF Flt3L Tpo	2–10d	Total CD34$^+$ cell numbers	↑		↓ p21 ↑ HoxB4 Activation of GSK3β-dependent signaling pathway	Bug *et al.*[45]
					CFU plating and replating activity	↑ ↑			
					CFU-S (d12)				
	M	LSK	IL-3		% engraftment	↑ 2.2			
	H	UCB HSC	SCF Flt-3L Tpo	14d	Total CD45$^+$CD34$^+$ cell numbers	↑ 2.0		↑ HoxB4	Seet *et al.*[46]
					% CD45$^+$CD34$^+$ cells in S-phase	↑			
					SRC frequency	↑ 6.0			
Chlamy docin	H	CD34$^+$ MPB	SCF Flt3L Tpo	24h	% engraftment (SRC)	↑ 4.0	↑ 2.0	↑ H4-Ac	Young *et al.*[47]
5aza +TSA	H	CD34$^+$ UCB	Serum SCF Flt3L MGDF IL-3	9d	Total CD34$^+$ cell numbers	↑ 2.5	↑ 5	↑ H4-Ac	Araki *et al.*[48]
					Total CD34$^+$CD90$^+$ cell numbers	↑ 6	↑ 12.5		
					CFU activity	↑ 8	↑ 9.8		
					CAFC activity	↑ 7	↑ 11.5		
					SRC frequency of CD34$^+$CD90$^+$ cells	↑ 40	↑ 9		

Continued

Table 1. *Continued*

Molecule	Species	Input cells	Supplements	Treatment time	Effect			Mechanism	Reference
					Assay	Fold over control	Fold over input		
BIO	H	CD34$^+$ UBC	Serum SCF Flt3L Tpo	5d	% cells in G_0/G_1 phase + Ki67^{dim}	↑ 28 + 1.5		↑ p57, Jagged-1, BMP8-b, RARRES2, Noggin	Ko *et al.*[49]
					CFU activity	↑ 2	↑ 30	↑β-catenin nuclear-location	
					CFU replating activity	↑ 6	↑ 15	↓ cyclin D1, CEBPδ	
					Engraftment of expanded cells	↑ 2	↔		
	H	CD34$^+$ UBC	Serum SCF Flt-3L Tpo	5d	Total CD34$^+$ cell number	↓		↑ β-catenin nuclear-location	Holmes *et al.*[50]
					CFU	↓		↑ c-myc, HoxB4	
					% engraftment	↔		↑ CXCR4 in stroma-adherent population	
					SRC frequency (6 weeks)	↑ 2.5			
SR1	H	CD34$^+$ MPB UBC	SCF Flt3L Tpo IL-6	3–5w	Total CD34$^+$ cell numbers	↑ 47		Antagonizing the aryl hydrocarbon receptor	Boitano *et al.*[51]
					CFU	↑ 65	↑ 9500		
					Engraftment 1 week p-tpx	↑ 24	↑ 6	↓ AHRR and CYP1B1	
					SRC frequency	↑ 14	↑ 17		
					Secondary SRC frequency	↑ 4	↑ 12		
PGE2	M	WBM		2h	CFU-S (d12)	↑ 3		↑ Survivin, CXCR4 ↓ caspase-3	North *et al.*[52]
	H	LSK			HSC frequency (CRU)	↑ 2-4	↑		Hoggatt *et al.*[53]
NR-101	H	CD34$^+$ CD34$^+$38$^-$ UCB		7d	Total CD34$^+$CD38$^-$ cell number	↑ 2.3		Activation of STAT5, but not of STAT3. Accumulation of HIF-1α and enhanced activation of its targets.	Nishino *et al.*[54]
					% CD34$^+$CD38$^-$ cells in G1/G0-phase	↑ 2.1			
					SRC frequency	↑ 2.3	↑ 2.9		

FGFs, fibroblast growth factors; IGF-2, insulin growth factors 2; Angplts, angiopoietin-like proteins; ATRA, all-*trans* retonic acid; TEPA, tetra-ethylenepentamine; VPA, valproic acid; TSA, trichostatin A; 5aza, 5-aza-2′-deoxycytidine; BIO, 6-bromoindirubin-3′-oxime; SR1, StemRegenin1; PGE$_2$, prostaglandin E$_2$; M, mouse; H, human; BM, bone marrow; SP, side population; UCB, umbilical cord blood; MPB, mobilized peripheral blood; SCF, stem cell factor; Tpo, thromopoietin; Flt3L, Ftl3 ligand; IL, interleukin; MGDF, megakaryocyte growth and development factor; w, week; d, day; CAFC, cobblestone area forming cell; CFU(-S), colony forming cell (spleen); CRU, competitive repopulating unit; SRC, NOD/SCID-repopulating cell; H4-AC, histone 4 acetylation.

translocation to the nucleus, leading to activation of Wnt target gene transcription.[65] Ectopic expression of a constitutively active form of β-catenin resulted in expansion of the stem cell pool.[59] In this study, upon activation of Wnt signaling, upregulation of HoxB4 and Notch1 were reported, suggesting a link between Wnt and Notch signaling pathways. Work of Duncan *et al.* supports cooperation of these two pathways in maintaining the HSC pool, thus demonstrating the importance of the Wnt pathway in cell proliferation and survival and a requirement of Notch for maintaining HSCs in an undifferentiated state.[66]

Notch pathway

Notch and Notch-ligands, such as Jagged-1, Jagged-2, and Delta-1, are expressed by HSCs and their microenvironment, indicating a potential role for Notch interactions in hematopoiesis.[67] Studies on the involvement of the Notch pathway in HSC fate regulation demonstrated that an immobilized form of Delta-1, but not its soluble form, can expand murine and human HSPCs in culture. Incubation of human UCB CD34$^+$ cells with Delta-1 and cytokines resulted in increased numbers of NOD/SCID repopulating cells with secondary transplantation ability. Similar to the dependence of HSCs on Wnt-protein levels, the Notch-mediated effects were suggested to be dose-dependent, since HSPC expansion was observed only with lower Delta-1 concentrations.[68,69] A subsequent clinical phase I trial showed rapid neutrophil recovery, enhanced myeloid engraftment, and no signs of graft-versus-host disease (GVHD) following transplantation with Delta-1–expanded human UCB cells,[70] although it remains unclear whether the UCB cells benefitted from the expansion protocol.

Shh/BMP signals

Cytokine-based expansion cultures of HSCs can also be improved by addition of other developmental factors, namely by soluble Sonic hedgehog proteins (Shh). Following Shh stimulation of human CD34$^+$CD38$^-$Lin$^-$ cells, enhanced cell proliferation and increased recovery of NOD/SCID repopulating cells were reported.[71] Additionally, Shh-induced HSPC expansion was suggested to be dependent on downstream bone morphogenic protein-4 (BMP-4) signaling, because inhibition of BMP-4 abrogated Shh-induced expansion.[71] Since human HSCs express BMP receptors, addition of soluble BMP-4 could also improve human HSPC proliferation and maintenance following *in vitro* culture.[72]

FGF signaling

The observation that murine HSCs express high levels of FGF receptors has led to investigation of the role of FGF signaling in HSPC fate regulation. Among this large protein family, FGF-1 and FGF-2 in particular were shown to maintain/expand multilineage, serially-transplantable, long-term repopulating HSCs when added to cytokine- and serum-free cultures of unfractionated mouse BM.[34,35] However, addition of classical cytokines to these cultures and stimulation of proliferation overruled FGF-dependent stem cell preservation. Moreover, FGFs could not support the maintenance of an undifferentiated state of purified murine HSCs in culture, and only appeared active when stromal elements were present, suggesting an indirect nature of the FGF effects.[35]

IGF and Angptls

Recently, several soluble proteins produced by the endothelium and fetal liver, including insulin-like growth factor 2 (IGF-2), IGF binding protein 2 (IGFBP-2), and angiopoietin-like proteins (Angptls), have been identified as growth factors that can enhance *ex vivo* expansion of HSCs. Zhang *et al.* reported that addition of IGF-2, a protein expressed in fetal liver cells, to serum-free cultures of fetal liver HSCs resulted in increased stem cell frequency in a transplantation setting.[36] Shortly afterward, it was reported that IGFBP-2 displayed similar or even better stem cell–supporting effects and that IGF-2 could be replaced by IGFBP-2 in *ex vivo* HSC expansion cultures.[73] These cultures also contained a combination of a developmental factor, FGF-1 (mentioned above), and two cytokines, SCF and Tpo. The efficiency of this expansion cocktail could be further enhanced by addition of angiopoietin-like proteins, which[37] include several members, such as Angplt2, Angplt3, Angplt5, and Angplt7. Culture of murine HSCs with these factors resulted in a 30-fold net expansion of long-term repopulating HSCs, one of the highest HSC amplifications achieved to date by extrinsic factors. Similar expansion levels have also been reported in human HSC cultures supplemented with the Lodish and Zhang cocktail.[37,73,74]

Pleiotrophin

Another novel microenvironmental factor, pleiotrophin, expressed by human brain endothelial cells, has recently been reported to be essential for HSC maintenance. Mouse HSCs cultured in the presence of pleiotrophin displayed a net increase in *in vivo*–measured stem cell frequency and enhanced secondary repopulation activity upon serial transplantation. Pleiotrophin was also shown to improve *ex vivo* expansion of human CD34$^+$CD38$^-$Lin$^-$ HSCs, albeit to a lesser extent compared with murine cells. Pleiotrophin-mediated expansion was abrogated by addition of specific PI3K or Notch inhibitors to the culture medium, suggesting involvement of these pathways in the pleiotrophin stem cell–enhancing effects.[38] Additionally, whereas pleiotrophin deficiency did not show any *in vivo* steady-state hematopoietic effects, repression of pleiotrophin in stromal cells resulted in defective hematopoiesis.[75]

The studies discussed above indicate the potential usefulness of exploiting the molecular pathways involved in stem cell self-renewal in improving *ex vivo* HSPC amplification cultures. The challenge remains to integrate the various cell-intrinsic and cell-extrinsic signals into the regulatory networks controlling HSC fate outcomes and to decipher how these factors are linked with each other. Likely, the balance of various coregulated stimuli, acting simultaneously or sequentially, is crucial for HSC fate outcomes, and the ability to mimic the orchestra of these signals may lie in the success of *in vitro* stem cell-expansion protocols.

Small molecules

In recent years, evidence has accumulated suggesting that in addition to natural cytokines/factors chemical compounds can also have potent effects on stem cell expansion protocols (Table 1). Studies were launched to assess the ability of such compounds to control stem cell fates in culture, where the effects ranged from enhancement of stem cell expansion to stimulation of lineage-specific differentiation, including effects on iPS reprogramming efficiency.[76–78]

ATRA

Retinoic acid (RA), a derivative of vitamin A (retinol) normally present in low physiological concentration in serum, represents a naturally occurring small molecule.[79] Retinoids have been shown to play important roles in development, differentiation, and homeostasis of a wide variety of cell types. The retinoic acid receptor agonist all-trans retinoic acid (ATRA) has a well-documented effect on hematopoietic cells in cancer treatment, where it can induce differentiation of leukemic cells.[80] Studies with RA receptor (RAR) deletion have revealed the importance of retinoids in normal hematopoiesis, since RAR-γ knock-out mice showed markedly reduced HSC numbers.[81] However, the utility of retinoids for *ex vivo* HSPC expansion is unclear. Purton *et al.* suggested that addition of ATRA to cultures of mouse HSPCs enhanced the maintenance of short- and long-term repopulating cells and augmented self-renewal in serial transplantation experiments.[39,40] Additionally, increased Notch1 and HoxB4 mRNA levels were observed upon ATRA stimulation, suggesting that ATRA effects could be at least partially mediated via known cell-intrinsic regulators of HSC self-renewal.[81] On the other hand, retinoic acid was also reported to negatively affect HSPC *ex vivo* expansion, since inhibition of retinoid signaling resulted in enhanced HSPC self-renewal *in vitro*.[82] Therefore, the effects of retinoids on *in vitro* HSPC self-renewal and differentiation need further investigation.

Copper chelator, TEPA

It has been shown that increased cellular copper levels can improve ATRA-stimulated differentiation of leukemic cells.[83] Additionally, ions such as magnesium, calcium, or copper (Cu) are known to play an important role in cell function, since disturbed homeostasis of these ions is associated with clinical symptoms. Copper malnutrition resulted in hematopoietic cell maturation arrest but did not affect progenitor cell development, which suggests that cellular Cu balance may play a role in HSPC proliferation and differentiation. Studies on modulation of cellular Cu content during *ex vivo* culture demonstrated that elevated copper levels accelerated cell maturation, whereas decreased Cu levels, achieved by addition of the Cu chelator tetraethylenepentamine (TEPA), attenuated *ex vivo* human HSPC differentiation, resulting in expansion of early progenitor cells. TEPA-supplemented, long-term cytokine cultures of $CD34^+$ cord blood cells significantly increased the number of $CD34^+CD38^-$ cells and enhanced NOD/SCID repopulating capacity.[41,42] A phase I/II clinical trial in which TEPA-cultured cord blood cells were coinfused with uncultured cord blood cells showed the safety of this approach.[43,84] Although engraftment of TEPA-treated cells could be observed in almost all study subjects, the time required for neutrophil and platelet recovery was not changed compared with uncultured cord blood cells.[84] The efficiency of TEPA-cultured HSPC transplantation product is currently under investigation in an ongoing phase II/III study.

HDIs

Since stem cells are believed to be characterized by a specific (transcriptionally permissive) epigenetic status, epigenetically active compounds could possibly modulate stem cell fates. Several small molecules have been identified that alter the epigenetic status of cells. Histone deacetylase inhibitors (HDIs) or demethylating agents are an example of such compounds. Several groups have studied the effects of valproic acid (VPA), an epigenetic drug known for its anticancer activity, on regulation of *in vitro* HSC fate determination. Addition of VPA to cultures of murine or human HSPCs preserved the expression of primitive markers and an HSC phenotype.[44,45,85] VPA treatment was shown to maintain CFU-S activity in murine HSCs and resulted in elevated chimerism levels upon transplantation into lethally irradiated recipients.[44–46] Similar effects on the *ex vivo* expansion of human-mobilized peripheral blood CD34$^+$ HSCs were observed upon treatment with another HDI, chlamydocin.[47] Additionally, a combination of two epigenetic drugs was shown to additively affect HSC maintenance and/or expansion in culture.[48,86] Sequential addition of a demethylating agent, 5-aza-2′-deoxycytidine methyltransferase (5azaD), and the HDI trichostatin A (TSA) to various cytokine cocktails has been shown to improve maintenance of stem cells in culture, compared with single compound treatment.[48,86] As expected, in these HSCs, HDI exposure increased histone acetylation, whereas 5aza decreased methylation.[48] HDI treatment has also been shown to result in upregulation of cell-intrinsic HSPC fate regulators such as HoxB4 and AC133, in activation of Wnt signaling, and in downregulation of p21. Nevertheless, the exact mechanism of HDI-stimulated HSC expansion remains unclear.[44–48]

BIO

Small molecules that specifically target signaling pathways that play an important role in HSC self-renewal and maintenance could be beneficial for *ex vivo* expansion protocols. The synthetic compound 6-bromoindirubin-3′-oxime (BIO) can modulate Wnt signaling activity by targeting GSK3α and β, a negative regulator of Wnt signals.[50] Stimulation of cord blood CD34$^+$ cells with BIO led to accumulation of slowly dividing cells and enhanced replating activity.[50] Additionally, the exposure of cultured CD34$^+$ cells to BIO significantly increased engraftment and chimerism levels of the NOD/SCID repopulating cells.[49] As a GSK3β inhibitor, BIO treatment resulted in accumulation of β-catenin and its nuclear localization; however, the expression of Wnt target genes was not changed upon treatment.[49,50] Therefore, the key mechanism of BIO, and how it is responsible for enhanced HSPC activity, remain unknown.

SR1

Recent reports have shown that unbiased screens to search for factors that are able to maintain/expand HSCs can identify novel compounds with HSC self-renewal stimulatory activity. Using high-throughput screening, Boitano *et al.* identified an acryl hydrocarbon receptor (AhR) antagonist, referred to as StemRegenin1 (SR1), that was capable of enhancing human CD34$^+$ HSC cell amplification.[51] Cord blood CD34$^+$ cells cultured in the presence of SR1 led to a high net increase in the number of CD34$^+$ cells. Importantly, SR1-treated human cells demonstrated a 17-fold increase in cell numbers that were capable of hematopoietic reconstitution of sublethally irradiated mice and showed enhanced multilineage, long-term potential. Although it has been shown that hHSCs express AhR, which has been implicated in pathways regulating hematopoiesis, pathways including HES-1, β-catenin, and CXCR4, the exact mechanism of SR1-induced HSC expansion is not known yet.[51]

PGE$_2$

Another novel compound, the small lipid mediator prostaglandin E$_2$ (PGE$_2$), was identified as a regulator of HSC self-renewal using high-throughput library screens in zebrafish.[52] Chemicals enhancing PGE$_2$ synthesis have been shown to increase HSC numbers in zebrafish and mouse embryos, whereas those blocking its synthesis decrease stem cell numbers, indicating that modulation of the prostaglandin pathway might affect stem cell characteristics.[52] Components of the prostaglandin pathway and PGE$_2$ receptors are present on both mouse and human stromal cells and HSCs.[52,53] The enhancing role of PGE$_2$ on adult HSCs has been demonstrated in competitive transplantation models, where short stimulation of mouse cells with PGE$_2$ prior to transplantation increased the frequency of short- and long-term repopulating cells.[52] Hoggatt *et al.* confirmed the stimulatory effects of PGE$_2$ on mouse HSCs and additionally reported that the induced competitive advantage of treated HSCs remained upon serial transplantation. Although the exact mechanism of PGE$_2$ is not known, studies have shown that the PGE$_2$ effects might be explained by stimulation of HSC survival, proliferation, and self-renewal associated with upregulation of survivin, an inhibitor of apoptosis, and with a reduction of the active intercellular form of caspase-3. Furthermore, PGE_2 may enhance HSPC homing, as suggested by upregulated expression of the CXCR4 gene.[53]

NR-101

The studies reported above have employed the use of chemical compounds added to cultures that also contain cytokines. Yet, small molecules have also been used as a full replacement of cytokines. Tpo is one of the most efficient cytokines supporting HSC proliferation and survival in culture and is widely used in HSC expansion protocols. Several small molecules have been chemically synthesized that are able to activate c-MPL, the receptor for Tpo. Nishino *et al.* screened more than 400 of these molecules for their HSC stimulatory role and found a novel small molecule c-MPL agonist, NR-101, that more effectively expanded HSCs *ex vivo* compared with Tpo. Following culture, NR-101 increased $CD34^+$ and $CD34^+CD38^-$ cell numbers and NOD/SCID-repopulating cell frequencies compared with Tpo or with fresh cord blood $CD34^+$ cells. NR-101 was shown to activate the major pathways of Tpo/c-MPL signaling, but displayed extended signal activation compared to Tpo.[54]

Finding that HSPC fate can be modulated chemically by addition of small molecules to the culture media is convenient in light of *ex vivo* HSC expansion and offers, next to cytokines and developmental factors, an additional level of control of HSC self-renewal and differentiation *in vitro*. The promising potential of chemically enhanced HSC self-renewal and its cooperation with other self-renewal factors should be carefully evaluated and tested for use in clinical stem cell expansion protocols.

Conclusions

Although HSCs can expand extensively *in vivo*, conditions that reliably induce robust HSC expansion *in vitro* have yet to be discovered. Expansion attempts with the use of hematopoietic cytokines were rather disappointing and therefore diminished the initial enthusiasm of *ex vivo* stem cell expansion. A potential role of developmental factors and the discovery that self-renewal can be controlled by chemical compounds has revived this hope. Today, thousands of molecules are screened for their effects on HSC fates, and the utility of these factors for *in vitro* stem cell expansion is being investigated. Experience gained from many years of research indicates that the most comprehensive approach to develop optimal HSC expansion conditions may involve a combination of methods. It is likely that these conditions would involve multiple biological and chemical compounds that act in concert to induce cell survival and division while simultaneously preventing stem cell differentiation. Accordingly, several signaling pathways and/or compounds have been reported to act additively or synergistically to modulate HSPC fates in culture.[85,87,88] Additionally, methods that improve HSC homing and survival after transplantation could also be combined with *ex vivo* HSC expansion to potentially further enhance the efficiency of HSPC engraftment. Difficulties in defining HSC expansion conditions result partly from insufficient understanding of the molecular mechanisms controlling HSC fate determination. Identification of key molecules and the interaction networks important for self-renewal could significantly improve expansion attempts and may also lead to development of specific compounds targeting these self-renewal factors. Importantly, the challenge remains to assess to what extent it will be possible to generate fully functional HSCs, and their derivatives, for future clinical regenerative medicine applications.

Acknowledgments

This study was supported by the Landsteiner Foundation for Blood Research (LSBR-0702), Dutch Platform for Tissue Engineering (DPTE; STW-GGT6727), and by the European Community's Seventh Framework Program (FP7/2007-2013) under Grant Agreement n° 222989 (StemExpand).

Conflicts of interest

The authors declare no conflicts of interest.

References

1. Barker, J.N. 2007. Umbilical cord blood (UCB) transplantation: an alternative to the use of unrelated volunteer donors? *ASH Education Program Book* **2007:** 55–61.
2. Wagner, J.E., J.N. Barker, T.E. DeFor, *et al.* 2002. Transplantation of unrelated donor umbilical cord blood in 102 patients with malignant and nonmalignant diseases: influence of CD34 cell dose and HLA disparity on treatment-related mortality and survival. *Blood* **100:** 1611–1618.
3. Brunstein, C.G., J.A. Gutman, D.J. Weisdorf, *et al.* 2010. Allogeneic hematopoietic cell transplantation for hematological malignancy: relative risks and benefits of double umbilical cord blood. *Blood* **116:** 4693–4699.

4. Pawliuk, R., C. Eaves & R. Humphries. 1996. Evidence of both ontogeny and transplant dose-regulated expansion of hematopoietic stem cells in vivo. *Blood* **88:** 2852–2858.
5. Iscove, N.N. & K. Nawa. 1997. Hematopoietic stem cells expand during serial transplantation in vivo without apparent exhaustion. *Curr. Biol.* **7:** 805–808.
6. Osawa, M., K. Hanada, H. Hamada, *et al.* 1996. Long-term lymphohematopoietic reconstitution by a single CD34-low/negative hematopoietic stem cell. *Science* **273:** 242–245.
7. Zhu, J. & S.G. Emerson. 2002. Hematopoietic cytokines, transcription factors and lineage commitment. *Oncogene* **21:** 3295–3313.
8. Akala, O.O. & M.F. Clarke. 2006. Hematopoietic stem cell self-renewal. *Curr. Opin. Genet. Dev.* **16:** 496–501.
9. Sauvageau, G., U. Thorsteinsdottir, C.J. Eaves, *et al.* 1995. Overexpression of HOXB4 in hematopoietic cells causes the selective expansion of more primitive populations in vitro and in vivo. *Genes Dev.* **9:** 1753–1765.
10. Antonchuk, J., G. Sauvageau & R.K. Humphries. 2002. HOXB4-induced expansion of adult hematopoietic stem cells ex vivo. *Cell* **109:** 39–45.
11. Krosl, J., P. Austin, N. Beslu, *et al.* 2003. In vitro expansion of hematopoietic stem cells by recombinant TAT-HOXB4 protein. *Nat. Med.* **9:** 1428–1432.
12. Fischbach, N.A., S. Rozenfeld, W. Shen, *et al.* 2005. HOXB6 overexpression in murine bone marrow immortalizes a myelomonocytic precursor in vitro and causes hematopoietic stem cell expansion and acute myeloid leukemia in vivo. *Blood* **105:** 1456–1466.
13. Lawrence, H.J., J. Christensen, S. Fong, *et al.* 2005. Loss of expression of the Hoxa-9 homeobox gene impairs the proliferation and repopulating ability of hematopoietic stem cells. *Blood* **106:** 3988–3994.
14. Thorsteinsdottir, U., A. Mamo, E. Kroon, *et al.* 2002. Overexpression of the myeloid leukemia-associated Hoxa9 gene in bone marrow cells induces stem cell expansion. *Blood* **99:** 121–129.
15. Fournier, M., C. Lebert-Ghali, G. Krosl, *et al.* 2012. HOXA4 induces expansion of hematopoietic stem cells in vitro and confers enhancement of Pro-B-cells in vivo. *Stem Cells and Dev.* **21:** 133–142.
16. Kamminga, L.M., L.V. Bystrykh, A. de Boer, *et al.* 2006. The polycomb group gene Ezh2 prevents hematopoietic stem cell exhaustion. *Blood* **107:** 2170–2179.
17. Iwama, A., H. Oguro, M. Negishi, *et al.* 2004. Enhanced self-renewal of hematopoietic stem cells mediated by the polycomb gene product Bmi-1. *Immunity* **21:** 843–851.
18. Rizo, A., B. Dontje, E. Vellenga, *et al.* 2008. Long-term maintenance of human hematopoietic stem/progenitor cells by expression of BMI1. *Blood* **111:** 2621–2630.
19. Han, Y., C.Y. Park, G. Bhagat, *et al.* 2010. microRNA-29a induces aberrant self-renewal capacity in hematopoietic progenitors, biased myeloid development, and acute myeloid leukemia. *J. Exp. Med.* **207:** 475–489.
20. Chen, C., L. Li, H.F. Lodish, *et al.* 2004. MicroRNAs modulate hematopoietic lineage differentiation. *Science* **303:** 83–86.
21. Guo, S., J. Lu, R. Schlanger, *et al.* 2010. MicroRNA miR-125a controls hematopoietic stem cell number. *Proc. Natl. Acad. Sci.* **107:** 14229–14234.
22. O'Connell, R.M., A.A. Chaudhuri, D.S. Rao, *et al.* 2010. MicroRNAs enriched in hematopoietic stem cells differentially regulate long-term hematopoietic output. *Proc. Natl. Acad. Sci.* **107:** 14235–14240.
23. Ooi, A.G.L., D. Sahoo, M. Adorno, *et al.* 2010. MicroRNA-125b expands hematopoietic stem cells and enriches for the lymphoid-balanced and lymphoid-biased subsets. *Proc. Natl. Acad. Sci.* **107:** 21505–21510.
24. Gerrits, A., M.A. Walasek, S. Olthof, *et al.* 2011. Genetic screen identifies microRNA cluster 99b/let-7e/125a as a regulator of primitive hematopoietic cells. *Blood* **119:** 377–387.
25. Bousquet, M., M.H. Harris, B. Zhou, *et al.* 2010. MicroRNA miR-125b causes leukemia. *Proc. Natl. Acad. Sci.* **107:** 21558–21563.
26. Ogawa, M. 1993. Differentiation and proliferation of hematopoietic stem cells. *Blood* **81:** 2844–2853.
27. Metcalf, D. 2008. Hematopoietic cytokines. *Blood* **111:** 485–491.
28. Sauvageau, G., N.N. Iscove & R.K. Humphries. 2004. In vitro and in vivo expansion of hematopoietic stem cells. *Oncogene* **23:** 7223–7232.
29. Goff, J.P., D.S. Shields & J.S. Greenberger. 1998. Influence of cytokines on the growth kinetics and immunophenotype of daughter cells resulting from the first division of single $CD34^{+}Thy\text{-}1^{+}lin^{-}$ cells. *Blood* **92:** 4098–4107.
30. Glimm, H. & C.J. Eaves. 1999. Direct evidence for multiple self-renewal divisions of human in vivo repopulating hematopoietic cells in short-term culture. *Blood* **94:** 2161–2168.
31. Nakauchi, H., K. Sudo & H. Ema. 2001. Quantitative assessment of the stem cell self-renewal capacity. *Ann. N. Y. Acad. Sci.* **938:** 18–25.
32. Uchida, N., B. Dykstra, K.J. Lyons, *et al.* 2003. Different in vivo repopulating activities of purified hematopoietic stem cells before and after being stimulated to divide in vitro with the same kinetics. *Exp. Hematol.* **31:** 1338–1347.
33. Ema, H., H. Takano, K. Sudo, *et al.* 2000. In vitro self-renewal division of hematopoietic stem cells. *J. Exp. Med.* **192:** 1281–1288.
34. de Haan, G., E. Weersing, B. Dontje, *et al.* 2003. In vitro generation of long-term repopulating hematopoietic stem cells by fibroblast growth factor-1. *Dev. Cell.* **4:** 241–251.
35. Yeoh, J.S.G., R. van Os, E. Weersing, *et al.* 2006. Fibroblast growth factor-1 and -2 preserve long-term repopulating ability of hematopoietic stem cells in serum-free cultures. *Stem Cells* **24:** 1564–1572.
36. Zhang, C.C. & H.F. Lodish. 2004. Insulin-like growth factor 2 expressed in a novel fetal liver cell population is a growth factor for hematopoietic stem cells. *Blood* **103:** 2513–2521.
37. Zhang, C.C., M. Kaba, G. Ge, *et al.* 2006. Angiopoietin-like proteins stimulate ex vivo expansion of hematopoietic stem cells. *Nat. Med.* **12:** 240–245.
38. Himburg, H.A., G.G. Muramoto, P. Daher, *et al.* 2010. Pleiotrophin regulates the expansion and regeneration of hematopoietic stem cells. *Nat. Med.* **16:** 475–482.
39. Purton, L.E., I.D. Bernstein & S.J. Collins. 1999. All-trans retinoic acid delays the differentiation of primitive hematopoietic precursors (lin–c-kit +Sca-1+) while

enhancing the terminal maturation of committed granulocyte/monocyte progenitors. *Blood* **94:** 483–495.

40. Purton, L.E., I.D. Bernstein & S.J. Collins. 2000. All-trans retinoic acid enhances the long-term repopulating activity of cultured hematopoietic stem cells. *Blood* **95:** 470–477.
41. Peled, T., E. Landau, J. Mandel, *et al.* 2004. Linear polyamine copper chelator tetraethylenepentamine augments long-term ex vivo expansion of cord blood-derived CD34+ cells and increases their engraftment potential in NOD/SCID mice. *Exp. Hematol.* **32:** 547–555.
42. Peled, T., E. Glukhman, N. Hasson, *et al.* 2005. Chelatable cellular copper modulates differentiation and self-renewal of cord blood-derived hematopoietic progenitor cells. *Exp. Hematol.* **33:** 1092–1100.
43. Peled, T., J. Mandel, R.N. Goudsmid, *et al.* 2004. Pre-clinical development of cord blood-derived progenitor cell graft expanded ex vivo with cytokines and the polyamine copper chelator tetraethylenepentamine. *Cytotherapy* **6:** 344–355.
44. De Felice, L., C. Tatarelli, M.G. Mascolo, *et al.* 2005. Histone deacetylase inhibitor valproic acid enhances the cytokine-induced expansion of human hematopoietic stem cells. *Cancer Res.* **65:** 1505–1513.
45. Bug, G., H. Gül, K. Schwarz, *et al.* 2005. Valproic acid stimulates proliferation and self-renewal of hematopoietic stem cells. *Cancer Res.* **65:** 2537–2541.
46. Seet, L., E. Teng, Y. Lai, *et al.* 2009. Valproic acid enhances the engraftability of human umbilical cord blood hematopoietic stem cells expanded under serum-free conditions. *Eur. J. Haematol.* **82:** 124–132.
47. Young, J.C., S. Wu, G. Hansteen, *et al.* 2004. Inhibitors of histone deacetylases promote hematopoietic stem cell self-renewal. *Cytotherapy* **6:** 328–336.
48. Araki, H., K. Yoshinaga, P. Boccuni, *et al.* 2007. Chromatin-modifying agents permit human hematopoietic stem cells to undergo multiple cell divisions while retaining their repopulating potential. *Blood* **109:** 3570–3578.
49. Ko, K., T. Holmes, P. Palladinetti, *et al.* 2011. GSK-3β inhibition promotes engraftment of ex vivo-expanded hematopoietic stem cells and modulates gene expression. *Stem Cells* **29:** 108–118.
50. Holmes, T., T.A. O'Brien, R. Knight, *et al.* 2008. Glycogen synthase kinase-3β inhibition preserves hematopoietic stem cell activity and inhibits leukemic cell growth. *Stem Cells* **26:** 1288–1297.
51. Boitano, A.E., J. Wang, R. Romeo, *et al.* 2010. Aryl hydrocarbon receptor antagonists promote the expansion of human hematopoietic stem cells. *Science* **329:** 1345–1348.
52. North, T.E., W. Goessling, C.R. Walkley, *et al.* 2007. Prostaglandin E2 regulates vertebrate haematopoietic stem cell homeostasis. *Nature* **447:** 1007–1011.
53. Hoggatt, J., P. Singh, J. Sampath, *et al.* 2009. Prostaglandin E2 enhances hematopoietic stem cell homing, survival, and proliferation. *Blood* **113:** 5444–5455.
54. Nishino, T., K. Miyaji, N. Ishiwata, *et al.* 2009. Ex vivo expansion of human hematopoietic stem cells by a small-molecule agonist of c-MPL. *Exp. Hem.* **37:** 1364–1377.
55. Audet, J., C.L. Miller, C.J. Eaves, *et al.* 2002. Common and distinct features of cytokine effects on hematopoietic stem and progenitor cells revealed by dose-response surface analysis. *Biotechnol. Bioeng.* **80:** 393–404.
56. Miller, C.L. & C.J. Eaves. 1997. Expansion in vitro of adult murine hematopoietic stem cells with transplantable lympho-myeloid reconstituting-ability. *Proc. Natl. Acad. Sci.* **94:** 13648–13653.
57. Bhatia, M., D. Bonnet, U. Kapp, *et al.* 1997. Quantitative analysis reveals expansion of human-hematopoietic repopulating cells after short-term ex vivo culture. *J. Exp. Med.* **186:** 619–624.
58. de Lau, W., N. Barker, T.Y. Low, *et al.* 2011. Lgr5 homologues associate with Wnt receptors and mediate R-spondin signalling. *Nature* **476:** 293–297.
59. Reya, T., A.W. Duncan, L. Ailles, *et al.* 2003. A role for Wnt signalling in self-renewal of haematopoietic stem cells. *Nature* **423:** 409–414.
60. Austin, T.W., G.P. Solar, F.C. Ziegler, *et al.* 1997. A role for the Wnt gene family in hematopoiesis: expansion of multilineage progenitor cells. *Blood* **89:** 3624–3635.
61. Willert, K., J.D. Brown, E. Danenberg, *et al.* 2003. Wnt proteins are lipid-modified and can act as stem cell growth factors. *Nature* **423:** 448–452.
62. Murdoch, B., K. Chadwick, M. Martin, *et al.* 2003. Wnt-5A augments repopulating capacity and primitive hematopoietic development of human blood stem cells in vivo. *Proc. Natl. Acad. Sci.* **100:** 3422–3427.
63. Van Den Berg, D.J., A.K. Sharma, E. Bruno, *et al.* 1998. Role of members of the Wnt gene family in human hematopoiesis. *Blood* **92:** 3189–3202.
64. Luis, T., B.E. Naber, P.C. Roozen, *et al.* 2011. Canonical Wnt signaling regulates hematopoiesis in a dosage-dependent fashion. *Cell Stem Cell.* **9:** 345–356.
65. MacDonald, B.T., K. Tamai & X. He. 2009. Wnt/β-catenin signaling: components, mechanisms, and diseases. *Dev Cell.* **17:** 9–26.
66. Duncan, A.W., F.M. Rattis, L.N. DiMascio, *et al.* 2005. Integration of notch and Wnt signaling in hematopoietic stem cell maintenance. *Nat. Immunol.* **6:** 314–322.
67. Varnum-Finney, B., L.E. Purton, M. Yu, *et al.* 1998. The Notch ligand, jagged-1, influences the development of primitive hematopoietic precursor cells. *Blood* **91:** 4084–4091.
68. Varnum-Finney, B., C. Brashem-Stein & I.D. Bernstein. 2003. Combined effects of Notch signaling and cytokines induce a multiple log increase in precursors with lymphoid and myeloid reconstituting ability. *Blood* **101:** 1784–1789.
69. Delaney, C., B. Varnum-Finney, K. Aoyama, *et al.* 2005. Dose-dependent effects of the Notch ligand Delta1 on ex vivo differentiation and in vivo marrow repopulating ability of cord blood cells. *Blood* **106:** 2693–2699.
70. Delaney, C., S. Heimfeld, C. Brashem-Stein, *et al.* 2010. Notch-mediated expansion of human cord blood progenitor cells capable of rapid myeloid reconstitution. *Nat. Med.* **16:** 232–236.
71. Bhardwaj, G., B. Murdoch, D. Wu, *et al.* 2001. Sonic hedgehog induces the proliferation of primitive human hematopoietic cells via BMP regulation. *Nat. Immunol.* **2:** 172–180.

72. Bhatia, M., D. Bonnet, D. Wu, *et al.* 1999. Bone morphogenetic proteins regulate the developmental program of human hematopoietic stem cells. *J. Exp. Med.* **189:** 1139–1148.
73. Huynh, H., S. Iizuka, M. Kaba, *et al.* 2008. Insulin-like growth factor-binding protein 2 secreted by a tumorigenic cell line supports ex vivo expansion of mouse hematopoietic stem cells. *Stem Cells* **26:** 1628–1635.
74. Zhang, C.C., M. Kaba, S. Iizuka, *et al.* 2008. Angiopoietin-like 5 and IGFBP2 stimulate ex vivo expansion of human cord blood hematopoietic stem cells as assayed by NOD/SCID transplantation. *Blood* **111:** 3415–3423.
75. Istvanffy, R., M. Kröger, C. Eckl, *et al.* 2011. Stromal pleiotrophin regulates repopulation behavior of hematopoietic stem cells. *Blood* **118:** 2712–2722.
76. Huangfu, D., R. Maehr, W. Guo, *et al.* 2008. Induction of pluripotent stem cells by defined factors is greatly improved by small-molecule compounds. *Nat. Biotech.* **26:** 795–797.
77. Li, W. & S. Ding. 2010. Small molecules that modulate embryonic stem cell fate and somatic cell reprogramming. *Trends Pharmacol. Sci.* **31:** 36–45.
78. Schugar, R.C., P.D. Robbins & B.M. Deasy. 2007. Small molecules in stem cell self-renewal and differentiation. *Gene Ther.* **15:** 126–135.
79. Collins, S.J. 2002. The role of retinoids and retinoic acid receptors in normal hematopoiesis. *Leukemia* **16:** 1896–1905.
80. Degos, L. & Z.Y. Wang. 2001. All *trans* retinoic acid in acute promyelocytic leukemia. *Oncogene* **20:** 7140–7145.
81. Purton, L.E., S. Dworkin, G.H. Olsen, *et al.* 2006. RARγ is critical for maintaining a balance between hematopoietic stem cell self-renewal and differentiation. *J. Exp. Med.* **203:** 1283–1293.
82. Chute, J.P., G.G. Muramoto, J. Whitesides, *et al.* 2006. Inhibition of aldehyde dehydrogenase and retinoid signaling induces the expansion of human hematopoietic stem cells. *Proc. Natl. Acad. Sci.* **103:** 11707–11712.
83. Bae, B. & S.S. Percival. 1993. Retinoic acid-induced HL-60 cell differentiation is augmented by copper supplementation. *J. Nut.* **123:** 997–1002.
84. de Lima, M., J. McMannis, A. Gee, *et al.* 2008. Transplantation of ex vivo expanded cord blood cells using the copper chelator tetraethylenepentamine: a phase I/II clinical trial. *Bone Marrow Transplant.* **41:** 771–778.
85. Walasek, M.A., L. Bystrykh, V. van den Boom, *et al.* 2012. The combination of valproic acid and lithium delay hematopoietic stem/progenitor cell differentiation. *Blood* **119:** 3050–3059. Epub 2012 Feb 10.
86. Milhem, M., N. Mahmud, D. Lavelle, *et al.* 2004. Modification of hematopoietic stem cell fate by 5aza 2'deoxycytidine and trichostatin A. *Blood* **103:** 4102–4110.
87. Watts, K.L., C. Delaney, R.K. Humphries, *et al.* 2010. Combination of HOXB4 and Delta-1 ligand improves expansion of cord blood cells. *Blood* **116:** 5859–5866.
88. Perry, J.M., X.C. He, R. Sugimura, *et al.* 2011. Cooperation between both Wnt/β-catenin and PTEN/PI3K/Akt signaling promotes primitive hematopoietic stem cell self-renewal and expansion. *Genes Dev.* **25:** 1928–1942.

Ann. N.Y. Acad. Sci. ISSN 0077-8923

ANNALS OF THE NEW YORK ACADEMY OF SCIENCES
Issue: *Hematopoietic Stem Cells VIII*

Enhancing engraftment of cord blood cells via insight into the biology of stem/progenitor cell function

Hal E. Broxmeyer
Department of Microbiology and Immunology, Indiana University School of Medicine, Indianapolis, Indiana

Address for correspondence: Hal E. Broxmeyer, Ph.D., Department of Microbiology and Immunology, Indiana University School of Medicine, 950 West Walnut Street, Indianapolis, IN 46202-5181. hbroxmey@iupui.edu

Cord blood (CB) transplantation has been used over the last 24 years to treat patients with malignant and nonmalignant disorders. CB has its advantages and disadvantages compared with other sources of hematopoietic stem cells (HSCs) and hematopoietic progenitor cells (HPCs) for transplantation. More knowledge of the cytokines and intracellular signaling molecules regulating HSCs and HPCs could be used to modulate these regulators for clinical benefit. This review provides information about the general field of CB transplantation and about studies from the author's laboratory that focus on regulation of HSCs and HPCs by CD26/DPPIV, SDF-1/CXCL12, the Rheb2-mTOR pathway, SIRT1, DEK, cyclin-dependent kinase inhibitors, and cytokines/growth factors. Cryopreservation of CB HSCs and HPCs is also briefly discussed.

Keywords: cord blood; stem and progenitor cells; intracellular signals; cytokines

Introduction

Cord blood (CB) has served as a transplantable source of hematopoietic stem cells (HSCs) and hematopoietic progenitor cells (HPCs) to treat malignant and nonmalignant disorders since our initial laboratory,[1–9] and clinical studies,[10–14] began over 20 years ago. There have now been over 25,000 CB transplants performed worldwide[15] since our initial clinical report of a child with Fanconi anemia who was successfully treated with CB from his human leukocyte antigen (HLA)-matched sister.[10] The field has moved rapidly, and advances in CB banking[16–22] and transplantation[23–44] have been encouraging, but there is still much to be learned to make CB transplantation a more efficient and efficacious procedure.[15,45]

There are advantages to using CB as a source of transplantable HSCs and HPCs over other sources. These include that CB is a readily available source of HLA-typed cells that are stored in CB banks; that CB had already been used to treat essentially all malignant and nonmalignant diseases that can be treated by bone marrow (BM) transplantation; and that CB transplantation induces a lower level of acute and chronic graft-versus-host disease (GVHD) when the CB is used as a single minimally manipulated unit. This latter characteristic of CB cells allows for more flexibility in related and unrelated donors (for example, by using partially HLA-mismatched CB) compared with BM transplantation. There are also disadvantages to using CB as a source of transplantable HSCs and HPCs. CB transplantation is associated with a slower time to neutrophil and platelet engraftment and to immune cell recovery than are BM and mobilized peripheral blood (mPB), differences in part due to the more limiting numbers of HSCs and HPCs in CB collections compared with BM and mPB. For BM and mPB one can collect many cells, but for CB what is collected at the birth of a baby is all that is available, although a means to collect greater numbers of CB-derived cells may be feasible in the future.[46] However, at present, the low number of HSCs and HPCs collected in CB is somewhat problematic for single CB unit transplantation of adults and higher-weight pediatric recipients; in addition, CB transplantation has been associated with enhanced graft failure.

A number of investigators, including from my own group, have worked on the means to enhance engraftment of limiting numbers of HSCs and HPCs under several types of conditions, including in a

doi: 10.1111/j.1749-6632.2012.06509.x

basic science laboratory setting, in preclinical animal studies, and in pilot phase I and other clinical studies. Examples of such attempts to foster a more potent CB transplant have been recently reviewed by experts in the field and include double CB transplantation,[40–43,47–50] *ex vivo* expansion of HSCs and HPCs,[51–54] intrabone transplantation,[55–57] and efforts to enhance the homing and engraftment of cells through *ex vivo* or *in vivo* inhibition of CD26, a dipeptidylpeptidase (DPPIV),[58–66] incubation of cells *ex vivo* with prostaglandin E,[67–70] or by fucosylation of donor cells *ex vivo*.[71,72]

Ultimately, a better understanding of the biology of HSCs and HPCs will allow for more innovative means to enhance the homing and engraftment of limiting numbers of HSCs and HPCs in CB. Our knowledge of the characteristics and functions of HSCs and HPCs is becoming clearer; we now know much more about the phenotypic[73] and functional[74,75] characteristics of human HSCs and HPCs, the cytokines that regulate these cells,[76,77] and the BM microenvironmental cells and factors that influence HSCs and HPCs *in vivo*.[78–80]

This review mainly focuses on some recent work from my laboratory that has evaluated intracellular and extracellular factors that influence HSC/HPC function, with the goal to eventually use this information to enhance the engrafting capability of HSCs and HPCs for clinical advantage. Through all of this work and with the aim of eventually translating of these studies to the clinic the goal has been: "The simpler, the better." The topics of research to be discussed include inhibition of CD26/DPPIV, intracellular molecule modulation of HSCs and HPCs, cytokine and growth factor effects on HSCs and HPCs, and cryopreservation of CB HSCs and HPCs.

CD26/DPPIV influence on the SDF-1/CXCL12-CXCR4 axis and homing/engraftment of HSCs and HPCs

Stromal cell-derived factor (SDF)-1/CXCL12 is a well-known chemokine that acts as a chemotactic (directed cell movement) agent for HSCs and HPCs through its action on the receptor CXCR4. SDF-1/CXCL12 has been implicated in the homing, survival, and nurturing of HSCs and HPCs.[76,80–82] CD26/DPPIV is an enzyme that cleaves dipeptides from the N-terminus of proteins after a proline or alanine. SDF-1/CXCL12 has a DPPIV cleavage site that CD26/DPPIV cleaves to produce a truncated SDF-1/CXCL12 molecule that is no longer chemotactic.[58] Moreover, the truncated SDF-1/CXCL12 blocks the chemotactic activity of full-length SDF-1/CXCL12.[58] Since SDF-1/CXCL12 acts as an *in vivo* homing molecule for HSCs and HPCs,[80,81] and since a number of cell types, including HSCs and HPCs, express CD26/DPPIV on their cell surface,[59] we reasoned that inhibition of CD26/DPPIV by small peptide molecules such as diprotin A (Ile-Pro-Ile) or Val-Pyr on the donor cells would enhance the homing and engrafting capability of limiting numbers of donor mouse BM cells in a congenic mouse model of competitive and noncompetitive HSC transplantation.[59] Our experiments and those of others confirmed and extended our initial hypothesis.[60–62] The engraftment-enhancing effects of CD26/DPPIV inhibition were extended to pretreatment of human CD34$^+$ donor CB[63,64] and mPB[65] cells transplanted into mice with a nonobese diabetic/severe combined immunodeficiency (NOD/SCID) genotype. *In vivo* inhibition of CD/DPPIV in recipient mice[65,66] effectively enhanced engraftment, an effect likely due to inhibition of DPPIV truncation of nonchemokine growth factors. We recently found that a number of colony-stimulating factors (CSFs), such as GM-CSF, G-CSF, IL-3, and EPO, contain DPPIV truncation sites, and that DPPIV inhibitors used to pretreat target cells that express CD26 enhance the detectable activity of the CSFs. Also, *in vivo* use of DPPIV inhibitors allows accelerated recovery of HSCs and HPCs after stress (e.g., chemotherapy and irradiation) (Broxmeyer *et al.*, unpublished observations; manuscript in revision). A Food and Drug Administration (FDA)-approved CD26 inhibitor is currently being evaluated in a pilot clinical study for its effects on enhancing the engrafting capability of single CB collections in adult patients with leukemia and lymphoma.[66]

Intracellular modulation of HSCs/HPCs

Understanding the intracellular signals involved in HSC and HPC function may lead to successful efforts to manipulate these signaling molecules for clinical benefit.

Effect of the Rheb2-mTOR pathway on HSC engraftment

Rheb is a member of the ras homologue enriched in the brain family of small ras-like GTPase molecules. Rheb cycles between active guanosine tripeptide (GTP) and inactive guanosine dipeptide (GDP)-bound forms. Both Rheb1 and Rheb2 are able to activate mammalian target of rapamycin (mTOR) signaling in mammalian cells. Since Rheb2 was found to be preferentially expressed in immature mouse HSCs compared with mature hematopoietic cells,[83] we evaluated effects of Rheb2 overexpression, by means of a MIEG3 bicistronic retroviral vector, on the function of mouse BM HSCs and HPCs.[84] In this study, we identified Rheb2 as a pathway important in expansion of immature progenitor cells *in vitro* and *in vivo*. However, this expansion was accompanied by a loss of HSC activity. We felt that regulating the activity of the Rheb-mTOR pathway might allow for effective expansion of cells without their loss of HSC repopulating ability. Toward this goal, we treated human $CD34^+$ CB cells *ex vivo* with a combination of HSC expansion cytokines (stem cell factor (SCF), Flt3-ligand (FL), and thrombopoietin (TPO)) in the presence and absence of rapamycin prior to assessing their engrafting capability in sublethally irradiated NOD/SCID IL-2R gamma chain null (NSG) mice.[85] *Ex vivo* rapamycin treatment of the $CD34^+$ CB cells in the presence of SCF, FL, and TPO greatly enhanced their engrafting capability. More mechanistic evaluation of these studies is warranted, as is further preclinical analysis for enhancing CB transplantation.

Tip110/p110nrb/SART3/p110 regulation of hematopoiesis

Tip110 is a Tat-interacting protein of 110 KDa that has been implicated in RNA metabolism and tumor-antigen presentation. Tip110 was found to regulate the transcription of HIV-1 and cellular genes.[86,87] It also functions as a general pre-mRNA splicing factor.[88] The various effects of Tip110 that impinged on transcription factors and cellular gene expression enticed us to evaluate a possible role for Tip110 in the regulation of hematopoiesis. Using Tip110-overexpressing transgenic mice, haploinsufficient ($Tip110^{+/-}$) mice, and means to up and down-regulate Tip110 expression in human cells through lentiviral gene transduction, we were able to demonstrate that Tip110 transgenic expression increased the numbers, cell cycling status, and survival of HPCs, while $Tip110^{+/-}$ mice manifested opposite effects of the Tip110 transgenic mice in terms of HPC function.[89] Also, $Tip110^{+/-}$ BM HPCs responded better and Tip110 transgenic BM HPCs worse than control mice to recovery from the cytotoxic effects of 5-flurouracil.[89] Mechanistically, Tip110 regulated expression of CMYC and GATA2, with Tip110 and CMYC regulating the expression of each other, thus linking Tip110 hematopoietic regulation to Tip110 reciprocal regulation of CMYC.[89] We also found that Tip110 was expressed in human embryonic stem (hES) cells, and that it was important for maintenance of expression of pluripotency factors such as NANOG, OCT4, and SOX2, and for pluripotency of hES cells.[90] How these Tip110 effects in hES cells are mediated and what their role in induced pluripotent stem (iPS) cells is remain to be investigated.

SIRT1 effects on hematopoiesis

SIRT1, a member of the sirtuin family encompassing seven proteins and histone deacytelases, has been conserved from bacteria to humans.[91] Mammalian sirtuins have been implicated in numerous cell functions, some with disease relevance.[92] SIRT1 is a human homologue of the Sir2 yeast protein. We showed that SIRT1 is involved in regulating apoptosis and Nanog expression in mouse embryonic stem (ES) cells when in the presence of leukemia inhibitory factor (LIF), at least in part by controlling the subcellular localization of the tumor suppressor p53.[93] We followed up that study by demonstrating the need for SIRT1 for differentiation of mouse ESCs in the absence of LIF but in the presence of 2-mercaptoethanol.[94] A mouse ES cell line deficient in SIRT1 formed few mature blast cell colonies, and the replated cells from these colonies were defective in hematopoietic potential. There was decreased primitive and definitive hematopoiesis associated with LIF removal-induced differentiation of the $SIRT1^{-/-}$ mouse ES cell line, and this corresponded to a delayed capacity to turn off expression of Oct4, nanog, and Fgr5, and to decrease expression of β-H1 globin, β-major globin, and Scl genes. $SIRT1^{-/-}$ mice had fewer yolk sac primitive erythroid precursors and manifested decreased embryogenesis. In adult mice, both $SIRT1^{-/-}$ and $SIRT1^{+/-}$ BM cells were decreased in number and cycling status compared with HPCs, an effect most apparent when cells were cultured *in vitro* under lower (5%), compared to normal (~20%), oxygen tension. Thus, these results link oxygen tension and SIRT1 activity. Most recently, others have suggested that cell-autonomous SIRT1 intracellular signaling may be dispensable for adult HSC functional maintenance in mice.[95]

How the above information on adult HSCs and HPCs may be used to modulate and influence engrafting capability remains to be determined. Of interest in this context is the observation that deficiency of SIRT1 in mouse ES cells causes down-regulation of the PTEN/JNK/FOXO1 pathway, with a concomitant block in reactive oxygen species (ROS)–induction of apoptosis.[96] Thus, at least in mouse ES cells, SIRT1 appears to play an important role in adjusting the PTEN/JNK/FOXO1 pathway for responsiveness to cellular ROS. Mitochondria and mitochondria-generated ROS are important for hematopoiesis,[97,98] and it is possible that SIRT1 may still play a role in asymmetric divisions of HSCs and self-renewal,[99] especially in cases of heavy stress-induced hematopoiesis. In this regard, energy metabolism is likely crucial to normal, and especially stress-induced hematopoiesis.

We recently assessed whether cellular energy homeostasis participates in the maintenance of pluripotency and self-renewal of mouse ES cells.[100] AMP-activated protein kinase (AMPK), one regulator of energy metabolism, is activated during stress that induces exhaustion of ATP. Using 5-aminoimidazole-4-carboxyamide ribonucleoside (AICAR), an AMPK activator, we demonstrated activation of the p53-p21^{cip1} pathway, decreased mouse ES cell proliferation, a G_1/S-phase cell cycle block, and decreased expression of the pluripotency markers NANOG and SSEA-1, without effects on Oct4 expression. These effects were associated with enhanced differentiation of erythroid cells from mES cells.

DEK regulation of hematopoiesis

DEK is a relatively newly defined molecule involved in a number of cellular activities, such as transcriptional repression and activation, processing of mRNA, and chromatin remodeling.[101,102] DEK is a unique molecule in that it can be found in the nucleus, yet it can also be secreted outside of the cell and influence other cell types;[101] for example, DEK can act as a chemoattractant for specific mature blood cells (CD8$^+$ T cells and natural killer cells) after being released from macrophages.[103] We identified a role for DEK in hematopoiesis.[104] DEK$^{-/-}$ mice have increased numbers of HPCs in BM and spleen; and purified recombinant DEK protein suppressed *in vitro* colony formation by wild-type mouse BM and human CB HPC. This negative effect of DEK on HPC proliferation *in vitro* and *in vivo* was associated with decreased long-term competitive and secondary mouse HSC repopulating capacity, suggesting a positive role for DEK in HSC functions such as engraftment. What these hematopoietic effects of DEK are due to requires further investigation, but this information may be of potential clinical relevance.

Role for cyclin-dependent kinase inhibitors in hematopoiesis

A number of cyclin-dependent kinase inhibitors (CDKIs) have been implicated in HSC and HPC function.[76,105–108] These include p21$^{cip1/waf1}$ (p21), p27^{kip1} (p27), and p18^{INK4c} (p18). However, other than a paper demonstrating that p18$^{-/-}$ counters the exhaustion of p21$^{-/-}$ HSCs after serial transplantation,[109] there is nothing else in the literature that links the different CDKIs and their networking interactions with HSC and HPC function. We have noted that CDKIs have different effects on HPC proliferation and that they differentially modulate the responsiveness of HPCs to synergistic combinations of cytokines such as a CSF plus SCF.[110] Deletion of p18 resulted in decreased numbers and proliferation of HPCs, effects similar to those previously reported for p21$^{-/-}$ mice.[77] These positive effects of p18 dominated over the negative effects of p27, where p27$^{-/-}$ was associated with enhanced HPC proliferation. The responsiveness of HPCs to suppression by certain chemokine family members was directly related to the ability of HPCs to respond to synergistic stimulation and cycling HPCs. Deletion of the p18 gene rescued the loss of chemokine suppression of the synergistic cell proliferation associated with deletion of the p21 gene. Thus, there is interplay among cell cycle regulators of HPC proliferation, and loss of one CDKI can sometimes be compensated for by another missing CDKI.

Cytokine/growth factor regulation of hematopoiesis

Numerous cytokines are known to regulate the proliferation and survival of HPCs.[76,77,108]

Role of immune cells in the regulation of hematopoiesis

T lymphocytes play a role in the proliferation of HPCs and their homeostasis,[111–113] in part through the transcription factors STAT4, STAT6, BCL-6, and BAZF, as well as the growth factor oncostatin M. We recently reported a role for STAT3-dependent IL-21 production from helper T cells in maintenance of HPC homeostasis.[114] We found decreased HPC proliferation in mice with a specific deficiency of STAT3 in T cells, and STAT3 expression was required for production of IL-21. Neutralization of IL-21, but not IL-22, resulted in a decrease in HPC number and cycling, similar to that seen in T cell–specific STAT3-deficient mice. Moreover, exogenous administration of IL-21 was able to rescue suppressed HPC proliferation in mice with STAT3$^{-/-}$ T cells.

Angiopoietin-like molecules-2 and -3 enhance survival of HPCs in CB

Angiopoietin-like (ANGPLT) molecules have been implicated in regulation of mouse fetal liver and BM HSCs and in human CB NOD/SCID-repopulating cells,[115–117] but no information was available on effects of ANGPTL molecules on HPCs. We identified the actions of ANGPTL-2 and -3 molecules on enhancement of the survival and replating ability of HPCs from human CB.[118] These activities of ANGPTL-2 and -3 were manifested through their coiled-coil domains. We did not detect functional activities of ANGPTL-4, -5, -6, or -7 on HPC survival or replating capacity, and none of the ANGPTL molecules tested influenced the proliferation of CB HPCs. The survival and replating effects of ANGPTL-2 and -3 on HPCs[118] and on expanding NOD/SCID-repopulating human HPCs[116] may be of future relevance to CB transplantation.

A role for neuronally active molecules on hematopoiesis

The nervous system has been implicated in microenvironmental control of hematopoiesis.[79] This opened up the possibility that other neuronally associated proteins and their receptors could play a role in hematopoiesis. We identified such a role for neurexophilin1.[119] Neurexophilins bind neurexin1α, and neurexin1α and dystroglycan are membrane receptors that serve as mutual ligands within the neuronal system. We found that neurexophilin1 was able to suppress the proliferation of HPCs and it acted through neurexin1α, an effect that could be counter-modulated by dystroglycan. The suppressive effect of neurexophilin1 on HPCs was direct, acting on the HPCs; and inhibition was apparent both *in vitro* on human CB and mouse BM HPCs, and *in vivo* after injection of recombinant neurexophilin1 into mice. Thus, a signaling axis in the hematopoietic system centers on neurexin1α and its modulation by neurexophilin1 and dystroglycan. Additional information on links between the nervous and hematopoietic systems may offer the opportunity to modulate one for the benefit of the other in a transplant setting.

Potential for modulation of intracellular signals by cytokines and growth factors and other means for enhanced CB transplantation

There are many transcription factors and other intracellular signaling molecules that impinge on HSC and HPC numbers and functions.[76,108] How all these different intracellular factors may interact with each other, if at all, will need to be better elucidated if they are to be modulated for clinical advantage with minimal side effects. Interconnected with these intracellular molecules are numerous cytokines and growth factors that can trigger/activate these intracellular molecules.[76,77] While much is known regarding cytokines and growth factor effects on HPC function, we still know very little of how these cytokines/growth factors influence HSC functions such as self-renewal, survival, and engraftment. Future information in these areas will likely have a positive influence on how we might be able to enhance HSC transplantation, especially with CB cells.

Cryopreservation of CB HSCs, HPCs, and other cells

CB transplantation is critically dependent on CB banking, which in turn is dependent on the capacity to adequately cryopreserve the HSCs and HPCs in CB and to maintain these cells in a frozen state. Many CB banks, both public and private, have been formed in order to supply cryopreserved CBs for transplantation purposes. However, how long such frozen CB units can be stored and then thawed for efficient recovery of HSCs and HPCs is critical for the success of CB banking and CB transplantation.

Over the last 23 plus years my laboratory reported on the cryopreservation and subsequent recovery of thawed cells.[1,7,120,121] Most recently, we demonstrated that we could recover functionally intact HPCs at high efficiency from cryopreserved CB after thawing of cells stored frozen for up to 21–23.5 years.[122] While there was a range of recoveries of HPCs from different CB units, 80–100% recovery was apparent for most samples, with maintenance of high proliferative and replating capability. Moreover, CB cryopreserved for up to 21 years could be thawed, the $CD34^+$ cells isolated, and these cells could be used to long-term repopulate primary and secondary immune-deficient mice, suggesting that HSCs with long-term marrow repopulating and self-renewing capacities were adequately cryopreserved and could be retrieved after thawing. From the long-term cryopreserved cells we were also able to retrieve functionally responsive $CD4^+$ and $CD8^+$ T cells and high-proliferative-potential endothelial colony-forming cells (i.e, endothelial progenitor cells); additionally, from the thawed and subsequently purified $CD34^+$ cell population we generated iPS cells that, *in vitro* and *in vivo*, could be differentiated into all three germ cell lineages.[122] Subsequent studies with the iPS cells showed that only a small percentage of these cells were fully reprogrammed. Efforts are underway to enhance the full reprogramming capacity of a larger percent of these iPS cells by modulating expression of the microRNA 302 cluster and its downstream targets. However, how effective such generated iPS cells may be for future regenerative medicine possibilities remains to be determined.

Concluding remarks

The intent of this article was to focus mainly on basic and translational research by my laboratory. There are many outstanding laboratories working on the means to gain better mechanistic insight into optimizing the functional activities of HSCs and HPCs. It is hoped that together this work will translate into clinical utility for health benefit, with one such benefit being enhanced efficacy of CB HSC/HPC transplantation.

Acknowledgments

Most of the studies in the reference list reported from the author's laboratory were funded through NIH R01 or NIH P01 grants. H. Broxmeyer is currently principal investigator (PI) on and supported by Public Health Service Grants R01 HL056416, R01 HL67384, R01 HL112669, and P01 DK090948 from the National Institutes of Health. He is also PI on NIH Training Grants T32 DK07519 and T32 HL07910.

Conflicts of interest

The author is on the medical scientific advisory board of CordUse, a cord blood banking company.

References

1. Broxmeyer, H.E., G.W. Douglas, G. Hangoc, *et al.* 1989. Human umbilical cord blood as a potential source of transplantable hematopoietic stem/progenitor cells. *Proc. Natl. Acad. Sci. USA* **86:** 3828–3832.
2. Broxmeyer, H.E., E. Gluckman, A. Auerbach, *et al.* 1990. Human umbilical cord blood: a clinically useful source of transplantable hematopoietic stem/progenitor cells. *Int. J. Cell Cloning* **8**(Suppl. 1): 76–91.
3. Auerbach, A.D., Q. Liu, R. Ghosh, *et al.* 1990. Prenatal identification of potential donors for umbilical cord blood transplantation for Fanconi Anemia. *Transfusion* **30:** 682–687.
4. Pollack, M.S., A.D. Auerbach, H.E. Broxmeyer, *et al.* 1991. The use of DNA amplification for DQ typing as an adjunct to serological prenatal HLA typing for the identification of potential donors for umbilical cord blood transplantation. *Human Immunol.* **30:** 45–49.
5. Broxmeyer, H.E., J. Kurtzberg, E. Gluckman, *et al.* 1991. Umbilical cord blood hematopoietic stem and repopulating cells in human clinical transplantation. *Blood Cells* **17:** 313–329.
6. Carow, C., G. Hangoc, S. Cooper, *et al.* 1991. Mast cell growth factor (c-kit ligand) supports the growth of human multipotential (CFU-GEMM) progenitor cells with a high replating potential. *Blood* **78:** 2216–2221.
7. Broxmeyer, H.E., G. Hangoc, S. Cooper, *et al.* 1992. Growth characteristics and expansion of human umbilical cord blood and estimation of its potential for transplantation of adults. *Proc. Natl. Acad. Sci. USA* **89:** 4109–4113.
8. Carow, C.E., G. Hangoc & H.E. Broxmeyer. 1993. Human multipotential progenitor cells (CFU-GEMM) have extensive replating capacity for secondary CFU-GEMM: an effect enhanced by cord blood plasma. *Blood* **81:** 942–949.
9. Lu, L., M. Xiao, R-N., Shen, *et al.* 1993. Enrichment, characterization and responsiveness of single primitive $CD34^{+++}$ human umbilical cord blood hematopoietic progenitor with high proliferative and replating potential. *Blood* **81:** 41–48.
10. Gluckman, E., H.E. Broxmeyer, A.D. Auerbach, *et al.* 1989. Hematopoietic reconstitution in a patient with Fanconi anemia by means of umbilical-cord blood from an HLA-identical sibling. *New Engl. J. Med.* **321:** 1174–1178.
11. Wagner, J.E., H.E. Broxmeyer, R.L. Byrd, *et al.* 1992. Transplantation of umbilical cord blood after myeloblative therapy: analysis of engraftment. *Blood* **79:** 1874–1881.
12. Kohli-Kumar, M., N.T. Shahidi, H.E. Broxmeyer, *et al.* 1993. Haematopoietic stem/progenitor cell transplant in Fanconi anemia using HLA-matched sibling umbilical cord blood cells. *Brit. J. Haematol.* **85:** 419–422.
13. Wagner, J.E., N.A. Kernan, M. Steinbuch, *et al.* 1995. Allogeneic sibling umbilical cord blood transplantation in forty-four children with malignant and non-malignant disease. *Lancet* **346:** 214–219.

14. Broxmeyer, H.E. & F.O. Smith. 2009. Cord blood hematopoietic cell transplantation. In *Thomas' Hematopoietic Cell Transplantation*, 4th ed. F.R. Appelbaum, S.J. Forman, R.S. Negrin & K.G. Blume, Eds.: 559–576. Wiley-Blackwell. West Sussex, UK.
15. Rocha, V. & H.E. Broxmeyer. 2009. New approaches for improving engraftment after cord blood transplantation. *Biol. Blood Marrow Transplant.* **16:** S126–S132.
16. Rubinstein, P., R.E. Rosenfield, J.W. Adamson & C.E. Stevens. 1993. Stored placental blood for unrelated bone marrow reconstitution. *Blood* **81:** 1679–1690.
17. Rubinstein, P., L. Dobrila, R.E. Rosenfield, *et al.* 1995. Processing and cryopreservation of placental/umbilical cord blood for unrelated bone marrow reconstitution. *Proc. Natl. Acad. Sci. USA* **92:** 10119–10122.
18. Guindi, E.S., T.F. Moss & M.T. Ernst. 2011. Public and private cord blood banking. In *Cord Blood Biology, Transplantation, Banking, and Regulation.* H.E. Broxmeyer, Ed.: 595–628. AABB Press. Bethesda, MD.
19. Regan, D.M. 2011. Cord blood banking: the development and application of cord blood banking processes, standards, and regulations. In *Cord Blood Biology, Transplantation, Banking, and Regulation.* H.E. Broxmeyer, Ed.: 633–644. AABB Press. Bethesda, MD.
20. Sniecinski, I. 2011. Cord blood banking in developing countries. In *Cord Blood Biology, Transplantation, Banking, and Regulation.* H.E. Broxmeyer, Ed.: 647–661. AABB Press. Bethesda, MD.
21. Boo, M., K. Welte & D. Confer. 2011. Accreditation and regulation of cord blood banking. In *Cord Blood Biology, Transplantation, Banking, and Regulation.* H.E. Broxmeyer, Ed.: 663–671. AABB Press. Bethesda, MD.
22. Welte, K., M. Boo & D. Confer. 2011. Worldwide searching and distribution of cord blood units. In *Cord Blood Biology, Transplantation, Banking, and Regulation.* H.E. Broxmeyer, Ed.: 673–684. AABB Press. Bethesda, MD.
23. Kurtzberg, J., M. Laughlin, M.L. Graham, *et al.* 1996. Placental blood as a source of hematopoietic stem cells for transplantation into unrelated recipients. *N. Engl. J. Med.* **335:** 157–166.
24. Wagner, J.E., J. Rosenthal, R. Sweetman, *et al.* 1996. Successful transplantation of HLA-matched and HLA-mismatched umbilical cord blood from unrelated donors: analysis of engraftment and acute graft-versus-host disease. *Blood* **88:** 795–802.
25. Gluckman, E., V. Rocha, A. Boyer-Chammard, *et al.* 1997. Outcome of cord-blood transplantation from related and unrelated donors. Eurocord Transplant Group and the European Blood and Marrow Transplantation Group. *N. Engl. J. Med.* **337:** 373–381.
26. Rubinstein, P., C. Carrier, A. Scaradavou, *et al.* 1998. Outcomes among 562 recipients of placental-blood transplants from unrelated donors. *N. Engl. J. Med.* **339:** 1565–1577.
27. Rocha, V., J.E. Wagner Jr., K.A. Sobocinski, *et al.* 2000. Graft-versus-host disease in children who have received a cord-blood or bone marrow transplant from an HLA-identical sibling. Eurocord and International Bone Marrow Transplant Registry Working Committee on Alternative Donor and Stem Cell Sources. *N. Engl. J. Med.* **342:** 1846–1854.
28. Laughlin, M.J., J. Barker, B. Bambach, *et al.* 2001. Hematopoietic engraftment and survival in adult recipients of umbilical-cord blood from unrelated donors. *N. Engl. J. Med.* **344:** 1815–1822.
29. Rocha, V., J. Cornish, E.L. Sievers, *et al.* 2001. Comparison of outcomes of unrelated bone marrow and umbilical cord blood transplants in children with acute leukemia. *Blood* **97:** 2962–2971.
30. Wagner, J.E., J.N. Barker, T.E. Defor, *et al.* 2002. Transplantation of unrelated donor umbilical cord blood in 102 patients with malignant and nonmalignant diseases: influence of CD34 cell dose and HLA disparity on treatment-related mortality and survival. *Blood* **100:** 1611–1618.
31. Barker, J.N., D.J. Weisdorf, T.E. Defor, *et al.* 2003. Rapid and complete donor chimerism in adult recipients of unrelated donor umbilical cord blood transplantation after reduced-intensity conditioning. *Blood* **102:** 1915–1919.
32. Rocha, V., M. Labopin, G. Sanz, *et al.* 2004. Transplants of umbilical-cord blood or bone marrow from unrelated donors in adults with acute leukemia. *N. Engl. J. Med.* **351:** 2276–2285.
33. Laughlin, M.J., M. Eapen, P. Rubinstein, *et al.* 2004. Outcomes after transplantation of cord blood or bone marrow from unrelated donors in adults with leukemia. *N. Engl. J. Med.* **351:** 2265–2275.
34. Brunstein, C.G., J.N. Barker, D.J. Weisdorf, *et al.* 2007. Umbilical cord blood transplantation after nonmyeloablative conditioning: impact on transplantation outcomes in 110 adults with hematologic disease. *Blood* **110:** 3064–3070.
35. Eapen, M., P. Rubinstein, M.J. Zhang, *et al.* 2007. Outcomes of transplantation of unrelated donor umbilical cord blood and bone marrow in children with acute leukaemia: a comparison study. *Lancet* **369:** 1947–1954.
36. Kurtzberg, J., V.K. Prasad, S.L. Carter, *et al.* 2008. Results of the cord blood transplantation study (COBLT): clinical outcomes of unrelated donor umbilical cord blood transplantation in pediatric patients with hematologic malignancies. *Blood* **112:** 4318–4327.
37. Eapen, M., V. Rocha, G. Sanz, *et al.* 2010. Effect of graft source on unrelated donor haemopoietic stem-cell transplantation in adults with acute leukaemia: a retrospective analysis. *Lancet Oncol.* **11:** 653–660.
38. Eapen, M. 2011. Cord blood transplantation for leukemia in children. In *Cord Blood Biology, Transplantation, Banking, and Regulation.* H.E. Broxmeyer, Ed.: 467–474. AABB Press. Bethesda, MD.
39. Horwitz, M.E. & N. Chao. 2011. Umbilical cord blood transplantation for treatment of nonmalignant disorders. In *Cord Blood Biology, Transplantation, Banking, and Regulation.* H.E. Broxmeyer, Ed.: 477–483. AABB Press. Bethesda, MD.
40. Gluckman, E., A. Ruggeri & V. Rocha. 2011. Extending the use of cord blood cells to adults. In *Cord Blood Biology, Transplantation, Banking, and Regulation.* H.E. Broxmeyer, Ed.: 487–499. AABB Press. Bethesda, MD.
41. Brunstein, C.G. & J.E. Wagner. 2011. Double umbilical cord blood transplantation for malignant disorders. In *Cord Blood Biology, Transplantation, Banking, and Regulation.* H.E. Broxmeyer, Ed.: 503–524. AABB Press. Bethesda, MD.

42. Ponce, D.M. & J.N. Barker. 2011. Determinants of engraftment after double-unit cord blood transplantation. In *Cord Blood Biology, Transplantation, Banking, and Regulation.* H.E. Broxmeyer, Ed.: 529–541. AABB Press. Bethesda, MD.
43. Chen, Y.-B., C. Cutler & K. Ballen. 2011. Reduced-intensity conditioning and double umbilical cord blood transplantation. In *Cord Blood Biology, Transplantation, Banking, and Regulation.* H.E. Broxmeyer, Ed.: 545–553. AABB Press. Bethesda, MD.
44. McCullough, J. & D. McKenna. 2011. Management of umbilical cord blood at the transplant center. In *Cord Blood Biology, Transplantation, Banking, and Regulation.* H.E. Broxmeyer, Ed.: 585–592. AABB Press. Bethesda, MD.
45. Petz, L.D., S.R. Spellman & L. Gragert. 2011. The underutilization of cord blood transplantation: extent of the problem, causes, and methods of improvement. In *Cord Blood Biology, Transplantation, Banking, and Regulation.* H.E. Broxmeyer, Ed.: 557–581. AABB Press. Bethesda, MD.
46. Broxmeyer, H.E., S. Cooper, D.M. Haas, *et al.* 2009. Experimental basis of cord blood transplantation. *Bone Marrow Transplantation* **44**(Special Issue): 627–633.
47. Barker, J.N., D.J. Weisdorf, T.E. Defor, *et al.* 2005. Transplantation of 2 partially HLA-matched umbilical cord blood units to enhance engraftment in adults with hematologic malignancy. *Blood* **105:** 1343–1347.
48. Ballen, K.K., T.R. Spitzer, B.Y. Yeap, *et al.* 2007. Double unrelated reduced-intensity umbilical cord blood transplantation in adults. *Biol. Blood Marrow Transplant.* **13:** 82–89.
49. Brunstein, C.G., J.A. Gutman, D.J. Weisdorf, *et al.* 2010. Allogeneic hematopoietic cell transplantation for hematologic malignancy: relative risks and benefits of double umbilical cord blood. *Blood* **116:** 4693–4699.
50. Cutler, C., K. Stevenson, H.T. Kim, *et al.* 2011. Double umbilical cord blood transplantation with reduced intensity conditioning and sirolimus-based GVHD prophylaxis. *Bone Marrow Transplant.* **46:** 659–667.
51. Shpall, E.J., R. Quinones, R. Giller, *et al.* 2002. Transplantation of *ex vivo* expanded cord blood. *Biol. Blood Marrow Transplant.* **8:** 368–376.
52. Jaroscak, J., K. Goltry, A. Smith, *et al.* 2003. Augmentation of umbilical cord blood (UCB) transplantation with *ex vivo*-expanded UCB cells: results of a phase 1 trial using the AastromReplicell System. *Blood* **101:** 5061–5067.
53. Delaney, C., S. Heimfeld, C. Brashem-Stein, *et al.* 2010. Notch-mediated expansion of human cord blood progenitor cells capable of rapid myeloid reconstitution. *Nat. Med.* **16:** 232–236.
54. Delaney, C. 2011. *Ex vivo* expansion of cord blood stem and progenitor cells. In *Cord Blood Biology, Transplantation, Banking, and Regulation.* H.E. Broxmeyer, Ed.: 215–224. AABB Press. Bethesda, MD.
55. Frassoni, F., F. Gualandi, M. Podesta, *et al.* 2008. Direct intrabone transplant of unrelated cord-blood cells in acute leukaemia: a phase I/II study. *Lancet Oncol.* **9:** 831–839.
56. Brunstein, C.G., J.N. Barker, D.J. Weisdorf, *et al.* 2009. Intra-BM injection to enhance engraftment after myeloablative umbilical cord blood transplantation with two partially HLA-matched units. *Bone Marrow Transplant.* **43:** 935–940.
57. Frassoni, F., M. Podesta, R. Varaldo, *et al.* 2011. Intrabone transplantation of cord blood cells and the journey of hematopoietic cells. In *Cord Blood Biology, Transplantation, Banking, and Regulation.* H.E. Broxmeyer, Ed.: 239–250. AABB Press. Bethesda, MD.
58. Christopherson, K.W., G. Hangoc & H.E. Broxmeyer. 2002. Cell surface peptidase CD26/DPPIV regulates CXCL12/SDF-1a mediated chemotaxis of human $CD34^+$ progenitor cells. *J. Immunol.* **169:** 7000–7008.
59. Christopherson, K.W. II, G. Hangoc, C. Mantel & H.E. Broxmeyer. 2004. Modulation of hematopoietic stem cell homing and engraftment by CD26. *Science* **305:** 1000–1003.
60. Tian, C., J. Bagley, D. Forman & J. Iacomini. 2006. Inhibition of CD26 peptidase activity significantly improves engraftment of retrovirally transduced hematopoietic progenitors. *Gene Ther.* **13:** 652–658.
61. Peranteau, W.H., M. Endo, O.O. Adibe, *et al.* 2006. CD26 inhibition enhances allogeneic donor-cell homing and engraftment after in utero hematopoietic-cell transplantation. *Blood* **108:** 4268–4274.
62. Wyss, B.K., A.F.W. Donnelly, D. Zhou, *et al.* 2009. Enhanced homing and engraftment of fresh but not *ex vivo* cultured murine marrow cells in submyeloablated hosts following CD26 inhibition by Diprotin A. *Exp. Hematol.* **37:** 814–823.
63. Campbell, T.B., G. Hangoc, Y. Liu, *et al.* 2007. Inhibition of CD26 in human cord blood $CD34^+$ cells enhances their engraftment of nonobese diabetic/severe combined immunodeficiency mice. *Stem Cells Develop.* **16:** 347–354.
64. Christopherson, K.W. II, L.A. Paganessi, S. Napier & N.K. Porecha. 2007. CD26 inhibition on $CD34^+$ or $lineage^-$ human umbilical cord blood donor hematopoietic stem cells/hematopoietic progenitor cells improves long-term engraftment into NOD/SCID/$beta2^{null}$ immunodeficient mice. *Stem Cells Dev.* **16:** 355–360.
65. Kawai, T., U. Choi, P.O. Liu, *et al.* 2007. Diprotin A infusion into nonobese diabetic/severe combined immunodeficiency mice markedly enhances engraftment of human mobilized $CD34^+$ peripheral blood cells. *Stem Cells Dev.* **16:** 361–370.
66. Farag, S.S. & H.E. Broxmeyer. 2011. CD26 inhibition to enhance cord blood engraftment. In *Cord Blood Biology, Transplantation, Banking, and Regulation.* H.E. Broxmeyer, Ed.: 203–210. AABB Press. Bethesda, MD.
67. North, T.E., W. Goessling, C.R. Walkley, *et al.* 2007. Prostaglandin E2 regulates vertebrate haematopoietic stem cell homeostasis. *Nature* **447:** 1007–1011.
68. Hoggatt, J., P. Singh, J. Sampath & L.M. Pelus. 2009. Prostaglandin E2 enhances hematopoietic stem cell homing, survival, and proliferation. *Blood* **113:** 5444–5455.
69. Durand, E.M. & L.I. Zon. 2010. Newly emerging roles for prostaglandin E2 regulation of hematopoiesis and hematopoietic stem cell engraftment. *Curr. Opin. Hematol.* **17:** 308–312.
70. Pelus, L.M., J. Speth & J. Hoggatt. 2011. Prostaglandin E_2 and other eicosanoid-based strategies to enhance engraftment. In *Cord Blood Biology, Transplantation, Banking, and Regulation.* H.E. Broxmeyer, Ed.: 187–197. AABB Press. Bethesda, MD.

71. Xia, L., J.M. Mcdaniel, T. Yago, *et al.* 2004. Surface fucosylation of human cord blood cells augments binding to P-selectin and E-selectin and enhances engraftment in bone marrow. *Blood* **104:** 3091–3096.
72. Robinson, S.N., P.J. Simmons, P.A. Zweidler-McKay & E.J. Shpall. 2011. Fucosylation for improved engraftment. In *Cord Blood Biology, Transplantation, Banking, and Regulation*. H.E. Broxmeyer, Ed.: 227–234. AABB Press. Bethesda, MD.
73. Chitteti, B.R., M.A. Kacena & E.F. Srour. 2011. Phenotypic characterization of hematopoietic stem cells. In *Cord Blood Biology, Transplantation, Banking, and Regulation*. H.E. Broxmeyer, Ed.: 75–83. AABB Press. Bethesda, MD.
74. Cai, S., H. Wang & K.E. Pollok. 2011. Immunodeficient mouse models for assessing human hematopoiesis. In *Cord Blood Biology, Transplantation, Banking, and Regulation*. H.E. Broxmeyer, Ed.: 87–106. AABB Press. Bethesda, MD.
75. Notta, F., S. Doulatov, E. Laurenti, *et al.* 2011. Isolation of single human hematopoietic stem cells capable of long-term multilineage engraftment. *Science* **333:** 218–221.
76. Shaheen, M. & H.E. Broxmeyer. 2009. The humoral regulation of hematopoiesis. In *Hematology: Basic Principles and Practice*, 5th ed. R. Hoffman, E.J. Benz Jr., S.J. Shattil, *et al.*, Eds.: 253–275. Elsevier Churchill Livingston. Philadelphia, PA.
77. Shaheen, M. & H.E. Broxmeyer. 2011. Hematopoietic cytokines and growth factors. In *Cord Blood Biology, Transplantation, Banking, and Regulation*. H.E. Broxmeyer, Ed.: 35–63. AABB Press. Bethesda, MD.
78. Chute. J. 2011. Regulation of hematopoiesis by the bone marrow vascular niche. In *Cord Blood Biology, Transplantation, Banking, and Regulation*. H.E. Broxmeyer, Ed.: 133–147. AABB Press. Bethesda, MD.
79. Lucas, D. & P.S. Frenette. 2011. Cellular components and regulation of the hematopoietic stem cell niche. In *Cord Blood Biology, Transplantation, Banking, and Regulation*. H.E. Broxmeyer, Ed.: 153–162. AABB Press. Bethesda, MD.
80. Kalinkovich, A., J. Canaani & T. Lapidot. 2011. The dynamic and modifiable nature of hematopoietic stem cells and their niches. In *Cord Blood Biology, Transplantation, Banking, and Regulation*. H.E. Broxmeyer, Ed.: 169–182. AABB Press. Bethesda, MD.
81. Peled, A., I. Petit, O. Kollet, *et al.* 1999. Dependence of human stem cell engraftment and repopulation of NOD/SCID mice on CXCR4. *Science* **283:** 845–848.
82. Broxmeyer, H.E., S. Cooper, L. Kohli, *et al.* 2003. Transgenic expression of stromal cell-derived factor-1/CXC chemokine ligand 12 enhances myeloid progenitor cell survival/antiapoptosis *in vitro* in response to growth factor withdrawal and enhances myelopoiesis *in vivo*. *J. Immunol.* **170:** 421–429.
83. Ivanova, N.B., J.T. Dimos, C. Schaniel, *et al.* 2002. A stem cell molecular signature. *Science* **298:** 601–604.
84. Campbell, T.B., S. Basu, G. Hangoc, *et al.* 2009. Overexpression of Rheb2 enhances mouse hematopoietic progenitor cell growth while impairing stem cell repopulation. *Blood* **114:** 3392–3401.
85. Rohrabaugh, S.L., T.B. Campbell, G. Hangoc & H.E. Broxmeyer. 2011. *Ex vivo* rapamycin treatment of human cord blood CD34^{+} cells enhances their engraftment of NSG mice. *Blood Cells Mol. Dis.* **46:** 318–320.
86. Liu, Y., J. Li, B.O. Kim, *et al.* 2002. HIV-1 Tat protein-mediated transactivation of the HIV-1 long terminal repeat promoter is potentiated by a novel nuclear Tat-interacting protein of 110 kDa, Tip110. *J. Biol. Chem.* **277:** 23854–23863.
87. Liu, Y., B.O. Kim, C. Kao, *et al.* 2004. Tip110, the human immunodeficiency virus type 1 (HIV-1) Tat-interacting protein of 110 kDa as a negative regulator of androgen receptor (AR) transcriptional activation. *J. Biol. Chem.* **279:** 21766–21773.
88. Bell, M., S. Schreiner, A. Damianov, *et al.* 2002. p110, a novel human U6 snRNP protein and U4/U6 snRNP recycling factor. *EMBO J.* **21:** 2724–2735.
89. Liu, Y., K. Timani, C. Mantel, *et al.* 2011. Tip110/p110^{nrb}/SART3/p110 regulation of hematopoiesis through c-Myc. *Blood*. **117:** 5643–5651.
90. Liu, Y., M-R. Lee, K. Timani, *et al.* 2012. Tip110 maintains expression of pluripotent factors in and pluripotency of human embryonic stem cells. *Stem Cells Dev.* **21**(6): 829–833.
91. Frye, R.A. 2000. Phylogenetic classification of prokaryotic and eukaryotic Sir2-like proteins. *Biochem. Biophys. Res. Comm.* **273:** 793–798.
92. Haigis, M.C. & D.A. Sinclair. 2010. Mammalian sirtuins: biological insights and disease relevance. *Annual Rev. Pathol.* **5:** 253–295.
93. Han, M.-K., E.K. Song, Y. Guo, *et al.* 2008. SIRT1 regulates apoptosis and *Nanog* expression in mouse embryonic stem cells by controlling p53 subcellular localization. *Cell Stem Cell* **2:** 241–251.
94. Ou, X., H-D. Chae, R-H. Wang, *et al.* 2011. SIRT1 deficiency compromises mouse embryonic stem cell hematopoietic differentiation, and embryonic and adult hematopoiesis in the mouse. *Blood* **117:** 440–450.
95. Leko, V., B. Varnum-Finney, H. Li, *et al.* 2012. SIRT1 is dispensable for function of hematopoietic stem cells in adult mice. *Blood* [Epub ahead of print January 4, 2012, doi: 10.1182/blood-2011-09-377077].
96. Chae, H-D. & H.E. Broxmeyer. 2011. SIRT1 deficiency downregulates PTEN/JNK/FOXO1 pathway to block reactive oxygen species-induced apoptosis in mouse embryonic stem cells. *Stem Cells Dev.* **20:** 1277–1285.
97. Mantel, C., S. Messina-Graham & H.E. Broxmeyer. 2010. Upregulation of nascent mitochondrial biogenesis in mouse hematopoietic stem cells parallels upregulation of CD34 and loss of pluripotency: a potential strategy for reducing oxidative risk in stem cells. *Cell Cycle* **9:** 1–10.
98. Mantel, C., S. Messina-Graham, & H.E. Broxmeyer. 2011. Superoxide flashes, ROS, and the mPTP: potential implications for hematopoietic stem cell function. *Current Opin. Hematopoiesis* **18:** 208–213.
99. Mantel, C.R., R-H. Wang, C. Deng & H.E. Broxmeyer. 2008. Sirt1, notch, and stem cell "age asymmetry." *Cell Cycle* **7:** 2821–2825.
100. Chae, H.D., M.R. Lee & H.E. Broxmeyer. 2012. AICAR induces G1/S arrest and enhances erythroid differentiation

in mouse embryonic stem cells by modulating p53-Nanog pathways. *Stem Cells* **30:** 140–149.
101. Waldmann, T., I. Scholten, F. Kappes, *et al.* 2004. The DEK protein-an abundant and ubiquitous constituent of mammalian chromatin. *Gene* **343:** 1–9.
102. Kappes, F., T. Waldmann, V. Mathew, *et al.* 2011. The DEK oncoprotein is a Su(var) that is essential to heterochromatin integrity. *Genes Dev.* **25:** 673–678.
103. Mor-Vaknin, N., A. Punturieri, K. Sitwala, *et al.* 2006. The DEK nuclear autoantigen is a secreted chemotactic factor. *Mole. Cell Biol.* **26:** 9484–9496.
104. Broxmeyer, H.E., F. Kappas, N. Mor-Vaknin, *et al.* 2011. DEK regulates hematopoietic stem cell engraftment and progenitor cell proliferation. *Stem Cells Dev* Oct 27 [Epub ahead of print, doi: 10.1089/scd.2011.0451].
105. Mantel, C., Z. Luo, J. Canfield, *et al.* 1996. Involvement of $p21^{cip-1}$ and $p27^{kip-1}$ in the molecular mechanisms of steel factor induced proliferative synergy *in vitro* and of $p21^{cip-1}$ in the maintenance of stem/progenitor cells *in vivo. Blood* **88:** 3710–3719.
106. Cheng, T., N. Rodrigues, H. Shen, *et al.* 2000. Hematopoietic stem cell quiescence maintained by p21cip1/waf1. *Science* **287:** 1804–1808.
107. Stier, S., T. Cheng, R. Forkert, *et al.* 2003. *Ex vivo* targeting of p21Cip1/Waf1 permits relative expansion of human hematopoietic stem cells. *Blood* **102:** 1260–1266.
108. Shaheen M. & H.E. Broxmeyer. In press. Principles of cytokine signaling. In *Hematology: Basic Principles and Practice*, 6th ed. R. Hoffman, E.J. Benz Jr, S.J. Shattil, *et al.*, Eds. Elsevier Churchill Livingston. Philadelphia, PA.
109. Yuan, Y., H. Shen, D.S. Franklin, *et al.* 2004. *In vivo* self-renewing divisions of haematopoietic stem cells are increased in the absence of the early G1-phase inhibitor, $p18^{INK4C}$. *Nat. Cell Biol.* **6:** 436–442.
110. Broxmeyer, H.E., D.S. Franklin, S. Cooper, *et al.* In press. Hematopoietic progenitor cell proliferation and responsiveness to cytokines/chemokines vis $p18^{INK4C}$, and its functional interactions with $p21^{CIP1/WAF1}$ and $p27^{KIP1}$. *Stem Cells Dev.*
111. Broxmeyer, H.E., H. Bruns, S. Zhang, *et al.* 2002. Th1 cells regulate hematopoietic progenitor cell homeostasis by production of oncostatin M. *Immunity* **16:** 815–825.
112. Kaplan, M.H., C-H. Chang, S. Cooper, *et al.* 2003. Distinct requirements for Stat4 and Stat6 in hematopoietic progenitor cell responses to growth factors and chemokines. *J. Hematotherapy Stem Cell Res.* **12:** 401–408.
113. Broxmeyer, H.E., S. Sehra, S. Cooper, *et al.* 2007. BAZF-deficient mice have unusual alterations in hematopoietic progenitor cell activity similar to that of BCL-6 deficient mice. *Mole Cell Biol.* **27:** 5275–5285.
114. Kaplan, M.H., N.L. Glosson, G.L. Stritesky, *et al.* 2011. STAT3-dependent IL-21 production from T cells regulates hematopoietic progenitor cell homeostasis. *Blood* **117:** 6198–6201.
115. Zhang, C.C., M. Kaba, G. Ge, *et al.* 2006. Angiopoietin-like proteins stimulate *ex vivo* expansion of hematopoietic stem cells. *Nat. Med.* **12:** 240–245.
116. Zhang, C.C., M. Kaba, S. Iizuka, *et al.* 2008. Angiopoietin-like 5 and IGFBP2 stimulate *ex vivo* expansion of human cord blood hematopoietic stem cells as assayed by NOD/SCID transplantation. *Blood* **111:** 3415–3423.
117. Zheng, J., H. Huynh, M. Umikawa, *et al.* 2011. Angiopoietin-like protein 3 supports the activity of hematopoietic stem cells in the bone marrow niche. *Blood* **117:** 470–479.
118. Broxmeyer, H.E., E.F. Srour, S. Cooper, *et al.* 2012. Angiopoietin-like-2 and -3 act through their coiled-coil domains to enhance survival and replating capacity of human cord blood hematopoietic progenitors. *Blood Cells Mol. Dis.* **48:** 25–29.
119. Kinzfogl, J., G. Hangoc & H.E. Broxmeyer. 2011. Neurexin1a/neurexophilin 1 function as a myelosuppressive axis in human cord blood and murine bone marrow hematopoiesis. *Blood* **118:** 565–575.
120. Broxmeyer, H.E. & S. Cooper. 1997. High efficiency recovery of immature hematopoietic progenitor cells with extensive proliferative capacity from human cord blood cryopreserved for ten years. *Clin. Exp. Immunol.* **107:** 45–53.
121. Broxmeyer, H.E., E.F. Srour, G. Hangoc, *et al.* 2003. High efficiency recovery of hematopoietic progenitor cells with extensive proliferative and *ex vivo* expansion activity and of hematopoietic stem cells with NOD/SCID mouse repopulation ability from human cord blood stored frozen for 15 years. *Proc. Natl. Acad. Sci. USA* **100:** 645–650.
122. Broxmeyer, H.E., M-R. Lee, G. Hangoc, *et al.* 2011. Hematopoietic stem/progenitor cells, generation of induced pluripotent stem cells, and isolation of endothelial progenitors from 21- to 23.5-year cryopreserved cord blood. *Blood* **117:** 4773–4777.

Ann. N.Y. Acad. Sci. ISSN 0077-8923

ANNALS OF THE NEW YORK ACADEMY OF SCIENCES
Issue: *Hematopoietic Stem Cells VIII*

Immune reconstitution and strategies for rebuilding the immune system after haploidentical stem cell transplantation

Lena Oevermann, Peter Lang, Tobias Feuchtinger, Michael Schumm, Heiko-Manuel Teltschik, Patrick Schlegel, and Rupert Handgretinger
Department of Hematology/Oncology, Children's University Hospital, Tübingen, Germany

Address for correspondence: Rupert Handgretinger, Children's University Hospital, Hoppe-Seyler-Strasse 1, 72076 Tübingen, Germany. Rupert.Handgretinger@med.uni-tuebingen.de

Haploidentical hematopoietic stem cell transplantation is a curative alternative option for patients without an otherwise suitable stem cell donor. In order to prevent graft-versus-host disease (GvHD), different *in vitro* and *in vivo* T cell–depletion strategies have been developed. A delayed immune reconstitution is common to all these strategies, and an impaired immune function after haploidentical transplantation with subsequent infections is a major cause of deaths in these patients. In addition to *in vitro* and *in vivo* T cell–depletion methods, posttransplant strategies to rapidly rebuild the immune system have been introduced in order to improve the outcome. Advances in *in vitro* and *in vivo* T cell–depletion methods, and adoptive transfer of immune cells of the innate and specific immune system, will contribute to reduce the risk of GvHD, lethal infections, and the risk of relapse of the underlying malignant disease.

Keywords: haploidentical transplantation; immune reconstitution; adoptive transfer; T cell recovery

Introduction

Hematopoietic stem cell transplantation (HSCT) offers a curative approach for patients with hematological malignancies and some nonmalignant diseases in children and adults for which no other curative standard therapies are available. After its introduction by Thomas *et al.*,[1] improvements in the conditioning regimens, supportive care, prophylaxis and treatment of graft-versus-host disease (GvHD), anti-infectious prophylaxis, and treatment of posttransplant infections have established HSCT as a major pillar in the treatment of otherwise untreatable diseases.[2]

Initially, the pool of stem cell donors was limited to human leukocyte antigen (HLA)–identical sibling donors and was later considerably expanded to matched unrelated volunteer donors by the establishment of national and international donor registries. Only 30% of the patients in need of a transplant have an HLA-identical sibling and the success of finding an appropriate donor in the world wide donor registries depends on HLA diversity and patients' ethnic background. A significant number of patients will have a progression of the underlying disease and die during the search time or reach a point of the disease in which HSCT has become impossible.

For those patients without an HLA-identical donor, two alternative strategies exist: transplantation with matched or mismatched umbilical cord blood (UCB),[3] or with haploidentical stem cells from either of the parents for younger patients, haploidentical adult children for their parents, haploidentical siblings of the patients, or haploidentical donors from the extended family.

However, delayed immune reconstitution is common to all strategies of HSCT and is more pronounced after UCB and haploidentical transplantation exposing the patients at high risk for severe and often lethal infections, among them aspergillosis, cytomegalovirus (CMV), and adenovirus (ADV) infections.[4]

While the first attempts of haploidentical transplantation with unmanipulated bone marrow (BM) were associated with severe and most often lethal

doi: 10.1111/j.1749-6632.2012.06606.x

side effects,[5] some progress has been made by using BM where the GvHD-causing T lymphocytes had been poorly depleted.[6] Thereafter, different *in vitro* graft manipulation strategies have been introduced in haploidentical HSCT to deplete the graft of T lymphocytes for prevention of GvHD, including $CD34^+$ selection,[7–10] CD3 depletion,[11] CD3/CD19 depletion,[12] and more recently, the depletion of αβ T lymphocytes (TcRαβ/CD19).[13] In addition, transplantation of T cell–replete BM or peripheral stem cells (PBSC) with intensive pharmacological prophylaxis for GvHD, or with the *in vivo* depletion of T cells using either polyclonal antithymocyte globulin (ATG) or monoclonal antibodies (alemtuzumab), has been performed.[14–16] Another approach is *in vivo* allodepletion using posttransplant high-dose cyclophosphamide (Cy) to induce tolerance.[17,18]

Common to all these described strategies in haploidentical transplantation is delayed immune reconstitution caused by *in vitro* or *in vivo* T cell depletion, HLA disparity, immunosuppressive regimens to prevent or treat GvHD, and decreased age-dependent thymic function after transplantation. Because of the persistent lack of a sufficient immune response, patients can develop severe and lethal infections and viral reactivation; bacterial and/or fungal infections are also common complications after haploidentical HSCT.

Therefore, any attempt to improve the outcome of haploidentical HSCT should focus on strategies to accelerate immune reconstitution, as this will not only have an influence on the risk of infections but also on the risk of relapse of the underlying malignant disease.

In this short review, immune reconstitution, and its influence on the infectious risk and the antimalignancy effect of the graft, will be addressed with regard to the various *in vitro* and *in vivo* T cell–depletion methods employed in haploidentical HSCT. Moreover, several clinical approaches to accelerate immune reconstitution, such as adoptive transfer of specific T cells, will be discussed.

Haploidentical HSCT within T cell depleted grafts

$CD34^+$ selection

Aversa *et al.* first reported in a seminal study the use of haploidentical highly-purified $CD34^+$ PBSC obtained by a two-step enrichment protocol consisting of soybean agglutination and two rounds of E-rosetting.[9] Forty-three patients with high-risk acute leukemia were transplanted. All patients engrafted, and no acute or chronic GvHD was observed in the evaluable patients; the transplant-related mortality (TRM) was 40%. Peripheral counts of natural killer (NK) cells returned to normal within two to four weeks after transplantation, whereas the number of $CD4^+$ helper T cells was below 100 and 200 cells per microliter for as long as 10 and 16 months, respectively. In addition, the number of T cells responsive to polyclonal activators was very low in the first 10 months posttransplant. Eleven patients died from infection, including bacterial (five patients), fungal (five patients), and viral (one patient) infections. Two patients died from Epstein-Barr virus (EBV)–associated posttransplantation lymphoproliferative disorder (PTLD). The selection of highly-purified $CD34^+$ hematopoietic stem cells using the Clinimacs system[7] was established in order to prevent GvHD and EBV-LPLD. Using this device, 39 children received large numbers of $CD34^+$ haploidentical stem cells with no pharmacological prophylaxis for GvHD.[8] No significant GvHD of EBV-PTLD was observed. The reconstitution was more rapid if the number of transplanted $CD34^+$ cells exceeded 20×10^6/kg recipient's body weight. Ten patients died from TRM, among them five patients with viral infections and two with fungal infections. In subsequent pediatric studies with 102 patients, a high rate of infectious complications and TRM were observed. However, there was a center effect; centers that performed more than nine haploidentical transplants per year had a better outcome, due to a lower TRM and less infectious complications.[19] In a subsequent study in adult patients, Aversa *et al.* transplanted 104 adult patients with acute leukemias using $CD34^+$ stem cells positively selected either by the Isolex or the Clinimacs method.[10] Thirty-eight patients died from TRM, among them 17 from viral infections, 5 from fungal infections, and 5 from bacterial infections, reflecting poor immune reconstitution in these patients. In subsequent studies with haploidentical $CD34^+$ stem cells and mainly in adult patients, a high TRM, mostly due to infections, was reported.[20,21]

CD3/19 depletion

There is a crucial difference between positively-selected $CD34^+$ stem cells and grafts negatively-depleted of T and B cells. While all $CD34^-$ cells are discarded with CD34$^+$ selection, the selective depletion of T and B cells retains large numbers of potential important immune cells in the graft, including myeloid and plasmacytoid dendritic cells, natural killer cells, and large numbers of monocytes and other myeloid cells mobilized with growth factors. Nevertheless, both strategies lead to an effective and sufficient depletion of T and B cells, with a slightly lower T cell–depletion rate for negative selection compared with CD34$^+$ selection.

A first study by Hale *et al.* consisted of 20 pediatric patients with hematological malignancies undergoing haploidentical HSCT with CD3/19-depleted PBSC, following a myeloablative total body irradiation (TBI)–based preparative regimen. All patients engrafted, six died from TRM and four relapsed; 10 patients are long-term survivors.[22] A following study included 25 patients with refractory disease ($n = 9$) or relapse after conventional HSCT ($n = 16$). Taking into account that all patients had already received extensive therapy, the preparative regimen was reduced to a non-TBI-based conditioning regimen with fludarabine, thiotepa, melphalan, and the CD3 antibody OKT-3.[23] Three patients did not engraft and two of them were rescued by another transplant from their original donor. One patient had early disease progression. The cumulative incidence of grade II–IV GvHD was 44% and 8% of grade III–IV GvHD; chronic GvHD was observed in 28% of the patients. Thirteen patients relapsed, four patients died of TRM, and eight patients are long-term survivors. It is noteworthy that none of the patients died from an infectious complication. A detailed analysis of the immune reconstitution of these two cohorts of patients showed a rapid T cell reconstitution after a non-TBI-based reduced conditioning regimen, with an earlier thymopoiesis and broader T cell receptor repertoire.[24]

Based on these promising results, similar studies were performed in adult[25] and pediatric patients.[26] In adult patients, T cell reconstitution was still delayed, with a median of 205 CD3$^+$ cells per microliter and 70 CD4$^+$ helper T cells per microliter at day 100. In addition, a skewed T cell receptor repertoire with oligoclonal T cell expansion at day 100 and normalization at day 200 posttransplant was seen. The engraftment of NK cells was rapid, with normal values at day 20. The TRM at day 100 and for the whole study period were 7% and 39%, respectively. Ten patients died from infectious complications. In the pediatric study, 38 patients received a haploidentical CD3/19-depleted graft. No lethal viral infections were seen and only one patient experienced a transplant-related death. A low TRM was subsequently reported in additional pediatric studies.[27,28] More recently, a high success rate of haploidentical transplantation with either CD34-positively selected or CD3/19-negatively depleted PBSC was reported in children with very high-risk leukemia.[29]

TcRαβ/CD19 depletion

A promising approach recently established is the negative depletion of T cell receptor $\alpha\beta^+$ (TcR$\alpha\beta^+$) T lymphocytes from mobilized peripheral stem cell grafts.[13] With this method, a 4- to 5-log depletion similar to CD34$^+$ selection and better than CD3/19 depletion can be achieved. NK cells, monocytes/myeloid cells, dendritic cells, and $\gamma\delta^+$ T lymphocytes are retained in the graft and cotransplanted together with the stem cells. The role of $\gamma\delta^+$ T cells in haploidentical transplantation is not clear. $\gamma\delta^+$ T lymphocytes are nonalloreactive T cells with potentially important antitumor and antileukemia effects, and might have a positive impact on the outcome after haploidentical HSCT. It has been shown in an eight-year follow-up study of haploidentical transplantation with poorly T cell–depleted BM that patients with an increased number of $\gamma\delta^+$ T cells following HSCT had a better event-free and overall survival compared with the patients with lower numbers of $\gamma\delta^+$ T lymphocytes.[30] It is noteworthy that there was no difference in the cumulative incidence of GvHD among the patients with high or low $\gamma\delta^+$ T cells. The first clinical study using TcR$\alpha\beta$-depleted grafts for haploidentical HSCT was performed at the Children's University Hospital in Tuebingen, Germany and at the Ospedale Bambino Gesu, Rome, Italy.[31] Initially, 23 patients were transplanted. The median number of transplanted $\alpha\beta^+$ T cells and $\gamma\delta^+$ T cells was 14×10^3/kg and 11.9×10^6/kg, respectively. An analysis of the immune reconstitution of the patients transplanted in Tübingen using a non-TBI-based conditioning regimen showed a rapid immune reconstitution with 350 (range 21–824) CD3$^+$ T cells per microliter, 66 (range 12–177) CD4$^+$ helper T cells per microliter, and 599 (range 227–1,390) CD56$^+$ NK cells per microliter at day 28 posttransplant. No transplant-related death was

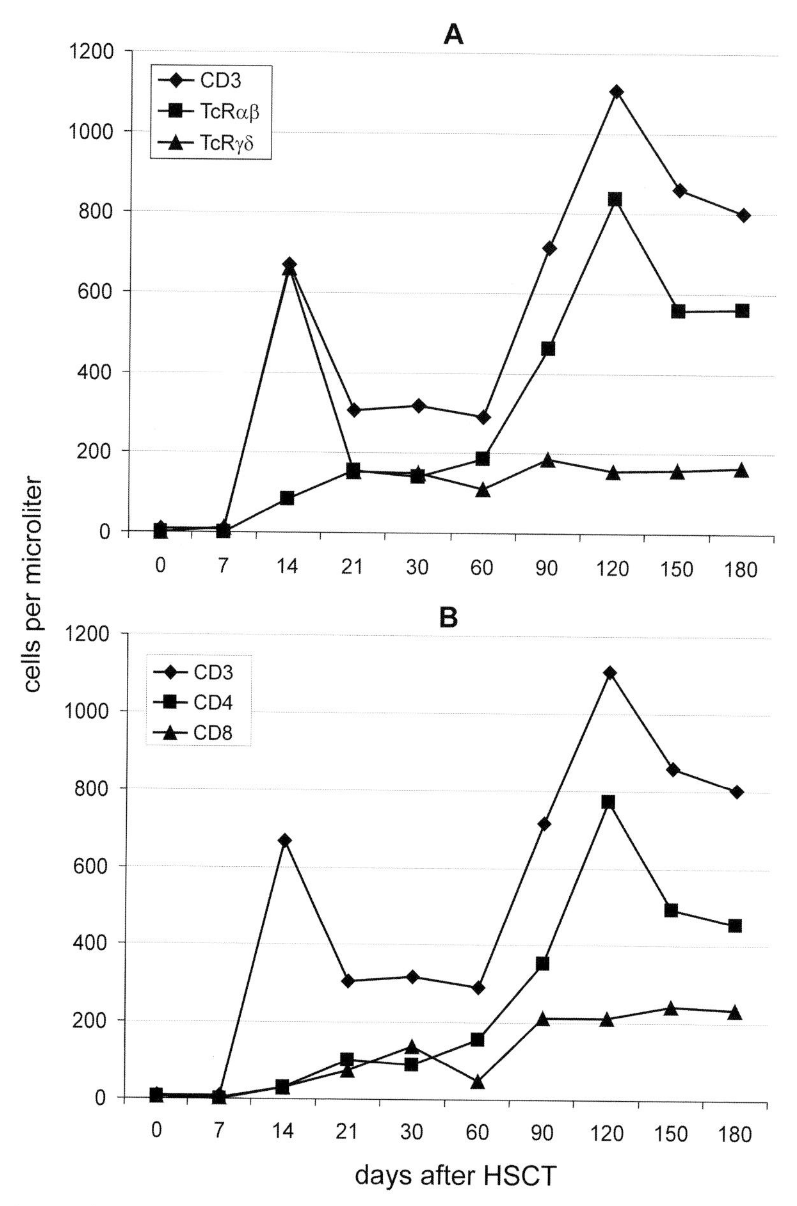

Figure 1. Reconstitution of $CD3^+$, $TcR\alpha\beta^+$, and $TcR\gamma\delta^+$ lymphocytes (A) and of $CD3^+$, $CD4^+$, and $CD8^+$ lymphocytes (B) after transplantation of $TcR\alpha\beta^+$-depleted haploidentical stem cells.

seen in this cohort. As of December 2011, a total of 17 patients have been transplanted in Tübingen with 1 transplant-related death from multiorgan failure (R. Handgretinger, unpublished data). In these pediatric patients, an early engraftment and expansion of donor-derived $\gamma\delta^+$ T cells, starting at around day 7 posttransplant followed by the expansion of donor-derived $\alpha\beta^+$ T cells starting at around day 14, was observed (Fig. 1A). In Figure 1B, the recovery of $CD3^+$, $CD4^+$, and $CD8^+$ lymphocytes is shown. The second cohort of patients transplanted in Rome received a mainly TBI-based myeloablative conditioning regimen. One patient died from lung aspergillosis and 10 patients are alive and disease free, with a short median follow-up of 4 months (range: 1–9 months).

In a retrospective comparison of the Tübingen cohorts receiving either CD3/19-depleted or

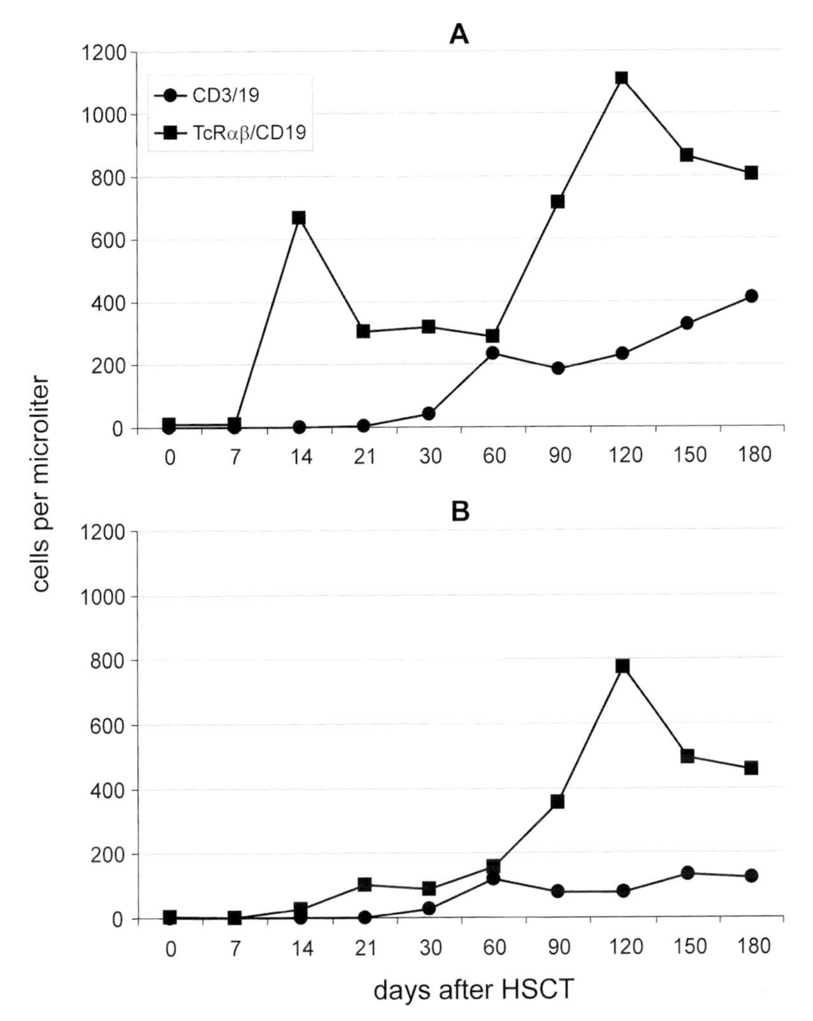

Figure 2. Retrospective analysis of the reconstitution of $CD3^+$ lymphocytes (A) and $CD4^+$ lymphocytes (B) after transplantation of either CD3/19-depleted or TcRαβ/CD19-depleted haploidentical stem cells.

TcRαβ-depleted PBSC using the same non-TBI-based conditioning regimen, a much more rapid reconstitution of the $CD3^+$ lymphocytes (Fig. 2A) and $CD4^+$ lymphocytes (Fig. 2B) could be observed. On the basis of these promising results, we are currently preparing a clinical study on pediatric and adult patients.

Haploidentical HSCT with unmanipulated grafts

In vivo *T cell depletion*

Promising results of unmanipulated HLA-mismatched/haploidentical blood and marrow transplantation using ATG for *in vivo* T cell depletion and posttransplant pharmacological GvHD prophylaxis were reported by Lu *et al.*[16] The outcome was similar to that achieved with HLA-identical sibling transplantation. The TRM for patients after mismatched transplantation was 22% and mainly caused by acute GvHD and pulmonary infections. Using alemtuzumab for *in vivo* T cell depletion, a reduced intensity conditioning regimen, and unmanipulated PBSC, Rizzieri *et al.* reported a TRM of 10% at day 100 and a two-year TRM of 40%.[32] In this study, an encouraging lymphocyte reconstitution and expansion of donor-derived T lymphocytes was observed after 3–6 months posttransplant.

Posttransplant infusion of cyclophosphamide

A more recently introduced approach is the application of high-dose cyclophosphamide (CY) posttransplant to perform allodepletion *in vivo*.[33] Here, donor-derived lymphocytes are exposed to host antigens and proliferating antigen-specific lymphocytes are generated. The well-timed application of CY is thought to deplete proliferating lymphocytes, while resting lymphocytes are spared. Luznik *et al.* performed a study with 68 patients suffering from hematological malignancies using T cell–replete grafts for haploidentical SCT after a nonmyeloablative conditioning regimen and applying CY (50 mg/kg) on day 3 or days 3 and 4 after transplantation. Graft failure occurred in 13% of the patients and the cumulative incidences of acute grade I–II and III–IV GvHD and chronic GvHD were 34%, 6%, and 22%, respectively. Overall survival and event-free survival two years after SCT was 36% and 26%, respectively.[17] A following study by the same group included 210 patients with hematological malignancies; 87% of patients have experienced sustained donor cell engraftment. The cumulative incidences of grades II–IV acute GVHD and chronic GVHD were 27% and 13%, respectively. Five-year cumulative incidence of nonrelapse mortality was 18%, relapse was 55%, and actuarial overall survival and event-free survivals was 35% and 27%, respectively.[34] Another group reported of 18 patients using this approach.[35] Immunosuppressive GvHD prophylaxis was discontinued between 60 and 340 days posttransplant (median 8.5 months). CMV reactivation occurred in 9 out of 14 seropositive patients. After a median follow-up of 12 months, 16 out of the 18 patients were alive and only 1 patient died from TRM.

Strategies to improve immune reconstitution

The development of strategies associated with an accelerated immune reconstitution is crucial for further improvement of the outcome after HSCT.[36] All currently known strategies have a more or less delayed immune reconstitution in common. Reconstitution of the different subsets of leukocytes occurs at different points after transplantation. The first 100 days are characterized by cellular immune deficiencies with a reduced number of cytotoxic T cells, NK cells, and T cells of the specific immune system.[4] Therefore, patients are susceptible to viral and fungal infections and CMV infection or CMV reactivation, and aspergillosis are among the major causes of death after haploidentical HSCT. Many different approaches exist to accelerate immune reconstitution. Noncellular approaches use factors, such as keratinocyte growth factor or growth hormone, but most strategies comprise cellular approaches, including adoptive transfer of immune cell subsets.

Noncellular approaches

Use of keratinocyte growth factor

Keratinocyte growth factor (KGF) has become important especially for prophylaxis and treatment of severe mucositis. Moreover, it mediates epithelial cell proliferation in different tissues, including thymic epithelial cells (TEC).[4] As TECs are targets of GvHD reactions after SCT, application of KGF should maintain TEC proliferation and their IL-7 production, which was shown in different nonhuman models.[37] Application in humans neither reduced the incidence of acute GvHD, nor CMV or fungal infections.[38]

Growth hormone

The use of growth hormone (GH) to strengthen thymic function has been investigated in patients undergoing UCBT.[4] Growth hormone–deficient mice showed atrophy of the thymus, which was reversible after administration of GH.[39] GH is thought to support thymus function by augmenting the number of lymphoid precursors homing to the thymus.[40]

Cytokines

Interleukin-2 (IL-2) exhibits pleiotropic effects on immune responses and is therefore thought to have beneficial effects after HSCT.[41] Ongoing studies are investigating the best IL-2 dosage and administration methods. IL-2 can enhance NK cell numbers and might augment the GvL effect, leading to reduced relapse rates after HSCT.[42] Application of IL-2 in combination with donor lymphocyte infusion (DLI) led to complete remissions in patients that did not respond to DLI alone.[43] IL-7, which mediates crucial pathways in thymic and nonthymic T cell development,[44] was found to increase T cell numbers in nontransplant cancer patients;[45] a possible positive effect after HSCT remains controversial. In a murine model, an increase of T cells without GvHD reduction was seen with IL-7,[46] whereas a different study found an augmented T cell number via peripheral expansion without *de novo* generation.[47] IL-15, a cytokine with a possible broad spectrum of action because it stimulates NK, T, and B cells, might play a role in tumor vaccination strategies,[48] and it might have a potential function on GvL effect, as it is known to stimulate NK cell function and survival.[49]

Cellular approaches

Adoptive transfer of NK cells

Alloreactive NK cells contribute to the graft versus leukemia reaction without causing GvHD.[50] Infusions of alloreactive NK cells after HSCT resulted in no major side effects in patients with different hematological malignancies.[51–53] Especially in patients with AML, NK cell infusions seem to have a beneficial effect on outcome in a nontransplant setting.[54]

Infusion of allodepleted donor T cells

Amrolia *et al.* could demonstrate an accelerated immune reconstitution in 16 patients who received adoptively transferred T cells that were allodepleted *in vitro* using an immunotoxin consisting of a CD25-specific antibody linked to the ricin α chain.[55] After 2 to 4 months, virus-specific T cells and a broad V_β T cell receptor repertoire could be observed, while the incidence of GvHD was low. In another 28 patients with advanced hematological malignancies (ALL = 24, AML = 4), CD8-depleted donor lymphocyte infusions were given.[56] The GvHD rate was 26%, two-year overall survival was 44%, and the two-year cumulative incidence of nonrelapse mortality was 26%.

Adoptive transfer of pathogen-specific T cells

Adoptive T cell transfer of CMV-specific T cells is a feasible method to treat patients with CMV disease or viremia after transplantation. In a study with 18 patients suffering from either CMV disease or viremia refractory to chemotherapy, CMV infection was cleared or significantly reduced in 33%, and viral control was reached in 12 of 16 patients. No induction of GvHD was found.[57] The safe transfer of ADV-specific T cells was also reported in a study with nine patients.[58] Recently, a method was established allowing the simultaneous generation of multivirus-specific and regulatory T cells for adoptive immunotherapy; clinical results with this method are pending.[59] Another method for adoptive T cell transfer is by photodynamic purging of alloreactive T cells.[60,61] Here, T cell allodepletion is reached by *in vitro* transfer of TH9402 dye into T cells followed by irradiation. Pathogen-specific T cells are retained, while GvHD-mediating T cells are eliminated, thus allowing safe infusion of T cells to improve donor immunity.

Adoptive transfer of $CD4^+CD25^+$ regulatory T cells

The use of $CD4^+CD25^+$ regulatory T cells (T_{reg} cells) was first described by Edinger *et al.* in mice.[62] In this pioneering work, it was shown that T_{reg} cells could prevent GvHD, while maintaining the graft-versus-tumor activity. These observations were subsequently translated into a clinical study by Di Ianni *et al.*[63] In a study including 28 patients with hematological malignancies undergoing haploidentical HSCT, rapid CD4/CD8 expansion and no occurrence of chronic GvHD after infusion of regulatory T cells, followed by the infusion of conventional T cells, were observed.

Conclusions

Advantages in *in vitro* T cell–depletion technologies, as well as the use of unmanipulated grafts with pharmacological *in vivo* T cell–depletion strategies, have allowed haploidentical transplantation with rapid and long-term engraftment, low rates of severe GvHD, and low incidence of TRM. Besides relapses of the underlying malignant diseases, poor immune reconstitution leading to severe and often lethal infectious complications is still the major reason for early and late mortality. Although tremendous progress has been achieved in the last decade, elimination of alloreactive T cells and preserving tumor- and pathogen-specific immunity will still be a major task to further improve the outcome after haploidentical transplantation. New technologies of T cell–depletion techniques, as well as new strategies of adoptive transfer of immune effector cells posttransplant, will be crucial for establishing haploidentical transplantation as a standard approach for patients in need of stem cell transplantation.

Conflicts of interest

The authors declare no conflicts of interest.

References

1. Thomas, E.D., C.D. Buckner, M. Banaji, *et al.* 1977. One hundred patients with acute leukemia treated by chemotherapy, total body irradiation, and allogeneic marrow transplantation. *Blood* **49:** 511–533.
2. Gyurkocza, B., A. Rezvani & R.F. Storb. 2010. Allogeneic hematopoietic cell transplantation: the state of the art. *Expert Rev. Hematol.* **3:** 285–299.

3. Wagner, J.E. & E. Gluckman. 2010. Umbilical cord blood transplantation: the first 20 years. *Semin. Hematol.* **47:** 3–12.
4. Seggewiss, R. & H. Einsele. 2010. Immune reconstitution after allogeneic transplantation and expanding options for immunomodulation: an update. *Blood* **115:** 3861–3868.
5. Powles, R.L., G.R. Morgenstern, H.E. Kay, *et al.* 1983. Mismatched family donors for bone-marrow transplantation as treatment for acute leukaemia. *Lancet* **1:** 612–615.
6. Mehta, J., S. Singhal, A.P. Gee, *et al.* 2004. Bone marrow transplantation from partially HLA-mismatched family donors for acute leukemia: single-center experience of 201 patients. *Bone Marrow Transplant.* **33:** 389–396.
7. Schumm, M., P. Lang, G. Taylor, *et al.* 1999. Isolation of highly purified autologous and allogeneic peripheral CD34+ cells using the CliniMACS device. *J. Hematother.* **8:** 209–218.
8. Handgretinger, R., T. Klingebiel, P. Lang, *et al.* 2001. Megadose transplantation of purified peripheral blood CD34(+) progenitor cells from HLA-mismatched parental donors in children. *Bone Marrow Transplant.* **27:** 777–783.
9. Aversa, F., A. Tabilio, A. Velardi, *et al.* 1998. Treatment of high-risk acute leukemia with T-cell-depleted stem cells from related donors with one fully mismatched HLA haplotype. *N. Engl. J. Med.* **339:** 1186–1193.
10. Aversa, F., A. Terenzi, A. Tabilio, *et al.* 2005. Full haplotype-mismatched hematopoietic stem-cell transplantation: a phase II study in patients with acute leukemia at high risk of relapse. *J. Clin. Oncol.* **23:** 3447–3454.
11. Gordon, P.R., T. Leimig, I. Mueller, *et al.* 2002. A large-scale method for T cell depletion: towards graft engineering of mobilized peripheral blood stem cells. *Bone Marrow Transplant.* **30:** 69–74.
12. Barfield, R.C., M. Otto, J. Houston, *et al.* 2004. A one-step large-scale method for T- and B-cell depletion of mobilized PBSC for allogeneic transplantation. *Cytotherapy* **6:** 1–6.
13. Chaleff, S., M. Otto, R.C. Barfield, *et al.* 2007. A large-scale method for the selective depletion of alphabeta T lymphocytes from PBSC for allogeneic transplantation. *Cytotherapy* **9:** 746–754.
14. Huang, X.J., D.H. Liu, K.Y. Liu, *et al.* 2006. Haploidentical hematopoietic stem cell transplantation without in vitro T-cell depletion for the treatment of hematological malignancies. *Bone Marrow Transplant.* **38:** 291–297.
15. Huang, X.J. & Y.J. Chang. 2011. Unmanipulated HLA-mismatched/haploidentical blood and marrow hematopoietic stem cell transplantation. *Biol. Blood Marrow Transplant* **17:** 197–204.
16. Lu, D.P., L. Dong, T. Wu, *et al.* 2006. Conditioning including antithymocyte globulin followed by unmanipulated HLA-mismatched/haploidentical blood and marrow transplantation can achieve comparable outcomes with HLA-identical sibling transplantation. *Blood* **107:** 3065–3073.
17. Luznik, L., P.V. O'Donnell, H.J. Symons, *et al.* 2008. HLA-haploidentical bone marrow transplantation for hematologic malignancies using nonmyeloablative conditioning and high-dose, posttransplantation cyclophosphamide. *Biol. Blood Marrow Transplant.* **14:** 641–650.
18. Luznik, L., J. Bolanos-Meade, M. Zahurak, *et al.* 2010. High-dose cyclophosphamide as single-agent, short-course prophylaxis of graft-versus-host disease. *Blood* **115:** 3224–3230.
19. Klingebiel, T., J. Cornish, M. Labopin, *et al.* 2010. Results and factors influencing outcome after fully haploidentical hematopoietic stem cell transplantation in children with very high-risk acute lymphoblastic leukemia: impact of center size: an analysis on behalf of the Acute Leukemia and Pediatric Disease Working Parties of the European Blood and Marrow Transplant group. *Blood* **115:** 3437–3446.
20. Ciceri, F., M. Labopin, F. Aversa, *et al.* 2008. A survey of fully haploidentical hematopoietic stem cell transplantation in adults with high-risk acute leukemia: a risk factor analysis of outcomes for patients in remission at transplantation. *Blood* **112:** 3574–3581.
21. Walker, I., N. Shehata, G. Cantin, *et al.* 2004. Canadian multicenter pilot trial of haploidentical donor transplantation. *Blood Cells Mol. Dis.* **33:** 222–226.
22. Hale, G.A., K.A. Kasow, K. Gan, *et al.* 2005. Haploidentical stem cell transplantation with CD3 depleted mobilized peripheral blood stem cell grafts for children with hematologic malignancies. ASH Annual Meeting Abstracts. **106:** 2910.
23. Hale, G.A., K.A. Kasow, R. Madden, *et al.* 2006. Mismatched family member donor transplantation for patients with refractory hematologic malignancies: long-term followup of a prospective clinical trial. ASH Annual Meeting Abstracts. **108:** 3137.
24. Chen, X., G.A. Hale, R. Barfield, *et al.* 2006. Rapid immune reconstitution after a reduced-intensity conditioning regimen and a CD3-depleted haploidentical stem cell graft for paediatric refractory haematological malignancies. *Br. J. Haematol.* **135:** 524–532.
25. Federmann, B., M. Hagele, M. Pfeiffer, *et al.* 2011. Immune reconstitution after haploidentical hematopoietic cell transplantation: impact of reduced intensity conditioning and CD3/CD19 depleted grafts. *Leukemia* **25:** 121–129.
26. Handgretinger, R., X. Chen, M. Pfeiffer, *et al.* 2007. Feasibility and outcome of reduced-intensity conditioning in haploidentical transplantation. *Ann. N.Y. Acad. Sci.* **1106:** 279–289.
27. Dufort, G., S. Pisano, A. Incoronato, *et al.* 2012. Feasibility and outcome of haploidentical SCT in pediatric high-risk hematologic malignancies and Fanconi anemia in Uruguay. *Bone Marrow Transplant* **47:** 663–668.
28. Palma, J., L. Salas, F. Carrion, *et al.* 2012. Haploidentical stem cell transplantation for children with high-risk leukemia. *Pediatr. Blood Cancer* [Epub ahead of print].
29. Leung, W., D. Campana, J. Yang, *et al.* 2011. High success rate of hematopoietic cell transplantation regardless of donor source in children with very high-risk leukemia. *Blood* **118:** 223–230.
30. Godder, K.T., P.J. Henslee-Downey, J. Mehta, *et al.* 2007. Long term disease-free survival in acute leukemia patients recovering with increased gammadelta T cells after partially mismatched related donor bone marrow transplantation. *Bone Marrow Transplant.* **39:** 751–757.
31. Handgretinger, R., P. Lang, T.F. Feuchtinger, *et al.* 2011. Transplantation of TcR{alpha}{beta}/CD19 depleted stem

cells from haploidentical donors: robust engraftment and rapid immune reconstitution in children with high risk leukemia. ASH Annual Meeting Abstracts. **118:** 1005.
32. Rizzieri, D.A., L.P. Koh, G.D. Long, *et al.* 2007. Partially matched, nonmyeloablative allogeneic transplantation: clinical outcomes and immune reconstitution. *J. Clin. Oncol.* **25:** 690–697.
33. Luznik, L., L.W. Engstrom, R. Iannone & E.J. Fuchs. 2002. Posttransplantation cyclophosphamide facilitates engraftment of major histocompatibility complex-identical allogeneic marrow in mice conditioned with low-dose total body irradiation. *Biol. Blood Marrow Transplant.* **8:** 131–138.
34. Munchel, A., C. Kesserwan, H.J. Symons, *et al.* 2011. Nonmyeloablative, HLA-haploidentical bone marrow transplantation with high dose, post-transplantation cyclophosphamide. *Pediatr. Rep.* **3**(Suppl 2):e15.
35. Tuve, S., J. Gayoso, C. Scheid, *et al.* 2011. Haploidentical bone marrow transplantation with post-grafting cyclophosphamide: multicenter experience with an alternative salvage strategy. *Leukemia* **25:** 880–883.
36. Fuji, S., M. Kapp & H. Einsele. 2012. Challenges to preventing infectious complications, decreasing re-hospitalizations, and reducing cost burden in long-term survivors after allogeneic hematopoietic stem cell transplantation. *Semin. Hematol.* **49:** 10–14.
37. Min, D., P.A. Taylor, A. Panoskaltsis-Mortari, *et al.* 2002. Protection from thymic epithelial cell injury by keratinocyte growth factor: a new approach to improve thymic and peripheral T-cell reconstitution after bone marrow transplantation. *Blood* **99:** 4592–4600.
38. Levine, J.E., B.R. Blazar, T. Defor, J.L. Ferrara & D.J. Weisdorf. 2008. Long-term follow-up of a phase I/II randomized, placebo-controlled trial of palifermin to prevent graft-versus-host disease (GVHD) after related donor allogeneic hematopoietic cell transplantation (HCT). *Biol. Blood Marrow Transplant.* **14:** 1017–1021.
39. Murphy, W.J., S.K. Durum, M.R. Anver & D.L. Longo. 1992. Immunologic and hematologic effects of neuroendocrine hormones. Studies on DW/J dwarf mice. *J. Immunol.* **148:** 3799–3805.
40. Knyszynski, A., S. Dler-Kunin & A. Globerson. 1992. Effects of growth hormone on thymocyte development from progenitor cells in the bone marrow. *Brain Behav. Immun.* **6:** 327–340.
41. Soiffer, R.J., C. Murray, K. Cochran, *et al.* 1992. Clinical and immunologic effects of prolonged infusion of low-dose recombinant interleukin-2 after autologous and T-cell-depleted allogeneic bone marrow transplantation. *Blood* **79:** 517–526.
42. Perillo, A., L. Pierelli, A. Battaglia, *et al.* 2002. Administration of low-dose interleukin-2 plus G-CSF/EPO early after autologous PBSC transplantation: effects on immune recovery and NK activity in a prospective study in women with breast and ovarian cancer. *Bone Marrow Transplant.* **30:** 571–578.
43. Nadal, E., A. Fowler, E. Kanfer, *et al.* 2004. Adjuvant interleukin-2 therapy for patients refractory to donor lymphocyte infusions. *Exp. Hematol.* **32:** 218–223.
44. Fry, T.J., E. Connick, J. Falloon, *et al.* 2001. A potential role for interleukin-7 in T-cell homeostasis. *Blood* **97:** 2983–2990.
45. Sportes, C., F.T. Hakim, S.A. Memon, *et al.* 2008. Administration of rhIL-7 in humans increases in vivo TCR repertoire diversity by preferential expansion of naive T cell subsets. *J. Exp. Med.* **205:** 1701–1714.
46. Alpdogan, O., S.J. Muriglan, J.M. Eng, *et al.* 2003. IL-7 enhances peripheral T cell reconstitution after allogeneic hematopoietic stem cell transplantation. *J. Clin. Invest.* **112:** 1095–1107.
47. Storek, J., T. Gillespy, III, H. Lu, *et al.* 2003. Interleukin-7 improves CD4 T-cell reconstitution after autologous CD34 cell transplantation in monkeys. *Blood* **101:** 4209–4218.
48. Rubinstein, M.P., A.N. Kadima, M.L. Salem, *et al.* 2002. Systemic administration of IL-15 augments the antigen-specific primary CD8+ T cell response following vaccination with peptide-pulsed dendritic cells. *J. Immunol.* **169:** 4928–4935.
49. Lin, S.J., P.J. Cheng, D.C. Yan, P.T. Lee & H.S. Hsaio. 2006. Effect of interleukin-15 on alloreactivity in umbilical cord blood. *Transpl. Immunol.* **16:** 112–116.
50. Leung, W. 2011. Use of NK cell activity in cure by transplant. *Br. J. Haematol.* **155:** 14–29.
51. Brehm, C., S. Huenecke, A. Quaiser, *et al.* 2011. IL-2 stimulated but not unstimulated NK cells induce selective disappearance of peripheral blood cells: concomitant results to a phase I/II study. *PLoS. One* **6:** e27351.
52. Rizzieri, D.A., R. Storms, D.F. Chen, *et al.* 2010. Natural killer cell-enriched donor lymphocyte infusions from A 3-6/6 HLA matched family member following nonmyeloablative allogeneic stem cell transplantation. *Biol. Blood Marrow Transplant.* **16:** 1107–1114.
53. Passweg, J.R., A. Tichelli, S. Meyer-Monard, *et al.* 2004. Purified donor NK-lymphocyte infusion to consolidate engraftment after haploidentical stem cell transplantation. *Leukemia* **18:** 1835–1838.
54. Rubnitz, J.E., H. Inaba, R.C. Ribeiro, *et al.* 2010. NKAML: a pilot study to determine the safety and feasibility of haploidentical natural killer cell transplantation in childhood acute myeloid leukemia. *J. Clin. Oncol.* **28:** 955–959.
55. Amrolia, P.J., G. Muccioli-Casadei, H. Huls, *et al.* 2006. Adoptive immunotherapy with allodepleted donor T-cells improves immune reconstitution after haploidentical stem cell transplantation. *Blood* **108:** 1797–1808.
56. Dodero, A., C. Carniti, A. Raganato, *et al.* 2009. Haploidentical stem cell transplantation after a reduced-intensity conditioning regimen for the treatment of advanced hematologic malignancies: posttransplantation CD8-depleted donor lymphocyte infusions contribute to improve T-cell recovery. *Blood* **113:** 4771–4779.
57. Feuchtinger, T., K. Opherk, W.A. Bethge, *et al.* 2010. Adoptive transfer of pp65-specific T cells for the treatment of chemorefractory cytomegalovirus disease or reactivation after haploidentical and matched unrelated stem cell transplantation. *Blood* **116:** 4360–4367.
58. Feuchtinger, T., S. Matthes-Martin, C. Richard, *et al.* 2006. Safe adoptive transfer of virus-specific T-cell immunity for the treatment of systemic adenovirus infection after allogeneic stem cell transplantation. *Br. J. Haematol.* **134:** 64–76.

59. Lugthart, G., S.J. Albon, I. Ricciardelli, *et al.* 2012. Simultaneous generation of multivirus-specific and regulatory T cells for adoptive immunotherapy. *J. Immunother.* **35:** 42–53.
60. Perruccio, K., F. Topini, A. Tosti, *et al.* 2008. Photodynamic purging of alloreactive T cells for adoptive immunotherapy after haploidentical stem cell transplantation. *Blood Cells Mol. Dis.* **40:** 76–83.
61. Mielke, S., R. Nunes, K. Rezvani, *et al.* 2008. A clinical-scale selective allodepletion approach for the treatment of HLA-mismatched and matched donor-recipient pairs using expanded T lymphocytes as antigen-presenting cells and a TH9402-based photodepletion technique. *Blood* **111:** 4392–4402.
62. Edinger, M., P. Hoffmann, J. Ermann, *et al.* 2003. CD4+CD25+ regulatory T cells preserve graft-versus-tumor activity while inhibiting graft-versus-host disease after bone marrow transplantation. *Nat. Med.* **9:** 1144–1150.
63. Di, I.M., F. Falzetti, A. Carotti, *et al.* 2011. Tregs prevent GVHD and promote immune reconstitution in HLA-haploidentical transplantation. *Blood* **117:** 3921–3928.

Ann. N.Y. Acad. Sci. ISSN 0077-8923

ANNALS OF THE NEW YORK ACADEMY OF SCIENCES
Issue: *Hematopoietic Stem Cells VIII*

Current insights into neutrophil homeostasis

Stefanie Bugl,[1*] Stefan Wirths,[1*] Martin R. Müller,[1] Markus P. Radsak,[2] and Hans-Georg Kopp[1]

[1]Department of Medical Oncology, Hematology, Immunology, Rheumatology and Pulmology, Department of Internal Medicine II, South West German Comprehensive Cancer Center, University Hospital of Tübingen, Tübingen, Germany. [2]Department of Internal Medicine III (Hematology, Oncology, Pneumology), Johannes Gutenberg University Medical Center, Mainz, Germany

Address for correspondence: Hans-Georg Kopp, M.D., Department of Hematology/Oncology, Eberhard-Karls University, Otfried-Mueller-Strasse 10, D-72076 Tübingen, Germany. hans-georg.kopp@med.uni-tuebingen.de

Neutrophil granulocytes represent the first immunologic barrier against invading pathogens, and neutropenia predisposes to infection. However, neutrophils may also cause significant collateral inflammatory damage. Therefore, neutrophil numbers are tightly regulated by an incompletely understood homeostatic feedback loop adjusting the marrow's supply to peripheral needs. Granulocyte colony-stimulating factor (G-CSF) is accepted to be the major determinant of neutrophil production, and G-CSF levels have, soon after its discovery, been described to be inversely correlated with neutrophil counts. A neutrophil sensor, or "neutrostat," has, therefore, been postulated. The prevailing feedback hypothesis was established in adhesion molecule–deficient mice; it includes macrophages and Th17 cells, which determine G-CSF levels in response to the number of peripherally transmigrated, apoptosing neutrophils. Recent work has deepened our understanding of homeostatic regulation of neutrophil granulopoiesis, but there are still inconsistent findings and unresolved questions when it comes to a plausible hypothesis, similar to the feedback control models of red cell or platelet homeostasis.

Keywords: granulopoiesis; neutropenia; homeostasis

Introduction

Neutrophils are the most abundant white blood cells in the human blood; they are produced at a rate of approximately 10^9 cells/kg bodyweight per day[1,2] and form part of the inborn immunity against bacteria and fungi. Febrile neutropenia remains to be one of the most important complications of cytotoxic chemotherapy. On the other hand, neutrophils are capable of tissue damage themselves; indeed, increased neutrophil counts are independently associated with mortality.[3] It is, therefore, obvious that neutrophil homeostasis is crucial. Granulocyte colony-stimulating factor (G-CSF), the most important cytokine-regulating neutrophil production and release from the marrow, is used in the clinic to shorten the duration of chemotherapy-induced neutropenia. However, in contrast to the regulation of thrombopoiesis or erythropoiesis, there is still no generally accepted hypothesis on the feedback mechanisms linking neutrophil demand and supply.

Peripheral blood neutrophil numbers are not only determined by production: in the steady state only 1–2% of mature murine neutrophils are found circulating;[4] egress of neutrophils from the marrow is regulated by CXCL12 and CXCR4 interaction; and stimulation by cytokines such as G-CSF prolongs neutrophil survival. Moreover, extravasation and clearance of apoptotic neutrophils itself appear to be critical processes in regulating neutrophil production.

Thus, the major determinants of peripheral blood neutrophil number are production, bone marrow egress, margination, and extravasation/clearance. Herein, we summarize available data on the regulation of neutrophil homeostasis and attempt to reconcile seemingly contradictory evidence.

*Both authors contributed equally to this work.

doi: 10.1111/j.1749-6632.2012.06607.x

G-CSF and the G-CSF receptor

G-CSF not only affects proliferation but also cell survival, differentiation commitment, maturation induction, and the functional stimulation of mature neutrophils.[5] With the discovery of G-CSF and its reciprocal correlation with neutrophil counts in patients after myelosuppressive chemotherapy and cyclic neutropenia, a feedback mechanism was postulated that included a hypothetical neutrostat that would sense peripheral neutrophil counts and adjust production by the marrow accordingly.[2,6] Indeed, G-CSF–deficient mice were later found to be neutropenic and to display a neutrophil mobilization defect as well as impaired immunity against *Listeria monocytogenes*. However, the mice still harbored between 20–30% of normal neutrophil numbers.[7,8] This finding suggests that further determinants of granulopoiesis are able to partly substitute for the G-CSF signal.

The most important source of G-CSF is generally accepted to be the bone marrow stromal compartment, although it remains unclear which culturable stromal cell type is the quantitatively most important contributor.[9,10]

In contrast to wild-type mice, adhesion molecule–deficient mice display a strong positive correlation between peripheral blood neutrophils and G-CSF levels.[11] In addition, transfusion of mature granulocytes into neutropenic mice did not reduce G-CSF protein levels.[12] These findings suggest that sensing of neutrophil numbers might occur at an extravascular rather than intravascular site, and receptor dependent scavenging by G-CSF receptor (G-CSF-R) positive cells may not play a major role.

Site of clearance of neutrophils as turnstile of granulopoiesis

The turnstile hypothesis implies that the site of neutrophil clearance would generate the feedback signal stimulating granulopoiesis. Current controversies, both on the most important anatomic site and on the cell type contributing to a positive feedback, may originate in the fact that different genetically modified mouse models have been used to evaluate these issues.

Turnover of neutrophil granulocytes is rapid; the postmitotic transit time estimated in mammals is about five days at the maximum.[13] Importantly, there is ongoing controversy regarding the correct methodology for determining neutrophil half-life, and a consensus on this issue has not been achieved.[14] The circulating time in peripheral blood, estimated by *ex vivo*–manipulated neutrophils, appears to be shorter than 12 hours.[15] Extravasation of senescent neutrophils is regulated by CXCR4,[16] and tracking studies demonstrate quantitatively equivalent clearance in liver, spleen, and bone marrow.[17] Senescent neutrophils undergo apoptosis,[18] which induces "find-me" and "eat-me" signals. Engulfment by phagocytes, in turn, initiates anti-inflammatory signals via nuclear peroxisome-proliferator–activated receptor γ (PPARγ) and liver X receptor (LXR) in macrophages.[19]

Indeed, interference with neutrophil find-me signaling by knocking out disintegrin and metalloproteinase domain–containing protein 10 (ADAM10) in mice creates a myeloproliferative syndrome with frank neutrophilia.[20,21] Moreover, genetic deficiency in the nuclear LXR ($LXR\alpha\beta^{-/-}$), which causes impaired clearance of apoptotic neutrophils by macrophages, results in neutrophilia.[22] Although the authors did not observe frank inflammation, anti-inflammatory effects of LXR are well characterized. Notably, LXR represses genes responsive to Toll-like receptor (TLR) signaling.[19,23] The importance of macrophages in the regulation of neutrophil granulopoiesis is underscored by data obtained in macrophage-deficient mice. Conditional deletion of the anti-apoptotic gene cellular FLICE-like inhibitory protein (c-FLIP) in lysozyme M (LysM)–expressing cells ($c\text{-}FLIP^{fl/fl}$ LysM-Cre) resulted in absent splenic and marrow macrophages.[24] Moreover, there was G-CSF–dependent extreme neutrophilia in the marrow, blood, peritoneum, and spleen concurrent with splenomegaly and extramedullary hematopoiesis. Interestingly, transplant studies showed that wild-type marrow chimerism was sufficient to suppress neutrophilia and splenomegaly in $c\text{-}FLIP^{fl/fl}$ LysM-Cre mice.[24] Dramatic neutrophilia in the context of a myeloproliferative syndrome was also observed in the CD11c:DTA mouse, which lacks conventional dendritic cells (cDC).[25] Moreover, specific inducible depletion of $CD169^{+}$ macrophages reduced marrow-expressed CXCL12.[26] In further experiments, the authors demonstrated this effect to be mediated by interplay with $Nestin^{+}$ bone marrow stromal cells. This study indicates that regulation of granulopoiesis might be influenced by macrophage-mediated regulation of bone marrow egress. In contrast to these data, which indicate that macrophage deficiency or dysfunction disinhibit neutrophil granulopoiesis, another group found that clearance of neutrophils by marrow-derived stromal macrophages *in vitro* stimulates the latter to produce G-CSF.[17]

Adhesion molecule (CD18, E/P-selectin, $\alpha_L\beta_2$ (LFA-1), core2-glucosaminyltransferase (Core-2), or E/P/L-selectin)–deficient mice in which neutrophils cannot transmigrate into peripheral tissues, display peripheral blood neutrophilia. Stark *et al.* established a feedback loop explaining this phenomenon.[27] Under normal conditions, transmigrated neutrophils are phagocytosed by macrophages and DC, which decrease the production of IL-23 upon phagocytosis of transmigrated apoptotic neutrophils. Macrophages and DC are the major source of IL-23,[28] which induces IL-17 in $\gamma\delta$ and $\alpha\beta$ T cells as well as NK and NKT cells.[29] Increased IL-17 levels lead to an increase of G-CSF followed by increased neutrophil granulopoiesis. Therefore, adhesion molecule deficiency, which renders neutrophils incapable of transmigration, leads to a constant IL-23, IL-17, and G-CSF–dependent overproduction of neutrophils. Accordingly, the authors propose a turnstile model where neutrophils transmigrate in the gastrointestinal tract.[27] The above-mentioned neutrophilic $LXR\alpha\beta^{-/-}$ mouse displayed increased levels of IL-23 and Th17 T cells, verifying this hypothesis.[22]

However, further data indicate redundant regulation of granulopoiesis: short-term antibody-mediated neutralization of IL-17 in wild-type mice does not alter neutrophil numbers.[17] Moreover, macrophage deficiency in c-FLIP$^{f/f}$ LysM-cre mice is associated with increased G-CSF and neutrophilia, but not elevated IL-17 levels.[24] In addition, increased granulopoiesis subsequent to antibody-induced neutropenia or alum-induced inflammation was found to be independent from IL-17.[30] Further complexity to the question of whether IL-17 and/or T cells were prerequisites for neutrophil homeostasis is added by the fact that IL-17 may come from sources other than Th17 T cells. In fact, IL-17 is not only produced by lymphocytes but also by intestinal Paneth cells and even by neutrophils.[31] Indeed, steady-state neutrophil numbers in T- and B-lymphopenic $RAG1^{-/-}$ mice are identical to wild-type controls, and Gr1-antibody–mediated neutropenia results in increased granulopoiesis indistinguishable from control mice.[30] This observation is confirmed in mice devoid of NK cells as well.[32]

A number of studies on the turnstile hypothesis have demonstrated that macrophages function as regulators of granulopoiesis. Of note, lack of macrophages or their proper function stimulates granulopoiesis. Thus, a negative regulating signal on granulopoiesis in these models is lost. In contrast, in adhesion-deficient mice, a positive signal is generated via IL-23 upon reduced phagocytosis of apoptotic neutrophils. In this context, it would be of interest whether macrophage malfunction in adhesion deficiency models contributed to deregulated granulopoiesis.

Effects of marrow hypocellularity and neutrophil egress

Inflammation induced by injection of the adjuvant alum as well as *Staphylococcus aureus* results in mobilization of neutrophils from the marrow and stimulates granulopoiesis.[30,33] The fact that antibody-induced neutropenia by itself is sufficient to trigger stimulation of granulopoiesis led to the conclusion that mobilization-induced bone marrow hypocellularity might stimulate granulopoiesis.[30,33]

Conflicting data arose in the LysM$^{cre/-}$ CXCR4$^{fl/-}$ mouse, whose neutrophils are prematurely released from the marrow because of a retention defect. However, these animals do not display increased proliferation of granulopoietic progenitors. Indeed, upon infection with *Listeria monocytogenes*, mobilization of neutrophils was absent, but no difference in infection control was observed.[34,35] Anchoring of neutrophils in the marrow is dependent on interactions between CXCR2, CXCR4, and their respective chemokine ligands.[36,37] The notion that egress from the marrow is not essential for stimulation of granulopoiesis was supported by experiments in which CXCR4 signaling was blocked during peritoneal infection; the respective mice showed stimulated granulopoiesis without neutrophil mobilization.[38] Moreover, regulation of CXCL12 is downstream of G-CSF, i.e., secondary to stimulation of granulopoiesis.[36]

Conditional gene knockout in neutrophils is hampered by the paucity of neutrophil-restricted gene constructs. LysM expression is shared between neutrophils and macrophages, causing difficulties in ascribing effects to either population. To induce experimental neutropenia, the RB6–8C5 antibody against Gr-1 is frequently used. As it binds both Ly6G and Ly6C, monocytes and lymphocyte subsets are also targeted. Ribechini *et al.* observed that RB6–8C5 effectively depleted peripheral blood neutrophils but was ineffective in reducing marrow neutrophil granulocytes. The authors ascribed this phenomenon to granulopoiesis stimulating effects of RB6–8C5.[39] Studies in neutropenic mice after administration of RB6–8C5 antibody are, therefore, criticized. The specific Ly6G antibody 1A8 does not suffer from the same limitations, however.[40]

Regulation of granulopoiesis by bone marrow stroma

Interestingly, genetic modification of the marrow stromal compartment is sufficient to induce chronic neutrophilia. Deficiency of the family member retinoic acid receptor γ (RARγ) in the stromal microenvironment but not in hematopoietic cells induced a phenotype of significantly increased granulocyte macrophage progenitors (GMPs) and neutrophils in bone marrow, spleen, and peripheral blood.[41] Similarly, loss of *IκBα* in hepatocytes,[42] deletion of *Dicer1* in osteoprogenitors,[43] and deficiency in *Retinoblastoma* in myeloid and stromal cells[44] leads to myeloproliferative syndromes or leukemia. Although different components of the bone marrow stroma seem to regulate neutrophil production, feedback into the stromal compartment is less clear.

Emergency versus steady-state granulopoiesis

The term *emergency granulopoiesis* was coined when neutropenic G-CSF–deficient mice challenged with microbial compounds through injection of infectious agents were found to be able to build up pronounced neutrophilia.[45] Because of this initial report, several groups have set out to characterize the underlying differences between G-CSF–independent neutrophilia in response to infection and G-CSF–dependent steady-state granulopoiesis.

CCAAT/enhancer-binding protein (C/EBP) transcriptions factors are important regulators of granulopoiesis. In noninfectious conditions neutrophil granulopoiesis in blood and bone marrow is dependent on C/EBPα. C/EBPα-deficient mice are completely deficient in granulocytes due to a differentation block at the common myeloid progenitor (CMP) level, suggesting C/EBPα is a master regulator of steady state granulopoiesis.[46,47] In contrast, emergency granulopoiesis has been described to be dependent on C/EBPβ, which promotes granulopoiesis in the presence of inflammatory growth factors.[48] These findings led to the commonly accepted conceptual distinction between steady-state and emergency granulopoiesis.

However, beyond the GMP stage, C/EBPα is not required for the development of mature granulocytes, and excess cytokines and danger signals can overcome the ascribed differentiation block. This effect has been shown to depend on C/EBPβ expression, whose complete loss does not influence steady-state granulopoiesis.[49] In addition, more recent work suggests that C/EBPβ is dispensable for inflammation-triggered granulopoiesis.[30] Together, these data suggest that the model of a single transcription factor determining separation of steady-state and emergency granulopoiesis may not fully reflect reality. Rather, fate determination of hematopoietic stem and progenitor cells (HSPCs) toward the granulocyte lineage can be corrected on demand, which, in case of innate immunity, occurs from the confrontation with microbial mediated danger.

Microbial components containing pathogen-associated molecular patterns (PAMPs) may be detected by pathogen recognition receptors (PRRs), which encompass membrane-bound TLRs and cytoplasmatic nucleotide-binding oligomerization domain–containing protein (NOD)-like receptors. HSPCs express TLRs and can be activated by TLR agonists such as LPS, *in vivo* and *in vitro*.[33,50] Although infection with *Escherichia coli* induces an increase of LSK (Lin^- $Sca\text{-}1^+$ $c\text{-}kit^+$ cells) in addition to increasing neutrophils, the mechanisms of activation of HSPCs by bacteria are not completely understood. Injection of LPS induced an eight- to nine-fold HSPC expansion in wild-type mice, while HSPC levels remained unchanged in C3H/HeJ mice, which harbor a loss-of-function mutation in the TLR4 gene.[33]

Colonization by a large and diverse bacterial flora is essential for development of the gut-associated immune system. Commensal bacteria–derived peptidoglycans were demonstrated to pass into circulation and induce maturation of neutrophils in a

NOD1-dependent manner.[51] Therefore, danger signaling by PRR is able to substitute cytokine signaling in emergency granulopoiesis and to invoke alternative signaling and transcription pathways. Commensal bacterial components may modulate innate immunity even in the steady state.

To establish the role of commensal germs in neutrophil homeostasis, we took advantage of germ-free mice.[12] These mice are maintained in special isolators and in an environment devoid of microorganisms or viruses.[52] The peripheral blood of germ-free mice was found to contain less than 10% of normal neutrophil levels, and plasma G-CSF was low.[12] These findings are consistent with results in germ-free rats, which have a 90% reduced number of neutrophils.[53] Thus, reduction of peripheral neutrophils in germ-free mice is even more pronounced than in G-CSF knockout mice, granulocyte macrophage colony-stimulating factor (GM-CSF) knockout mice, macrophage colony-stimulating factor (M-CSF) knockout mice, or triple knockout mice.[54] These findings suggest that microbiota may force the organism to adapt granulopoiesis to peripheral needs along a continuum, rather than a threshold effect that distinguishes steady-state from emergency situations.

Germ-free mice do not display increased levels of IL-17 or G-CSF. Still these mice have intact feedback loops, as their granulopoiesis is efficiently stimulated upon antibody-induced neutropenia, accompanied by elevation of G-CSF.[12] It is tempting to speculate that feedback is mediated by intrinsic damage-associated molecular pattern molecules (DAMPs) or minute amounts of PAMPs ingested by food.

Discussion

More than 25 years after G-CSF and its receptor were discovered there is still no generally accepted model of neutrophil homeostatic regulation, although neutropenia remains an important and costly clinical problem. Recently, several studies in preclinical models have substantially deepened our insight into neutrophil homeostasis. All original data regarding mechanisms of neutrophil homeostasis cited in this review have been generated in mice. Our knowledge of human neutrophil granulopoiesis is limited to correlative data, and further work will be necessary to gain more insight into human neutrophil physiology. However, there seem to be conflicting results in different mouse models of neutropenia and neutrophilia, respectively. Neutrophilia in adhesion molecule–deficient mice is due to disinhibited IL-23–, IL-17–, and G-CSF–dependent feedback loops,[27] while neutrophilia in splenic and marrow macrophage–deficient mice is caused by strongly increased G-CSF secreted by an extramedullary source and independent of IL-17.[24] Results by Furze *et al.* suggest that G-CSF is directly produced in response to engulfment of senescent, apoptosing neutrophils in the marrow by macrophages.[17] Antibody (RB6–8C5)-induced neutropenia results in HSPC expansion and neutrophil progenitor accumulation, which has been associated with a sensor of marrow hypocellularity.[30,33] However, each set of data has its shortcomings, for example, the recently described hematopoietic progenitor cell (HPC) stimulatory effect of RB6–8C5 antibody that leads to increased marrow cellularity and marrow neutrophil content, which may be indistinguishable from feedback effects on granulopoiesis.[39] We suggest that neutrophil depletion models in different mouse strains utilizing the specific anti-Ly6G antibody 1A8 should be performed in different wild-type and genetically modified mouse strains to compare the effects at the HSPC and peripheral blood levels. These results may help to reconcile inconsistencies in different mouse models employing different methodologies.

When it comes to the neutrostat or neutrophil turnstile, the location of neutrophil clearance may tell us about the location of the sensor: different sensor macrophages phagocytosing neutrophils in different organs may respond differently to G-CSF production.[55] When it comes to the debate about the most important anatomic site of neutrophil homing for clearance, methodological modifications of neutrophil reinfusion experiments may bring about interobserver discrepancies, as both maturation and activation states of reinfused neutrophils determine the location of degradation.[56]

Although marrow stromal macrophages phagocytosing senescent neutrophils enhance G-CSF secretion, engulfment of transmigrated apoptotic neutrophils by peripheral tissue macrophages leads to a direct or indirect (IL-23– and IL-17–mediated) suppression of G-CSF production.[17,24,27] Such an organ-specific, differential response is not only plausible from a teleological standpoint, but it is also underscored by the finding that different, organ-specific macrophages not only display differential expression of receptors but also respond in a subset-specific manner by cytokine expression after phagocytosis.[57–59] Therefore, seemingly contradictory findings in adhesion molecule–deficient mice compared with macrophage-deficient models may be reconciled by the fact that the influence of one macrophage subset or the other may prevail in a given model.

In conclusion, though there are still inconsistencies as to mechanism, it has become increasingly clear that the identity of the neutrostat may be related to phagocytes that are involved in the removal of senescent or apoptosing neutrophils, such as tissue-resident macrophages and DCs. These cells respond in a cell subset–specific manner. Moreover, they generate a signal that depends on the form of cell death to which the engulfed cell succumbed.[60]

There is substantial evidence that neutrophil numbers are determined by environmental rather than by solely genetic factors: although genetic differences in neutrophil counts are well known, the low neutrophil number in benign ethnic neutropenia, for example, may not be due to altered, genetically different neutrophil production but to pooling and margination effects.[15,61] However, we and others have shown that germ-free animals have baseline G-CSF levels, as well as strongly decreased neutrophils. In addition, the interesting fact that neutrophil counts in the U.S. population have been on a steady decline from 1958 to 2002, independent of age, gender, race, smoking, body mass index, and physical activity, may in part be an effect of improved hygiene.[3] Highly conserved innate immune system–signaling pathways may in fact represent the neutrostat, and may be responsible for increases of granulopoiesis for any need—may it be drastic bacteremia, food contaminated with LPS, or TLR-signaling by endogenous ligands—in excess of very low baseline neutrophil production.

In summary, a review of the literature suggests that the previous dogmatic separation of emergency versus steady-state granulopoiesis may represent different ends of the same spectrum of signaling prerequisites that contribute to the final common pathway of G-CSF–mediated neutrophil granulopoiesis. Further work will be required to identify the underlying mechanisms.

References

1. Dancey, J.T. *et al.* 1976. Neutrophil kinetics in man. *J. Clin. Invest.* **58:** 705–715.
2. Demetri, G.D. & J.D. Griffin. 1991. Granulocyte colony-stimulating factor and its receptor. *Blood* **78:** 2791–2808.
3. Ruggiero, C. *et al.* 2007. White blood cell count and mortality in the Baltimore longitudinal study of aging. *J. Am. Coll. Cardiol.* **49:** 1841–1850.
4. Semerad, C.L. *et al.* 2002. G-CSF is an essential regulator of neutrophil trafficking from the bone marrow to the blood. *Immunity* **17:** 413–423.
5. Metcalf, D. 2008. Hematopoietic cytokines. *Blood* **111:** 485–491.
6. Layton, J.E. *et al.* 1989. Evidence for a novel in vivo control mechanism of granulopoiesis: mature cell-related control of a regulatory growth factor. *Blood* **74:** 1303–1307.
7. Lieschke, G.J. *et al.* 1994. Mice lacking granulocyte colony-stimulating factor have chronic neutropenia, granulocyte and macrophage progenitor cell deficiency, and impaired neutrophil mobilization. *Blood* **84:** 1737–1746.
8. Liu, F. *et al.* 1996. Impaired production and increased apoptosis of neutrophils in granulocyte colony-stimulating factor receptor-deficient mice. *Immunity* **5:** 491–501.
9. Panopoulos, A.D. & S.S. Watowich. 2008. Granulocyte colony-stimulating factor: molecular mechanisms of action during steady state and 'emergency' hematopoiesis. *Cytokine* **42:** 277–288.
10. Watari, K. *et al.* 1994. Production of human granulocyte colony stimulating factor by various kinds of stromal cells in vitro detected by enzyme immunoassay and in situ hybridization. *Stem Cells* **12:** 416–423.
11. Forlow, S.B. *et al.* 2001. Increased granulopoiesis through interleukin-17 and granulocyte colony-stimulating factor in leukocyte adhesion molecule-deficient mice. *Blood* **98:** 3309–3314.
12. Bugl, S. *et al.* 2011. Neutropenia-induced feedback G-CSF production is regulated on the transcriptional level independent from the presence of commensal germs. *Onkologie* **34:** 116.
13. Pillay, J. *et al.* 2010. In vivo labeling with 2H2O reveals a human neutrophil lifespan of 5.4 days. *Blood* **116:** 625–627.
14. Tofts, P.S. *et al.* 2011. Doubts concerning the recently reported human neutrophil lifespan of 5.4 days. *Blood* **117:** 6050–6052.
15. von Vietinghoff, S. & K. Ley. 2008. Homeostatic regulation of blood neutrophil counts. *J. Immunol.* **181:** 5183–5188.
16. Martin, C. *et al.* 2003. Chemokines acting via CXCR2 and CXCR4 control the release of neutrophils from the bone marrow and their return following senescence. *Immunity* **19:** 583–593.
17. Furze, R.C. & S.M. Rankin. 2008. The role of the bone marrow in neutrophil clearance under homeostatic conditions in the mouse. *FASEB J.* **22:** 3111–3119.
18. Luo, H.R. & F. Loison. 2008. Constitutive neutrophil apoptosis: mechanisms and regulation. *Am. J. Hematol.* **83:** 288–295.

19. Ogawa, S. *et al.* 2005. Molecular determinants of crosstalk between nuclear receptors and toll-like receptors. *Cell* **122:** 707–721.
20. Blume, K.E. *et al.* 2012. Cleavage of annexin A1 by ADAM10 during secondary necrosis generates a monocytic "Find-Me" signal. *J. Immunol.* **188:** 135–145.
21. Yoda, M. *et al.* 2011. Dual functions of cell-autonomous and non-cell-autonomous ADAM10 activity in granulopoiesis. *Blood* **118:** 6939–6942.
22. Hong, C. *et al.* 2012. Coordinate regulation of neutrophil homeostasis by liver X receptors in mice. *J. Clin. Invest* **122:** 337–347.
23. Joseph, S.B. *et al.* 2003. Reciprocal regulation of inflammation and lipid metabolism by liver X receptors. *Nat. Med.* **9:** 213–219.
24. Gordy, C. *et al.* 2011. Regulation of steady-state neutrophil homeostasis by macrophages. *Blood* **117:** 618–629.
25. Birnberg, T. *et al.* 2008. Lack of conventional dendritic cells is compatible with normal development and T cell homeostasis, but causes myeloid proliferative syndrome. *Immunity* **29:** 986–997.
26. Chow, A. *et al.* 2011. Bone marrow CD169+ macrophages promote the retention of hematopoietic stem and progenitor cells in the mesenchymal stem cell niche. *J. Exp. Med.* **208:** 261–271.
27. Stark, M.A. *et al.* 2005. Phagocytosis of apoptotic neutrophils regulates granulopoiesis via IL-23 and IL-17. *Immunity* **22:** 285–294.
28. Oppmann, B. *et al.* 2000. Novel p19 protein engages IL-12p40 to form a cytokine, IL-23, with biological activities similar as well as distinct from IL-12. *Immunity* **13:** 715–725.
29. Steinman, L. 2007. A brief history of T(H)17, the first major revision in the T(H)1/T(H)2 hypothesis of T cell-mediated tissue damage. *Nat. Med.* **13:** 139–145.
30. Cain, D.W. *et al.* 2011. Inflammation triggers emergency granulopoiesis through a density-dependent feedback mechanism. *PLoS.One* **6:** e19957.
31. Cua, D.J. & C.M. Tato. 2010. Innate IL-17-producing cells: the sentinels of the immune system. *Nat. Rev. Immunol.* **10:** 479–489.
32. Bugl, S. *et al.* 2010. Lymphocytes are dispensable in neutrophil homeostasis. *Blood* **116:** 1080.
33. Scumpia, P.O. *et al.* 2010. Cutting edge: bacterial infection induces hematopoietic stem and progenitor cell expansion in the absence of TLR signaling. *J. Immunol.* **184:** 2247–2251.
34. Eash, K.J. *et al.* 2009. CXCR4 is a key regulator of neutrophil release from the bone marrow under basal and stress granulopoiesis conditions. *Blood* **113:** 4711–4719.
35. Ma, Q., D. Jones & T.A. Springer. 1999. The chemokine receptor CXCR4 is required for the retention of B lineage and granulocytic precursors within the bone marrow microenvironment. *Immunity* **10:** 463–471.
36. Petit, I. *et al.* 2002. G-CSF induces stem cell mobilization by decreasing bone marrow SDF-1 and up-regulating CXCR4. *Nat. Immunol.* **3:** 687–694.
37. Eash, K.J. *et al.* 2010. CXCR2 and CXCR4 antagonistically regulate neutrophil trafficking from murine bone marrow. *J. Clin. Invest* **120:** 2423–2431.
38. Delano, M.J. *et al.* 2011. Neutrophil mobilization from the bone marrow during polymicrobial sepsis is dependent on CXCL12 signaling. *J. Immunol.* **187:** 911–918.
39. Ribechini, E., P.J. Leenen & M.B. Lutz. 2009. Gr-1 antibody induces STAT signaling, macrophage marker expression and abrogation of myeloid-derived suppressor cell activity in BM cells. *Eur. J. Immunol.* **39:** 3538–3551.
40. Daley, J.M. *et al.* 2008. Use of Ly6G-specific monoclonal antibody to deplete neutrophils in mice. *J. Leukoc. Biol.* **83:** 64–70.
41. Walkley, C.R. *et al.* 2007. A microenvironment-induced myeloproliferative syndrome caused by retinoic acid receptor gamma deficiency. *Cell* **129:** 1097–1110.
42. Rupec, R.A. *et al.* 2005. Stroma-mediated dysregulation of myelopoiesis in mice lacking I kappa B alpha. *Immunity* **22:** 479–491.
43. Raaijmakers, M.H. *et al.* 2010. Bone progenitor dysfunction induces myelodysplasia and secondary leukaemia. *Nature* **464:** 852–857.
44. Walkley, C.R. *et al.* 2007. Rb regulates interactions between hematopoietic stem cells and their bone marrow microenvironment. *Cell* **129:** 1081–1095.
45. Basu, S. *et al.* 2000. "Emergency" granulopoiesis in G-CSF-deficient mice in response to Candida albicans infection. *Blood* **95:** 3725–3733.
46. Zhang, D.E. *et al.* 1997. Absence of granulocyte colony-stimulating factor signaling and neutrophil development in CCAAT enhancer binding protein alpha-deficient mice. *Proc. Natl. Acad. Sci. USA* **94:** 569–574.
47. Zhang, P. *et al.* 2004. Enhancement of hematopoietic stem cell repopulating capacity and self-renewal in the absence of the transcription factor C/EBP alpha. *Immunity* **21:** 853–863.
48. Hirai, H. *et al.* 2006. C/EBPbeta is required for 'emergency' granulopoiesis. *Nat. Immunol.* **7:** 732–739.
49. Screpanti, I. *et al.* 1995. Lymphoproliferative disorder and imbalanced T-helper response in C/EBP beta-deficient mice. *EMBO J.* **14:** 1932–1941.
50. Takizawa, H. *et al.* 2011. Dynamic variation in cycling of hematopoietic stem cells in steady state and inflammation. *J. Exp. Med.* **208:** 273–284.
51. Clarke, T.B. *et al.* 2010. Recognition of peptidoglycan from the microbiota by Nod1 enhances systemic innate immunity. *Nat. Med.* **16:** 228–231.
52. Inzunza, J. *et al.* 2005. Germfree status of mice obtained by embryo transfer in an isolator environment. *Lab. Anim.* **39:** 421–427.
53. Ohkubo, T. *et al.* 1999. Peripheral blood neutrophils of germ-free rats modified by in vivo granulocyte-colony-stimulating factor and exposure to natural environment. *Scand. J. Immunol.* **49:** 73–77.
54. Hibbs, M.L. *et al.* 2007. Mice lacking three myeloid colony-stimulating factors (G-CSF, GM-CSF, and M-CSF) still produce macrophages and granulocytes and mount an inflammatory response in a sterile model of peritonitis. *J. Immunol.* **178:** 6435–6443.
55. Bratton, D.L. & P.M. Henson. 2011. Neutrophil clearance: when the party is over, clean-up begins. *Trends Immunol.* **32:** 350–357.

56. Suratt, B.T. *et al.* 2001. Neutrophil maturation and activation determine anatomic site of clearance from circulation. *Am. J. Physiol Lung Cell Mol. Physiol.* **281:** L913-L921.
57. Fadok, V.A. *et al.* 1992. Different populations of macrophages use either the vitronectin receptor or the phosphatidylserine receptor to recognize and remove apoptotic cells. *J. Immunol.* **149:** 4029–4035.
58. Furze, R.C. & S.M. Rankin. 2008. Neutrophil mobilization and clearance in the bone marrow. *Immunology* **125:** 281–288.
59. Lucas, M. *et al.* 2003. Apoptotic cells and innate immune stimuli combine to regulate macrophage cytokine secretion. *J. Immunol.* **171:** 2610–2615.
60. Brereton, C.F. & J.M. Blander. 2011. The unexpected link between infection-induced apoptosis and a TH17 immune response. *J. Leukoc. Biol.* **89:** 565–576.
61. Bain, B.J. *et al.* 2000. Investigation of the effect of marathon running on leucocyte counts of subjects of different ethnic origins: relevance to the aetiology of ethnic neutropenia. *Br. J. Haematol.* **108:** 483–487.

Ann. N.Y. Acad. Sci. ISSN 0077-8923

ANNALS OF THE NEW YORK ACADEMY OF SCIENCES
Issue: *Hematopoietic Stem Cells VIII*

Prospects and challenges of induced pluripotent stem cells as a source of hematopoietic stem cells

Dirk W. van Bekkum and Harald M.M. Mikkers

Department of Molecular Cell Biology, Regenerative Medicine Program, Leiden University Medical Center, Leiden, the Netherlands

Address for correspondence: Harald M.M. Mikkers, Department of Molecular Cell Biology, Leiden University Medical Center, Postal zone S1P, PO Box 9600, 2300RC Leiden, the Netherlands. h.mikkers@lumc.nl

Many life-threatening hematological diseases are now treated by bone marrow transplantations, i.e., infusion of hematopoietic stem cells (HSCs). HSC transplantations are a valid option for the treatment of a variety of metabolic disorders, and even for solid tumors and some refractory severe autoimmune diseases. Unfortunately, the frequency and outcome of HSC transplantations are limited by a shortage of suitable donors. Induced pluripotent stem cells (iPSCs)—somatic cells that have acquired pluripotent stem cell characteristics by the ectopic expression of pluripotency-inducing factors—have been proposed as an alternative source of HSCs. Possible applications include cells of autologous, of autologous and genetically modified, or of allogeneic origin. Here, we provide a perspective on the distinct opportunities of iPSCs and discuss the challenges that lie ahead.

Keywords: induced pluripotent stem cells; hematopoietic stem cells; transplantation; banking; therapy

Hematopoietic stem cell transplantations as treatment of hematopoietic and other disorders

The majority of disorders commonly treated with hematopoietic stem cell (HSC) transplants are hematological malignancies and a variety of inherited and acquired hematopoietic diseases. In addition, lysosomal storage disorders, solid tumors, and autoimmune diseases have also been successfully treated in a more experimental setting. Congenital hematological disorders are usually caused either by mutations leading to aberrant hematopoietic differentiation or by mutations yielding non- or partially functioning cells. For example, severe combined immunodeficiency (SCID) patients lack mature T cells, sometimes in combination with the absence of other cells of lymphoid origin (either NK cells or B cells, or both), due to a block in lymphoid development. In α-thalassemia, too little functional hemoglobin is generated, and in sickle-cell disease a mutant and less efficient form of hemoglobin is produced.

Transplantation protocols of inherited disorders obviously exploit allogeneic donor HSCs. Donors are matched for HLA-A, -B, and -DR (6/6 match) or HLA-A, -B, -C, -DR, and -DQ (10/10 match), depending on the transplantation center. Allogeneic HSCs are isolated from bone marrow, peripheral blood (after the mobilization of HSCs), or umbilical cord blood. However, allogeneic transplantations are risky because of transplantation-related morbidity and mortality due to toxic conditioning regimens, take failures, infections, and graft-versus-host disease (GVHD).

Autologous bone marrow transplantations do not entail the risk of take failure and severe GVHD. Accordingly, gene-modified autologous HSCs are a very attractive alternative for the treatment of all congenital disorders that respond favorably to allogeneic HSC transplants, as well as for the treatment of aplastic anemia, which is caused by an acquired destruction of HSCs.

In the treatment of hematological malignancies, the primary function of the HSC graft is to rescue the patient's hematopoiesis following high-dose cytotoxic chemo- and radiation therapy. In certain subgroups of leukemias, allogeneic HSC transplants are preferred because they result in a lower incidence

doi: 10.1111/j.1749-6632.2012.06629.x
Ann. N.Y. Acad. Sci. 1266 (2012) 179–188 © 2012 New York Academy of Sciences.

of relapse than transplants of autologous HSCs. This difference is ascribed to the GVH reaction in general, as well as to a selective graft-versus-leukemia reaction (which, like GVH, kills residual leukemic cells), and, potentially, to the presence of residual tumor cells in the autologous graft. However, in some instances the advantage of a decreased relapse rate is offset by an increase in transplantation-related mortality due to GVHD. The overall success rate of allogeneic transplantations is relatively high, averaging approximately 50%. Survival depends on the disease treated, the donor cell source, haplotype match, and conditioning regimens.

In about 25% of the allogeneic transplantations, HLA-matched siblings provide the donors cells. The remainder of the patients is dependent on the worldwide bone marrow donor registry (BMDR) or, when no match is available, on donated umbilical cord blood cells. In spite of the constant increase in the size of donor registries and the number of banked umbilical cord cells, currently available donor material is insufficient to provide all patients in need with HLA-matched cells. For example, in the BMDR, a match can be found for about 80% of Caucasian Americans, whereas it is approximately 50% for African Americans. As populations are becoming more and more multiracial, chances of finding matched donors will not get better. In the following, we discuss ways that a new type of pluripotent stem cell—induced pluripotent stem cells (iPSCs)—may contribute to alleviating the scarcity of suitable HSC supply.

PSCs as an alternative source of HSCs?

One general solution to the donor shortage is the use of cells that *give rise to* hematopoietic stem cells and can be efficiently expanded in culture. Candidate cells are *self-renewing* PSCs. These cells expand in culture without losing their pluripotency features, meaning that they have the ability to generate all of the >200 cell types found in mammals, including cells belonging to the hematopoietic lineage.[1] *Embryonic stem cells* (ESCs), the best-characterized PSCs, are derived from blastocysts, in which a transient population of pluripotent cells resides.[2] ESCs have an unlimited lifespan, indicating that they can be maintained in culture infinitely. Thus, in theory, ESCs represent an unlimited source of HSCs, qualifying them as an ideal source of cells to be banked for transplantation purposes. However, the derivation of human ESC (hESC) lines is controversial for ethical reasons. As a result, regulation concerning the use and derivation of novel hESC lines varies between countries, effectively disqualifying them as a universal solution to the lack of suitable HSC donors. Because of ethical dilemmas and the failure of being able to establish hESC lines from cloned blastocysts, alternative ways of creating genotype-specific PSCs have been extensively explored.

In 2007, pioneering work performed in the laboratories of Yamanaka and Thompson demonstrated that human skin fibroblasts could be efficiently turned into PSCs by the expression of OCT4, SOX2, and either KLF4 and cMYC[3] or NANOG and LIN28.[4] These cells were named *induced pluripotent stem cells* (iPSCs). As their name suggests, iPSCs can give rise to cells of endodermal, mesodermal, and ectodermal origin. In fact they are very similar to ESCs.[5] Since these cells can be generated from almost every individual, possibilities that could only be dreamed of have come within reach (Fig. 1). In the short term, the most realistic iPSC applications include disease modeling, which facilitates the research on human disease pathogenesis, and drug discovery.[6] Future applications may also include cell-based therapies, such as HSC transplantation, as iPSC-derived HSCs would offer several advantages (Table 1).

Autologous iPSCs versus endogenous HSCs for the treatment of congenital disorders

Congenital disorders can be treated by the transplantation of allogenic HSCs from normal donors. However, as pointed out before, allogeneic grafts carry serious risks. Autologous HSCs are a valid option provided that the genetic defect can be efficiently corrected *ex vivo*. The preferred way to restore a genetic defect is by repair of the endogenous alleles through homologous recombination (HR) because the HSC genome is only altered at the position of the mutation. In addition, repair by HR preserves the endogenous control of the gene, creating a phenotype identical to that of heterozygous or hemizygous cells. Unfortunately, HR is a very inefficient process, requiring a selection procedure to identify the cells that have undergone correct HR. In addition, the cells that have undergone proper HR must be expanded to achieve the HSC numbers required

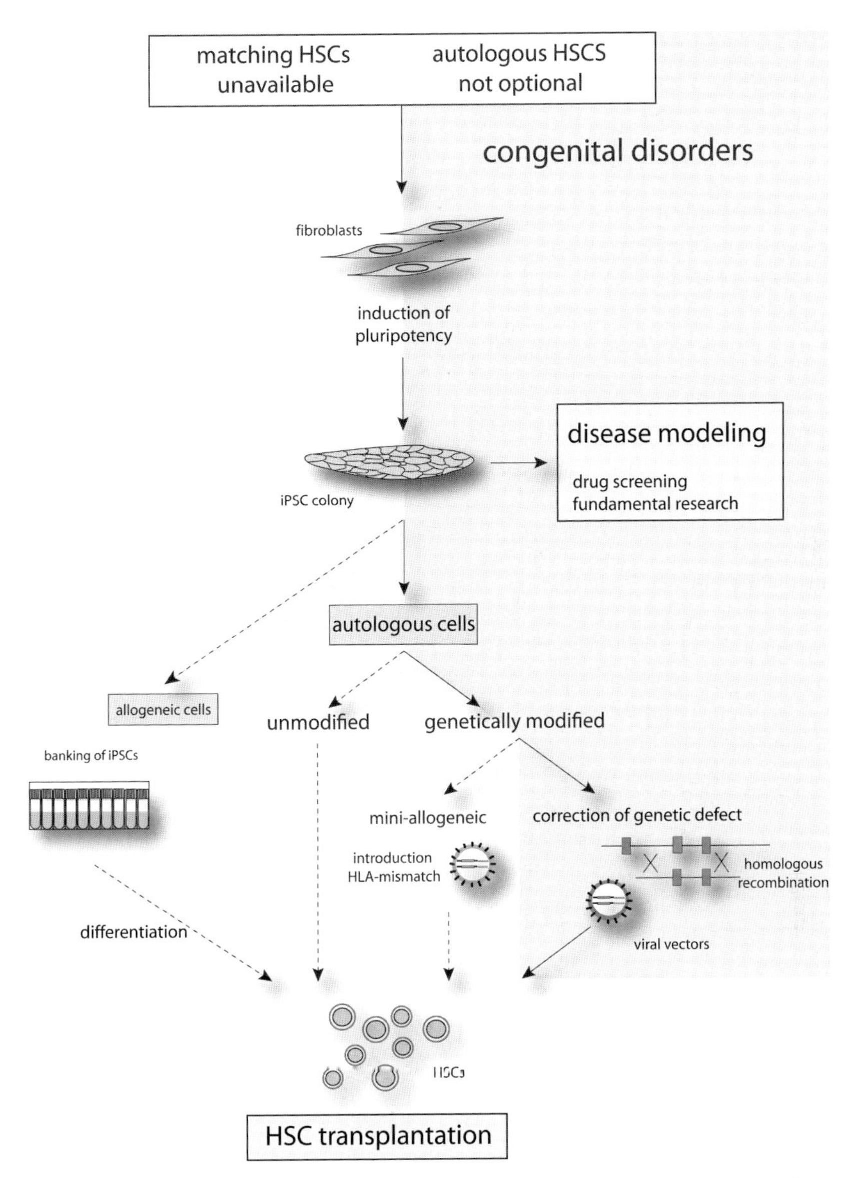

Figure 1. iPSC-based opportunities for HSC transplantation and the research on hematopoietic diseases. Left: iPSCs from healthy individuals can be used to make HLA-specific iPSC banks for transplantation purposes. Autologous iPSCs are an alternative for allogeneic HSCs to treat patients lacking endogenous HSCs. In addition, autologous iPSCs genetically modified (here by gene therapy stratgies) to induce a mild GVHD by HLA-mismatching can be used to treat certain forms of leukemia. Right: iPSCs from different patient-specific cell sources can be exploited for the generation of disease-specific cells (disease modeling) aiding fundamental research and drug screening, or for the treatment of patients suffering from congenital disorders.

for successful transplantation. Since the expansion of HSCs *in vitro* is very limited,[7] HSCs are currently unsuited to be repaired by HR, even despite the recent development of novel techniques that enhance homologous recombination via the induction of locus-specific double-strand breaks by zinc-finger nucleases (ZFN) or TAL effector nucleases (TALENs).[8,9] The only currently available means of long-term restoration is the use of retroviral gene therapy vectors that deliver a functional copy of the mutated gene by insertion into the genome of the HSC.

Table 1. Advantages and concerns of iPSC-based HSC transplantations in comparison with endogenous HSCs

Endogenous HSCs	HSCs from iPSCs
• Limited source	• Unlimited source of HSCs
• Contaminated with T cells (GVHD)	• Grafts devoid of T cells (less severe GVHD)
• HLA-matched grafts dependent on donor registries	• Theoretically always HLA-matched graft
• Autologous cells for some disorders not available	• Autologous cells always available
• Genetic modifications by viral vectors	• Genetic modification by viral vector or homologous recombination
Concerns about HSCs from iPSCs	
• Long-term repopulating iPSC-derived HSCs have not been generated	
• Safety has not been demonstrated	
• Immunogenic?	

Initial results of this form of gene therapy in X-linked SCID and ADA-SCID have been promising.[10,11] Eight out of ten treated ADA-SCID patients did not need enzyme-replacement therapy any longer, and only 3 suffered from serious adverse affects. For the SCID-X1 trial, 18 out of 19 patients survived the treatment and generated functional T cells. However, five SCID-X1 patients developed T cell leukemia due to retroviral vector–mediated insertional mutagenesis (LTFU 54–144 months), and the number of B and NK cells has since dropped to baseline levels in all patients. With improved gene therapy vectors, it is expected that observed adverse effects will be reduced. Nevertheless, HR-based strategies will always remain advantageous for the treatment of most disorders (Fig. 2); in particular, for those diseases where the disease gene consists of a very large open reading frame where extremely high transcript levels are needed, or where a very cell-specific and tightly-regulated transcription is required. Because iPSCs have an unlimited self-renewal capacity *in vitro* (as opposed to endogenous HSCs) the identification and selection of the correctly repaired cells by HR is feasible for iPSCs (Fig. 3).[12] HR efficiencies observed in iPSCs vary from 0.5% to 100% of the selected clones and depend on the length of homology between the targeting vector and the target locus, the method of delivery of the targeting vector (nonintegrating viral vectors or electroporation), the HR strategy (with or without nucleases), and on certain features of the disease locus, such as the presence of repetitive DNA and transcriptional activity.[13]

The application of autologous or autologous-mismatched iPSCs

For the treatment of noninherited blood disorders for which there is a lack of endogenous HSCs, such as the acquired forms of aplastic anemia and myelofibrosis, autologous iPSC-derived HSCs may have advantages over allogeneic grafts. For these purposes, genetic repair is not needed, and therefore the time required for the generation of the HSC transplants from the patient's fibroblasts is the only limiting factor.

The majority of acquired blood disorders requiring stem cell transplants include the different forms of leukemia and lymphoma. For some of these diseases, transplantations with autologous HSCs are effective, but, in general, allogeneic transfers are preferred for two reasons. First, the GVH response may eliminate residual tumor cells that survive the conditioning regimen. In some forms of leukemia, the occurrence of clinical GVH reduces the relapse incidence, which led to the postulation of a more specific *graft-versus-leukemia* reaction. Second, autologous HSCs, even when collected during remission and subjected to purging, may still contain malignant cells that theoretically could contribute to a relapse.[14,15] The latter problem would not be an issue if iPSC-derived cells were used. However, because iPSC-derived HSCs are autologous, they are not capable of inducing GVH reactions. To overcome this, it is technically possible to introduce one or more mismatches in either the major or minor histocompatibility genes, either by HR or with gene

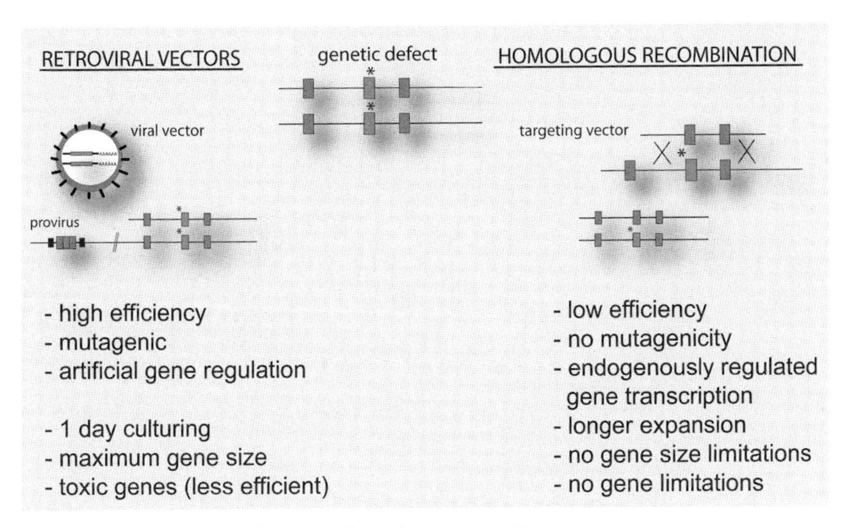

Figure 2. Current gene restoration approaches versus homologous recombination.

therapy vectors. A candidate would be HLA-DP, as HLA-DP mismatches have been associated with a decreased risk of relapse in HLA-matched (10/10) transplants.[16,17] It is obvious that such an approach should first be extensively tested in experimental models, but it remains an attractive option for those patients for whom a suitable donor cannot be found in current registries.

iPSC banking initiatives for HSC transplantations?

ESC and iPSC banking initiatives have been discussed by a number of people.[18–20] The majority of these reports discuss the number of lines required for tissue regeneration on the basis of the HLA-A, -B, and -DR alleles. It was estimated that an unlimited supply of 50 homozygous stem cell donors was sufficient to provide 80–90% of the Japanese population with an HLA-matched transplant.[20] Unfortunately, this number relates to *tissue* transplantations, which do not require the same degree of HLA-matching as HSC transplantations; the infusion of homozygous HSCs will lead to severe GVH in most instances. Nevertheless it remains an interesting question how many iPSC lines should be generated to cover a large proportion of a country's population. The current worldwide registry includes approximately 19 million donors, which is sufficient to supply beyond 80% of the Caucasian population with matched HSCs. However, receiving the matched donor cells at the right time in the right place is not as straightforward as it seems, and national or state iPSC banks may be helpful. The establishment of iPSC banks covering the majority of haplotypes is simply not realistic. However, banking the most frequent haplotypes may be an option to be considered when the problems of large scale production of iPSCs has been worked out. Furthermore, the establishment of iPSC banks covering the more rare HLA-alleles may be beneficial for patients who cannot be helped with current registries. It would be very informative to have detailed information on the haplotypes of the patients that cannot be provided with matched donor cells using current registries and donated cord blood units. These data would enable the estimation of the number of iPSC lines required to provide that proportion of the population with a rare haplotype. To the best of our knowledge, these data are not available to date, and therefore iPSC banking is not discussed in more detail.

Hematopoietic differentiation of iPSCs

The possibilities discussed above illustrate the promise of iPSC-based HSC transplantations. But what is the status of the hematopoietic differentiation protocols for iPSCs?

Thus far several differentiation approaches have been evaluated in terms of directing human iPSCs or ESCs toward different hematopoietic cell types. Current strategies invariably exploit the use of growth factors, of mouse or human stromal cells that exhibit hematopoietic instruction/maintenance potential,[21,22] or of embryoid bodies—or a combination of these.[23–25] Protocols are available for the

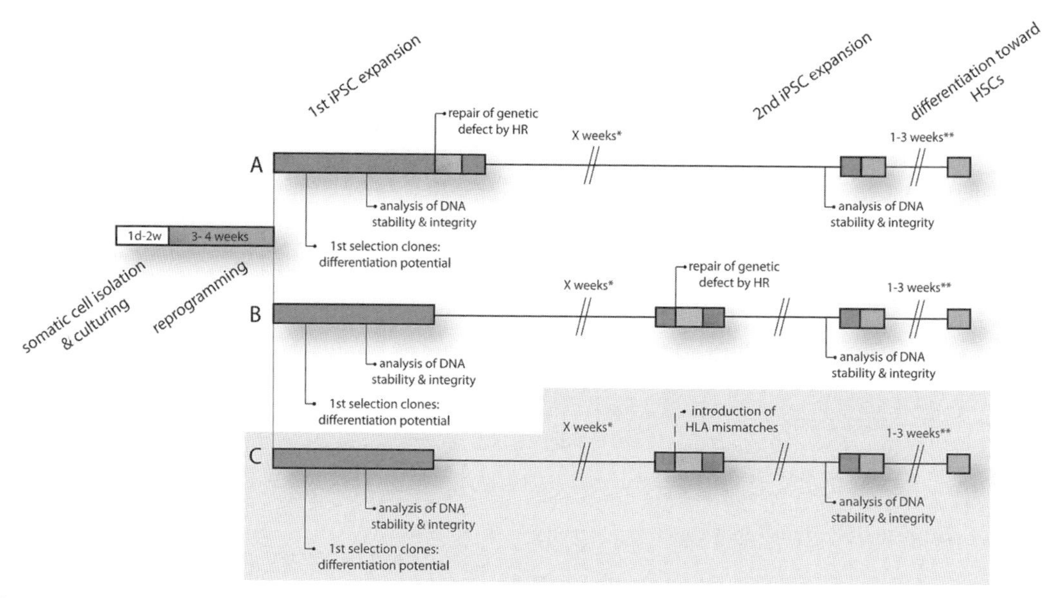

Figure 3. Three putative scenarios to generate genetically modified iPSCs from patients. After completion of the reprogramming phase that will take approximately four to six weeks, the generated iPSC clones are characterized and tested for their differentiation potential. The iPSC lines that make hematopoietic cells very efficiently are selected for further expansion, and genomic abnormalities are mapped. Selected lines are corrected by homologous recombination (HR), which can be performed after the initial expansion or near the end of the expansion phase. *Time needed for expansion depends on the time frame required to achieve genomically stabilized iPSC lines, and the number of HSCs needed for the transplantation. As protocols for the generation of long-term repopulating HSCs from PSCs have not been developed, it is currently not possible to provide an estimate of the time needed from patient transplantation. **On the basis of current protocols, the differentiation period is expected to last one to three weeks. (A) The defect is repaired by HR recombination during the initial expansion phase. (B) HR is performed at a later stage after genomic stabilization. (C) Genetic modifications by gene therapy approaches may also be performed once genomic stability has been achieved.

efficient differentiation *in vitro* of ESCs toward functional erythroid-megakaryocytic cells,[26–29] myeloid cells,[30,31] and NK cells.[32] For example, it is possible to generate 10–100 billion red blood cells from a single six-well plate of ESCs; 16% of such red blood cells express the adult beta-globin gene, indicating potential functionality.[27] In contrast to the generation of myeloid and erythroid cells, the *in vitro* generation of B and T lymphocytes has been more challenging.[33] Yet, from human ESCs it is now also possible to generate lymphoid cells with T cell receptor (TCR) rearrangements.[34]

The hematopoietic differentiation protocols developed for ESCs are expected to be valid for iPSCs as well. However, the directed differentiation toward fully differentiated cell types is only useful for *in vitro* disease modeling, and perhaps (in the future) for the large-scale production of red blood cells and thrombocytes to replace donor transfusions. For iPSC-based transplantation strategies, the generated HSCs should be capable of homing to the endogenous HSC niches in the bone marrow and to give rise to functional hematopoietic cells over the life span of an individual. Alternatively, long-lived progenitors may be exploited. Until now, data demonstrating the repopulation of the hematopoietic system by hESC-derived cells have been limited. In the majority of xenotransplantation studies performed with hESC- or iPSC-derived cells, human $CD45^+$ cells were absent or only present at low frequencies (<1%) for a short period of time after the infusion or intrafemoral injection of PSC-derived $CD34^+$ cells.[35–37] Only one study has demonstrated the presence of >1% of human $CD45^+$ cells in hematopoietic tissues upon secondary transplantation of ESC-derived $CD34^+$ cells.[38] However, this study failed to show functional B and T cell reconstitution, indicating that the differentiation ability of the repopulating cells was limited.

Recent experiments in which hiPSC-derived $CD34^+$ cells were transplanted also failed to show the long-term presence of donor type cells in the blood.[37,39] Upon isolation of the human CD34$^+$ cells from the bone marrow, iPSC-derived cells could be differentiated into distinct hematopoietic cell types *in vitro*, indicating a differentiation block *in vivo*.[37] However, these are results of only two studies, and it cannot be excluded that the iPSC lines and hematopoietic differentiation protocols affected the outcome. Similar repopulating failures have been encountered in experiments using HSC-like cells derived from mouse ESCs. These problems were eventually overcome by the transient overexpression of HoxB4 in mouse ESC-derived HSCs,[40] but HoxB4 is not effective in human ESC-derived HSC-like cells,[41] indicating important differences between the murine and human hematopoietic systems.

Thus, it remains enigmatic whether functional long-term repopulating cells can be generated from human PSCs. Although current evidence from *in vitro* experiments supports the notion that the HSC-like cells from human PSCs have the same differentiation potential as endogenous HSCs, their differentiation is deficient *in vivo*. *In vitro*, PSCs go through a hemangioblast-like stage, during which the cell is not yet committed to the endothelial or hematopoietic cell lineage.[42] This suggests that the imperfect transplantation behavior of the HSC-like cells derived from human PSCs have a too-immature phenotype and resemble HSCs of primitive hematopoiesis rather than long-term repopulating HSCs. It will be crucial for the full usage of iPSCs for hematopoietic transplantation purposes to mature these cells further into functional long term repopulating HSCs.

Safety issues of PSCs

If cells are considered for clinical purposes, it is crucial that a number of safety criteria be met. Extra care is required if the cells are transplanted as proliferative entities for establishing a complete tissue and to maintain the homeostasis of this tissue for the rest of a patient's life, as is the case for HSCs. Since the turnover rate of blood is high, transplanted HSCs and their transit amplifying progeny undergo many population doublings, increasing the likelihood of *selection*, which has been demonstrated by transplantation of retrovirally-marked mouse HSCs.[43] One of the initial concerns over the use of iPSCs was the requirement of retro- or lentiviral vectors to stably deliver pluripotency-inducing genes to the cells. These viral vectors contain promoter, enhancer, splicing, and polyadenylation sequences, and insert into the host genome, making them potent insertional mutagens.[44] In addition, the majority of the reprogramming genes have been linked to different kinds of cancers. For example, *MYC* is the most notorious, and has been found to be amplified in several types of leukemia. Not unexpectedly, mice originating from retrovirally-induced iPS cells developed tumors in a number of studies.[45,46] To circumvent the use of retroviral vector systems, better clinical proof strategies, such mRNA transfection,[47,48] miRNAs,[49] nonintegrating vector technologies,[50–52] and recombinant membrane-crossing proteins have been developed.[53] Also, the induction and expansion of iPSCs under current good manufacturing practice (cGMP) conditions is not a major hurdle any longer.[54] The huge interest in iPSCs has accelerated the development of cGMP-compliant iPSC derivation and culture protocols, enabling many laboratories worldwide to generate and expand cGMP iPSCs.[55]

With the technology for generating and expanding clinical-grade iPSCs at hand, the first steps toward the application of iPSCs for clinical purposes have been taken. Much can be learned from phase I clinical studies using human ESCs, which are very similar to those for iPSCs. One study involved the transplantation of ESC-derived retinal pigment epithelium to treat Stargardt's macular dystrophy and dry age-related macular degeneration.[56] In addition to efficacy, safety must be meticulously demonstrated in laboratory animals, but preferably also in syngeneic animal systems better resembling humans. Important safety issues include the risks of teratoma formation (undifferentiated ESCs have the capacity to form a benign tumor in which tissues of the three germ layers are present), the dissemination of PSC-derived cells into other organs, and genomic stability and integrity of the ESCs. To standardize the quality of the cells, ESC lines are extensively screened for genomic abnormalities, and master cell banks have been generated.

Similar procedures will apply to iPSCs before these cells can be applied in the clinic. Genome-wide analyses of several human iPSC lines by various labs demonstrated that human iPSCs have the tendency to acquire genetic aberrations during the reprogramming phase and/or first passages of culture. The sequencing of the majority of protein coding exons revealed that the iPSC lines invariably contained single point mutations. This was irrespective of the method of derivation. The observed mutation load was 2–15 mutations per iPSC line, of which a large proportion was nonsilent.[57] Approximately 50% of the mutations were originally present in the parental fibroblast lines, whereas the remainder was acquired during the first passages of culture. However, once established, the mutation load in iPSCs appears to stay constant over time, and was found to be independent of subsequent single-cell cloning steps.[56]

In addition to point mutations, copy number variations (CNVs) have been observed in several independently generated iPSC lines.[58,59] During the first culture periods, the CNVs increased but they were counterselected upon prolonged culture, leading to a genetic state resembling that of hESCs. These data indicated that the generation of iPSCs, and their subsequent expansion, might select for genetic mutations that are hazardous upon transplantation of their progeny. Therefore, improved derivation and culture conditions are necessary. However, these studies also underscore that an iPSC genotype is stable once an iPSC line has been maintained in culture for a prolonged period of time. This is very similar to what has been described for human ESCs. Since two human ESC lines have entered the clinic as part of a phase I trial, carefully selected iPSC lines may be useful for certain cell replacement therapies as well. However, evidence supporting the feasibility of iPSCs for regenerative medicine purposes is still missing, as the effects of the observed genetic aberrations have not yet been investigated in cell replacement experiments.

Current status on iPSC-mediated HSC transplantations

Two studies on inherited forms of anemia have already demonstrated the potential of autologous iPSCs for treatment of congenital blood disorders. One study was performed in a humanized mouse model of sickle cell anemia. Sickle iPSCs derived from tail-tip fibroblasts were corrected by HR. Transplantation of the corrected iPSC-derived HSCs in lethally irradiated mice yielded enhanced α-globin levels and the absence of sickle cells in the peripheral blood.[60] In the second study, iPSCs were generated from a β-thalassemia major (Cooley's syndrome) patient who lacked β-globin expression.[61] The mutant *HBB* gene was genetically corrected in the iPSC line by HR. Differentiation experiments using the repaired iPSCs demonstrated the presence of β-globin in erythroid colonies generated *in vitro* from bone marrow (CFU-E assay) and the presence of human β-globin in the peripheral blood of sublethally irradiated SCID mice six weeks after they received repaired $CD34^+$ cells. Both studies illustrate that a pathological phenotype can be eliminated upon repair of the mutant gene in iPSCs, and subsequent iPSC-derived HSC transplantation. However, restoration was shown for only a short period of time, not for the full lifespan of the transplanted animal. Therefore, full proof-of-principle of the use of repaired iPSCs to treat congenital blood disorders still has to be provided.

Prospects

It is evident that the iPSC technology will contribute to our understanding of disease pathogenesis and to the development of novel drugs. To what extent iPSC-derived HSCs, either based on iPSC banking initiatives or on (modified) autologous cells, will make their way into the clinic remains to be seen. The first iPSC-based hematological applications will likely be in the blood transfusion field, as was recently discussed elsewhere.[62] The generation of safe red blood cells and thrombocytes seems less complicated than that of transplantable HSCs. Transplantations of HSCs demand the highest quality of cells because they must remain active for the rest of a patient's life. Therefore, the presence of any tumor-prone cells must be excluded.

iPSCs derived and cultured under current state-of-the-art techniques undergo selection, as indicated by point mutations and CNVs. It will be essential to know the effects of every single genomic aberration in transplantation models. The generation of iPSCs devoid of mutations would be a preferred alternative, but since reprogramming and cell culture involve strong selection pressures, it is uncertain whether the long-term risk of leukemogenesis is avoidable. Nevertheless, the iPSC technology holds promise for HSC transplantation purposes, where suitable donors are scarce or unavailable.

The need for alternatives to current bone marrow supplies is expected to grow due to the expansion of multiracial populations throughout the world. However, formal proof for the efficient and safe long-term reconstitution of the blood-forming system by PSC-derived HSC-like cells has to be provided. In view of the likelihood that some factors involved in the *in vivo* behavior of iPSC-derived HSCs are species-specific, experiments involving subhuman primates are recommended as part of the translational research.

Acknowledgments

The authors thank Dr. R.C. Hoeben for comments. This work was supported by the Landsteiner Foundation for Research on Blood Transfusions (LSBR) (H.M.M.).

Conflicts of interest

The authors declare no conflicts of interest.

References

1. Jaenisch, R. & R. Young. 2008. Stem cells, the molecular circuitry of pluripotency and nuclear reprogramming. *Cell* **132:** 567–582.
2. Thomson, J. A. *et al.* 1998. Embryonic stem cell lines derived from human blastocysts. *Science* **282:** 1145–1147.
3. Takahashi, K. *et al.* 2007. Induction of pluripotent stem cells from adult human fibroblasts by defined factors. *Cell* **131:** 861–872.
4. Yu, J. *et al.* 2007. Induced pluripotent stem cell lines derived from human somatic cells. *Science* **318:** 1917–1920.
5. Puri, M.C. & A. Nagy. 2012. Concise review: embryonic stem cells versus induced pluripotent stem cells: the game is on. *Stem Cells* **30:** 10–14.
6. Inoue, H. & S. Yamanaka. 2011. The use of induced pluripotent stem cells in drug development. *Clin Pharmacol. Ther.* **89:** 655–661.
7. Sauvageau, G., N.N. Iscove & R.K. Humphries. 2004. In vitro and in vivo expansion of hematopoietic stem cells. *Oncogene* **23:** 7223–7232.
8. Cathomen, T. & J.K. Joung. 2008. Zinc-finger nucleases: the next generation emerges. *Mol.Ther.* **16:** 1200–1207.
9. Moscou, M.J. & A.J. Bogdanove. 2009. A simple cipher governs DNA recognition by TAL effectors. *Science* **326:** 1501.
10. Aiuti, A. *et al.* 2009. Hematopoietic stem cell gene therapy for adenosine deaminase deficient-SCID. *Immunol. Res.* **44:** 150–159.
11. Baum, C. 2011. Gene therapy for SCID-X1: focus on clinical data. *Mol. Ther.* **19:** 2103–2104.
12. Nakayama, M. 2010. Homologous recombination in human iPS and ES cells for use in gene correction therapy. *Drug Discov. Today* **15:** 198–202.
13. Zou, J., R. Cochran & L. Cheng. 2010. Double knockouts in human embryonic stem cells. *Cell Res.* **20:** 250–252.
14. Brenner, M.K. *et al.* 1993. Gene-marking to trace origin of relapse after autologous bone-marrow transplantation. *Lancet* **341:** 85–86.
15. Miller, C.B. *et al.* 2001. The effect of graft purging with 4-hydroperoxycyclophosphamide in autologous bone marrow transplantation for acute myelogenous leukemia. *Exp. Hematol.* **29:** 1336–1346.
16. Petersdorf, E.W. *et al.* 2001. The biological significance of HLA-DP gene variation in haematopoietic cell transplantation. *Br. J. Haematol.* **112:** 988–994.
17. Shaw, B.E. *et al.* 2007. The importance of HLA-DPB1 in unrelated donor hematopoietic cell transplantation. *Blood* **110:** 4560–4566.
18. Taylor, C.J. *et al.* 2005. Banking on human embryonic stem cells: estimating the number of donor cell lines needed for HLA matching. *Lancet* **366:** 2019–2025.
19. Nakatsuji, N., F. Nakajima & K. Tokunaga. 2008. HLA-haplotype banking and iPS cells. *Nature Biotechnol.* **26:** 739–740.
20. Nakatsuji, N. 2010. Banking human pluripotent stem cell lines for clinical application? *J. Dent. Res.* **89:** 757–758.
21. Kaufman, D.S. *et al.* 2001. Hematopoietic colony-forming cells derived from human embryonic stem cells. *Proc. Natl. Acad. Sci. USA* **98:** 10716–10721.
22. Vodyanik, M.A. *et al.* 2005. Human embryonic stem cell-derived CD34+ cells: efficient production in the coculture with OP9 stromal cells and analysis of lymphohematopoietic potential. *Blood* **105:** 617–626.
23. Chadwick, K. *et al.* 2003. Cytokines and BMP-4 promote hematopoietic differentiation of human embryonic stem cells. *Blood* **102:** 906–915.
24. Ng, E.S. *et al.* 2005. Forced aggregation of defined numbers of human embryonic stem cells into embryoid bodies fosters robust, reproducible hematopoietic differentiation. *Blood* **106:** 1601–1603.
25. Zambidis, E.T. *et al.* 2005. Hematopoietic differentiation of human embryonic stem cells progresses through sequential hematoendothelial, primitive, and definitive stages resembling human yolk sac development. *Blood* **106:** 860–870.
26. Giarratana, M.C. *et al.* 2011. Proof of principle for transfusion of in vitro-generated red blood cells. *Blood* **118:** 5071–5079.
27. Lu, S.J. *et al.* 2008. Biologic properties and enucleation of red blood cells from human embryonic stem cells. *Blood* **112:** 4475–4484.
28. Lu, S.J. *et al.* 2011. Platelets generated from human embryonic stem cells are functional in vitro and in the microcirculation of living mice. *Cell Res.* **21:** 530–545.
29. Takayama, N. *et al.* 2008. Generation of functional platelets from human embryonic stem cells in vitro via ES-sacs, VEGF-promoted structures that concentrate hematopoietic progenitors. *Blood* **111:** 5298–5306.
30. Yokoyama, Y. *et al.* 2009. Derivation of functional mature neutrophils from human embryonic stem cells. *Blood* **113:** 6584–6592.
31. Choi, K.D., M.A. Vodyanik & I.I. Slukvin. 2009. Generation of mature human myelomonocytic cells through expansion and differentiation of pluripotent stem cell-derived lin-CD34+CD43+CD45+ progenitors. *J. Clin. Invest.* **119:** 2818–2829.
32. Woll, P.S. *et al.* 2009. Human embryonic stem cells differentiate into a homogeneous population of natural killer cells with potent in vivo antitumor activity. *Blood* **113:** 6094–6101.
33. Martin, C.H. *et al.* 2008. Differences in lymphocyte developmental potential between human embryonic stem cell

and umbilical cord blood-derived hematopoietic progenitor cells. *Blood* **112:** 2730–2737.
34. Timmermans, F. *et al.* 2009. Generation of T cells from human embryonic stem cell-derived hematopoietic zones. *J. Immunol.* **182:** 6879–6888.
35. Wang, L. *et al.* 2005. Generation of hematopoietic repopulating cells from human embryonic stem cells independent of ectopic HOXB4 expression. *J. Exp. Med.* **201:** 1603–1614.
36. Narayan, A.D. *et al.* 2006. Human embryonic stem cell-derived hematopoietic cells are capable of engrafting primary as well as secondary fetal sheep recipients. *Blood* **107:** 2180–2183.
37. Risueno, R.M. *et al.* 2012. Inability of human induced pluripotent stem cell-hematopoietic derivatives to downregulate microRNAs in vivo reveals a block in xenograft hematopoietic regeneration. *Stem Cells* **30:** 131–139.
38. Ledran, M.H. *et al.* 2008. Efficient hematopoietic differentiation of human embryonic stem cells on stromal cells derived from hematopoietic niches. *Cell Stem Cell* **3:** 85–98.
39. Woods, N.B. *et al.* 2011. Brief report: efficient generation of hematopoietic precursors and progenitors from human pluripotent stem cell lines. *Stem Cells* **29:** 1158–1164.
40. Kyba, M., R.C. Perlingeiro & G.Q. Daley. 2002. HoxB4 confers definitive lymphoid-myeloid engraftment potential on embryonic stem cell and yolk sac hematopoietic progenitors. *Cell* **109:** 29–37.
41. Wang, L. *et al.* 2005. Generation of hematopoietic repopulating cells from human embryonic stem cells independent of ectopic HOXB4 expression. *J. Exp. Med.* **201:** 1603–1614.
42. Wang, L. *et al.* 2004. Endothelial and hematopoietic cell fate of human embryonic stem cells originates from primitive endothelium with hemangioblastic properties. *Immunity* **21:** 31–41.
43. Gerrits, A. *et al.* 2010. Cellular barcoding tool for clonal analysis in the hematopoietic system. *Blood* **115:** 2610–2618.
44. Mikkers, H. & A. Berns. 2003. Retroviral insertional mutagenesis: tagging cancer pathways. *Adv. Cancer Res.* **88:** 53–99.
45. Okita, K., T. Ichisaka & S. Yamanaka. 2007. Generation of germline-competent induced pluripotent stem cells. *Nature* **448:** 313–317.
46. Duinsbergen, D. *et al.* 2009. Tumors originating from induced pluripotent stem cells and methods for their prevention. *Ann. N.Y. Acad. Sci.* **1176:** 197–204.
47. Plews, J.R. *et al.* 2010. Activation of pluripotency genes in human fibroblast cells by a novel mRNA based approach. *PLoS ONE.* **5:** e14397.
48. Warren, L. *et al.* 2010. Highly efficient reprogramming to pluripotency and directed differentiation of human cells with synthetic modified mRNA. *Cell Stem Cell* **7:** 618–630.
49. Anokye-Danso, F. *et al.* 2011. Highly efficient miRNA-mediated reprogramming of mouse and human somatic cells to pluripotency. *Cell Stem Cell* **8:** 376–388.
50. Stadtfeld, M. *et al.* 2008. Induced pluripotent stem cells generated without viral integration. *Science* **322:** 945–949.
51. Ban, H. *et al.* 2011. Efficient generation of transgene-free human induced pluripotent stem cells (iPSCs) by temperature-sensitive Sendai virus vectors. *Proc. Natl. Acad. Sci. USA* **108:** 14234–14239.
52. Yu, J. *et al.* 2009. Human induced pluripotent stem cells free of vector and transgene sequences. *Science* **324:** 797–801.
53. Kim, D. *et al.* 2009. Generation of human induced pluripotent stem cells by direct delivery of reprogramming proteins. *Cell Stem Cell* **4:** 472–476.
54. Rodriguez-Piza, I. *et al.* 2010. Reprogramming of human fibroblasts to induced pluripotent stem cells under xeno-free conditions. *Stem Cells* **28:** 36–44.
55. Ausubel, L.J., P.M. Lopez & L.A. Couture. 2011. GMP scale-up and banking of pluripotent stem cells for cellular therapy applications. *Methods Mol. Biol.* **767:** 147–159.
56. Schwartz, S.D. *et al.* 2012. Embryonic stem cell trials for macular degeneration: a preliminary report. *Lancet* **379:** 713–720.
57. Gore, A. *et al.* 2011. Somatic coding mutations in human induced pluripotent stem cells. *Nature* **471:** 63–67.
58. Laurent, L.C. *et al.* 2011. Dynamic changes in the copy number of pluripotency and cell proliferation genes in human ESCs and iPSCs during reprogramming and time in culture. *Cell Stem Cell* **8:** 106–118.
59. Hussein, S.M. *et al.* 2011. Copy number variation and selection during reprogramming to pluripotency. *Nature* **471:** 58–62.
60. Hanna, J. *et al.* 2007. Treatment of sickle cell anemia mouse model with iPS cells generated from autologous skin. *Science* **318:** 1920–1923.
61. Wang, Y. *et al.* 2012. Genetic correction of beta-thalassemia patient-specific iPS cells and its use in improving hemoglobin production in irradiated SCID mice. *Cell Res.* **22:** 637–648.
62. Migliaccio, A.R. *et al.* 2012. The potential of stem cells as an in vitro source of red blood cells for transfusion. *Cell Stem Cell.* **10:** 115–119.